Lead in Glassy Materials in Cultural Heritage

SCIENCES

Physics of Condensed Matter,
Field Directors – Roland Pellenq and Pierre Levitz

Cultural and Industrial Heritage Materials,
Subject Head – Philippe Sciau

Lead in Glassy Materials in Cultural Heritage

Coordinated by
Anne Bouquillon
Patrice Lehuédé

ISTE

WILEY

First published 2023 in Great Britain and the United States by ISTE Ltd and John Wiley & Sons, Inc.

ISTE Ltd
27-37 St George's Road
London SW19 4EU
UK

www.iste.co.uk

John Wiley & Sons, Inc.
111 River Street
Hoboken, NJ 07030
USA

www.wiley.com

Library of Congress Control Number: 2022947641

British Library Cataloguing-in-Publication Data
A CIP record for this book is available from the British Library
ISBN 978-1-78945-076-7

ERC code:
PE3 Condensed Matter Physics
PE3_13 Structure and dynamics of disordered systems: soft matter (gels, colloids, liquid crystals, etc.), liquids, glasses, defects, etc.
SH5 Cultures and Cultural Production
SH5_6 History of art and architecture, arts-based research

Contents

Foreword by Isabelle Pallot-Frossard

Isabelle PALLOT-FROSSARD
Foundation for Cultural Heritage Sciences, Cergy-Pontoise, France

Glass, a material that man has synthesized for over 5,000 years, has a high heritage value, acquired through successive or simultaneous technical innovations required for its implementation, the multiple uses that civilizations have given to it, and its ubiquity in public collections. Its transparency, brightness, wide color variety and sometimes sonority properties have been explored by man to this day, in order to beautify his prestige objects, items dedicated to liturgical rites, and also those of daily life, to attire oneself, look at his reflection or still protect himself from the surrounding environment by closing up his living space or place of worship without darkening the inside ambiance. This polymorphic material, both resistant and brittle, is justifiably explored by a rich literature in materials science, art history and in history of techniques.

However, this book approaches a subject that is less well known, and which has not until present been the object of a historical and scientific synthesis: that of the combination of glass and lead, in a wide variety of uses and formulations, from glazes on ceramic to lead crystal glass, going through enamels on metal or glass, fake gems, Venetian blown glass, mosaic tesserae and certain pieces of stained glass windows. Due to its multiple properties, lead is omnipresent in various forms in heritage objects, sought-for as coloring agent, opacifier, siccative for oils, appreciated by alchemists and painters, used in pharmacopoeia and at the same time, paradoxically, dreaded since ancient times for its harmful effects on health.

Lead in Glassy Materials in Cultural Heritage,
coordinated by Anne BOUQUILLON and Patrice LEHUÉDÉ. © ISTE Ltd 2023.

The “marriage” referred to here is a “fusional” one, which integrates lead into the vitreous matrix, in the form of oxides, in order to give it color or on the contrary absence of color for the crystal glass, as fluxing agent to lower its melting temperature, modify its physical and optical properties or enhance its durability. The variety of resulting properties is huge during time and depends on uses. This book highlights the complexity of mechanisms involved, which required of the craftsmen who manufactured these objects relentless experimentations, based on know-how accumulated and transmitted in the workshops. These technical advances are recorded in many technical treatises, whose understanding nowadays is both necessary and difficult for reasons of vocabulary and cultural context, and sometimes of willingness from the authors to describe everything while protecting manufacturing secrets.

This book does not cover the “marriage of convenience”, a term that can be used to qualify the association of colored, cut and painted glass and a lattice of lead cames, whose plasticity was used to closer embed the glass pieces, which is the historical technique of stained glass. Another subject that is not covered here is the “forced marriage” between glass and lead, resulting from external contributions that frequently pollute the surface of glass exposed to the atmospheric environment, to the burial medium or to the close proximity of lead works.

However, the end of the volume deals with the effects of the “divorce” between glass and lead generated by leaching phenomena that alter the objects, particularly the lead glazes having been subjected to burial periods, and that continue in the museum collections. Moreover, the methods implemented for following up the progression of this weathering, in order to prevent it and slow it down or even to stop it, will also be described. The work conducted in the crystal glass industry will thus be understood, which will contribute to a better understanding and prevention of lead leaching in the crystal containers in contact with certain acid drinks, susceptible of harmful effects on human health.

Knowledge of lead glass may be at first sight perceived as a very specialized and limited subject. It is in reality an extremely wide field of research, and this work coordinated by Anne Bouquillon and Patrice Lehuédé is a true invitation to a time and space journey, in the quest for the most varied objects, knowledge of which requires the contribution of the finest techniques of modern chemistry and the history of techniques.

Foreword by Daniel R. Neuville

Daniel R. NEUVILLE[1,2]
[1] *IPGP, CNRS, University of Paris, Paris, France*
[2] *USTV, Paris, France*

Since ancient times, lead has been a mythical element whose use evolved with human civilizations. It was by turns praised and proscribed: praised due to its abundance, malleability, low melting temperature and its contribution to improving the properties of several materials (glass, ceramic), and also proscribed because of its toxicity (lead poisoning) known since ancient times.

Lead is an abundant element, relatively well distributed at the level of the Earth's crust. It is generally associated with copper, zinc or silver and therefore relatively easy to find. Lead was used by the first Sumerians who had developed the means to extract and purify it and who used it to color and enamel ceramics, seal amphorae, or produce cosmetics. In summary, lead has been present since ancient times in the common objects of our daily life. In parallel to its daily use, lead is related to alchemy; it is important to mention the works of Gilles de Rais, a figure who is at the origin of Barbe-Bleue. It is both used and feared by pharmacopoeia.

In this book dedicated to lead in vitreous materials of heritage, Anne Bouquillon and Patrice Lehuédé offer a new complete and updated perspective on the role of this element in glasses and glazes. Their book retraces the history and the manufacturing of the first lead glasses, which were in fact glazes covering pottery or opaque massive glasses. This book covers also the famous flintglass or lead crystal, which contributed to the significant progress of optics and astronomy. The more recent history of lead crystal is also approached from the perspective of its risks essentially related to the alteration and lixiviation of crystal.

Lead in Glassy Materials in Cultural Heritage,
coordinated by Anne BOUQUILLON and Patrice LEHUÉDÉ.

The reader will find a consideration on the origin of lead used in glasses, the difficulties encountered in making glasses and glazes (Part 1), and also the history and evolution of compositions of lead glasses throughout time and world. This journey of lead in glass leads from the very heart of China to the Vosges forests passing through the Arab world and ancient Rome. This scientific and technical journey is brightened up by a color chase, in which colorless crystal spans across the entire visible spectrum as an actual rainbow, in relation with chromium, copper, iron and all the other transition metals it contains. Part 2 presents a synthesis of current scientific works on the role of lead in glasses. The structural aspects related to the changes of coordination number of lead and their effects on the properties are amply described and commented upon. Part 3 traces back the history of lead glasses throughout the centuries and details the many glass recipes, clandestinely published, for making "proper" lead glass. This part takes us on a voyage across ages and throughout the world in search for the transparency and reflection of lead glass. Part 4 approaches the subject of history, implementation and evolution of lead glazes. These lead-rich glazes and their evolution have marked the history of ceramic and at the same time represent a technological and scientific challenge that is particularly well detailed in this book. As a conclusion, Part 5 approaches the difficult subjects of alteration, leaching and release of lead from the crystal. What is the role of lead escaping from glass in relation with the environment and health? These essential questions, which are at the root of the main critiques brought to lead crystal, are unambiguously explored and discussed here.

This book is the result of a considerable amount of work on lead in amorphous materials, glasses, enamels and glazes. The reader will find here both the historical aspects and the scientific and technical aspects of the use of this highly controversial element.

Introduction

Patrice LEHUÉDÉ and Anne BOUQUILLON
C2RMF, IRCP, Paris, France

Glass is a material that has exerted fascination since its invention, due to its exceptional qualities: transparency, shine, coloration and its multiple technical applications. Since the second millennium BC, humans have used this material for increasingly varied purposes: prestigious adornment and decoration elements initially intended for the elites, then objects of everyday life such as window panes and stained glass, mosaics, ceramic decorations (glazes) and metal decorations (enamels).

The composition of vitreous materials may include a very large number of elements, therefore their physical and chemical properties may vary widely. Hence, the introduction in the first century BC of lead for glazes and even earlier for massive glass was a major technological and esthetic progress, which was our intention to highlight. The bibliography dedicated to the history of heritage glass, enamels and glazes is plethoric, as is that concerning the characteristics and properties of the material "glass". On the other hand, the texts exclusively related to vitreous materials containing lead are much scarcer. This is why the first volume of the "Cultural and Industrial Heritage Materials" collection is dedicated to them. Its approach is strongly multidisciplinary, in an intent to offer the reader complementary perspectives. Lead glasses and glazes are at the intersection of several "industries" of the arts of fire: glass-makers, ceramists and metallurgists constitute an association of trades in which the discoveries made by some of them contribute to advancing the know-how and technologies of the others, in which the by-products of the activity of some are recovered and used by the others. The core of this book reviews the current knowledge on the intertwined history of these materials in various civilizations.

Lead in Glassy Materials in Cultural Heritage,
coordinated by Anne BOUQUILLON and Patrice LEHUÉDÉ. © ISTE Ltd 2023.

However, in our opinion, such a study cannot be approached in the absence of a precise idea of what a lead glass is and how it resists alterations over time.

The book we propose is therefore written by experts in the field of heritage sciences and material sciences. A first part briefly describes the advantages brought by lead in the glass-making matrix and the ensuing difficulties, as well as the contributions of isotopic analyses to the research of lead sources. This is followed by a more theoretical part related to the structure of lead glasses and its influence on glass color, particularly if it contains copper and iron. The third part focuses on the history of lead glass since its origin until present; it relies on the published ancient recipes and details the knowledge of Western medieval and Renaissance enamels and of Islamic lead glasses, whose significance can be better appreciated nowadays. The history of lead-glazed ceramics will be the object of a fourth part, with a specific review of the interactions between glaze and ceramic, on the one hand, and with the burial medium, on the other hand. Finally, considering the importance attached to environmental and public health problems associated with the use of lead, a fifth part will be dedicated to the interactions between lead glass and the liquid with which it is susceptible to get in contact, and to the standards that are nowadays applicable to the fabrication of these objects.

Our fullest gratitude goes to Philippe Sciau for his trust in us. We warmly thank Michel Dubus for our fruitful discussions and René Gy (SGR) and Alain Meunier (University of Poitiers) for having attentively read through certain parts of this book. We also thank the entire department of archives and new information technologies of C2RMF for their support.

PART 1

Overview and Specific Techniques for the Analysis of Lead Glasses and Glaze

1

Overview

Patrice LEHUÉDÉ
C2RMF, IRCP, Paris, France

Glass is often defined by contrast with crystal, as illustrated by the following definition formulated by Zarzycki (1982): "Glass is a non-crystallized solid that exhibits glass transition"[1]. This definition is however too restrictive for the heritage domain: many glasses involve in fact a crystallized phase, for example in the form of pigment or opacifier. It should therefore be considered as a glass a product that involves a significant glass phase and that has many points in common with glasses in the strict sense. The same difficulty arises for glazes, which certainly involve a significant glass phase, but may also include crystallized phases. Finally, the definition of lead glass is also debatable: modern instruments of characterization indicate the presence of lead in practically all glasses, often in minute traces. The fact that lead was deliberately added may be considered a criterion, which is difficult to prove, and it seemed to us that a 5% oxide weight threshold, as suggested by the researchers for the mosaic tesserae (Verità et al. 2009), was realistic, and we therefore adopted it.

1.1. Advantages brought by lead

The presence of lead is absolutely not essential for glass- or glaze-making, but in the history of glass a significant content (over 5% by weight of PbO) of lead can

1. Glass transition: passage from the solid state to the viscous liquid state. T_G: temperature at which this transition occurs.

Lead in Glassy Materials in Cultural Heritage,
coordinated by Anne BOUQUILLON and Patrice LEHUÉDÉ.

sometimes be found in silicate glasses, occasionally at least since 1500 BC (Turner 1956; Charleston 1960; Shortland and Eremin 2006) (Chapter 6).

Similarly, the first actual lead glazes (with high lead content) probably date from the Roman period (Tite et al. 1998). But a lead content above 5% can be found in certain glazes, partly in the form of lead antimonate $Pb_2Sb_2O_7$, since 1500 BC (Mecking 2013; Molina et al. 2014) (Chapter 10).

The fact that the introduction of lead in the vitreous matrix for certain applications appeared as a necessity is justified by the specific properties it brings. This chapter gives an overview of these properties, some of which will be extensively developed in the following chapters.

1.1.1. *Lead oxide as fluxing agent*

Lead oxide acts as a very efficient fluxing agent in lead glasses, in the same way as, for example, Na_2O or K_2O or B_2O_3, allowing elaboration of glasses and glazes at relatively low temperature. The binary diagram PbO-SiO_2 is a good illustration of this effect (see Figure 3.6): in the 60–100% molar PbO range, the mixture is fully molten for temperatures ranging between 700 and 800°C, which shows that the use of an additional fluxing agent is not necessary for glazes that are very rich in PbO for an implementation at a maximum temperature of about 1,000°C. The same does not apply for lower lead contents, for example those that can be observed in lead crystal glasses (25–33% PbO in oxide weight): it is important to add another fluxing agent such as K_2O to be able to elaborate the glass at a reasonable temperature of less than 1,170°C (Geller and Bunting 1936). But even at lower concentration, the fluxing agent role of lead oxide is prominent. It also increases the dissolution rate of impurities such as the fragments of refractory materials, so that the quality of the resulting glass is better than with traditional fluxing agents such as Na_2O (Leiser 1963; Shelby 2005, p. 125) and bubbles are more easily removed (Eppler and Eppler 2000, p. 215).

1.1.2. *Influence of lead oxide on viscosity*

PbO decreases the viscosity of glass, rendering it easier to process. Indeed, viscosity is the essential factor in all the glass hot forming processes, each process (blow-forming, fiber drawing, thermoforming, etc.) being optimized for a value of viscosity, and the obvious interest is to obtain this value of viscosity at a moderate temperature (Robertshaw et al. 2010, p. 369). Lead glasses are often qualified as "longer" than traditional soda-lime silica glasses, meaning that they have a lower viscosity variation rate with temperature (see Figure 1.1). This is an advantage, since forming is then less sensitive to temperature variation. However, similar to many other properties, viscosity depends on the overall composition of glass, and not only on its PbO and SiO_2 contents. Consequently, accurate comparisons of the behavior of various glasses must integrate all their compounds, making them more difficult.

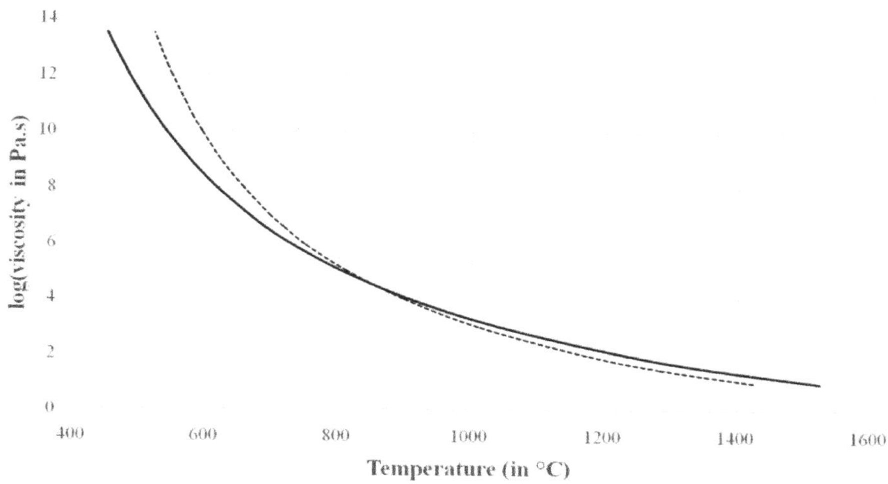

Figure 1.1. *Variation of the logarithm of viscosity (in Pa·s) as a function of temperature (in °C). Solid line: lead crystal; dotted line: soda-lime silica glass of float type. Curves calculated using Fluegel modeling (Fluegel 2007)*

The viscosity curve can be often estimated based on the composition of glass (see, for example, the works of Fluegel (2007)); unfortunately these works cover relatively few of the lead glasses and therefore they are less reliable for this type of glasses than, for example, for the soda-lime silica glasses.

1.1.3. *Influence of lead oxide on the expansion coefficient*

This influence is often of little interest for massive glasses, but it is essential for glazes. Indeed, during the cooling of a ceramic object covered with a glaze, if the expansion coefficient of the glaze is clearly different from that of the ceramic, stresses occur at the glaze/ceramic interface and they are often at the origin of the defects of the finished object: compression stresses that may lead to the burst of the glaze when the expansion coefficient of the glaze is lower than the one of the ceramic, extension stresses responsible for the appearance of cracks (crazing) in the opposite case. The generally admitted preference is for a glaze with an expansion coefficient that is slightly below that of ceramic by 5–15% (Tite et al. 1998). There are tables or even software for the calculation of the expansion coefficient of a glass based on its composition (Eppler and Eppler 2000, p. 251): they show that if Na_2O is replaced by PbO in a glass composition, the expansion coefficient decreases (the contribution of PbO to the expansion coefficient is of about one-fifth of that of the Na_2O for the same oxide weight concentration). The whole set of elements should however be taken into account in the estimation of the expansion coefficient. As the coefficient is normally above that of the ceramic, it is therefore interesting to use a lead glaze with low alkaline content in order to limit the crazing problems. Moreover, it is important to note that the presence of crystals in the glaze complicates particularly the estimation of the expansion coefficient based on the average composition, as well as the diffusion of elements of the glaze into the ceramic (or the opposite), or the volatilization of certain elements during the heat treatment.

1.1.4. *Influence of lead oxide on surface energy*

Bansal and Doremus (1986, p. 106, 111, 130) compiled the values of surface tension for many glass compositions. It systematically appears that the introduction of lead in the vitreous matrix clearly decreases its surface energy. This effect is explained by the high polarizability of Pb^{2+} ions (Scholze 1974, p. 224). The result is an easier wetting, which is a particularly interesting property in the case of glazes deposited on ceramic, since glaze adhesion is then better. Tite et al. (1998) gives 230 mJm^{-2} for a glaze with high lead content compared to 280 mJm^{-2} for an alkaline glaze. It is also this property that may explain the easier digestion of grains in the vitreous matrix containing lead. Lead presence also enhances the dispersion of opacifiers (Paynter and Jackson 2018). Finally, a lower surface energy has another advantage in the case of glazes: the latter have the natural tendency to form a smooth surface (Lehman 2002). Therefore, if the glaze contains bubbles, they tend to move up to the surface due to gravity and burst, leaving a hollow surface defect; this hollow, which significantly affects the aspect of the surface, is much easier smoothed down, at constant viscosity, if the surface energy of the glaze is low (Eppler and Eppler 2000, p. 216).

1.1.5. *Influence of lead oxide on the color*

The introduction of PbO in a colorless glass does not bring color as long as the content is moderate (crystal glass is renowned for its absence of color). But when the PbO content exceeds 30% mol, a yellow brown color may appear (Cohen et al.

1973), whose origin will be discussed in Chapter 4. On the other hand, the shade brought by certain coloring ions in solution in the glass may vary depending on the lead content of the matrix in which they are (Lehman 2002, p. 107). Hence, copper in the Cu^{2+} form gives a blue color to a soda-lime silica glass, of turquoise shade when the PbO content is below 40%, and emerald green when the lead content is higher (La Delfa et al. 2008) (Chapter 4).

Finally, some crystals such as lead antimonates $Sb_2O_5 \cdot 2PbO$, lead stannates $SnO_2 \cdot 2PbO$ and cuprite Cu_2O bring by their own color a shade to the glass (yellow for the first ones, red for the last one) and play the role of pigment. It can be noted that when coloring is done with lead antimonate pigments, the overall analysis of the glaze or glass usually leads to a mean PbO concentration much higher than the value calculated by stoichiometry based on the concentration of Sb_2O_5: the PbO/Sb_2O_5 ratio is often 2 or 3, and even more, when the stoichiometric ratio is 1.38, which means that a part of the lead is in the vitreous matrix (Kaczmarczyk and Hedge 1983).

In the case of opaque red glasses that are colored using cuprite Cu_2O crystals, for example those referred to as "sealing wax", there are often significant contents of PbO (20 % or more in oxide weight) and it is admitted that the presence of PbO in the glass promotes the crystallization of cuprite (Barber et al. 2009, pp. 115–128) (Chapter 6).

A further advantage of lead glazes in terms of color is that since they are elaborated at relatively low temperature, certain pigments that are unstable at high temperature can be readily used, such as cadmium sulfoselenides (Eppler and Eppler 2000, p. 34).

1.1.6. *Influence of lead oxide on devitrification*

Glasses maintained at a temperature below T_L[2] and above T_G tend to crystallize. The formation of such crystals utterly modifies the properties of the glass, and first of all their transparency. Glasses are always molten at a temperature above T_L: during cooling (and forming), it is therefore essential to rapidly cross the critical temperature range between T_L and T_G. The nature of the crystals forming during devitrification depends on the composition of glass and on temperature (e.g. α and β wollastonite $CaSiO_3$, diopside $CaMgSi_2O_6$ in the case of soda-lime silica glass) while the growth rate of these crystals depends on viscosity and therefore on temperature. This growth rate also varies significantly with the composition of

2. T_L: liquidus temperature; temperature above which there are no more crystals in the glass.

glasses; hence when one part of the lime is replaced by magnesia in the composition of a soda-lime silica glass, the liquidus temperature deceases, but also the devitrification rate (Barton and Guillemet 2005, p. 77). The same applies for the introduction of lead into the composition (Nordyke 1984; Hynes and Jonson 1997, p. 137; Lehman 2002, p. 17; Barber et al. 2009). Lead crystal glasses (close to $8SiO_2$-K_2O-PbO) have low sensitivity to devitrification, and the crystals that may form at the surface are silica, cristobalite and tridymite crystals, generated by a surface volatilization of lead (Volf 1984, p. 451). On the other hand, these glasses are not subjected to demixing (which is a separation of phases, similar to the case of devitrification, but the phase that precipitates is not crystallized), at least as long as the PbO content remains below 80% in oxide weight (see Chapter 3).

1.1.7. *Influence of lead oxide on glass redox*

In glasses and glazes, there are usually elements that are susceptible of being present under various degrees of oxidation. The most frequent is iron. It is responsible for a coloring that varies a lot depending on its oxidation state: Fe^{2+} or Fe^{3+}. It is therefore very important to characterize this degree of oxidation; this is done using the glass "redox", which is defined by Fe^{2+}/Fe_{total} ratio. Hence, for the usual soda-lime silica glasses, the redox varies between 0.9 for the very reduced glasses and 0.1 for oxidized glasses (Chopinet et al. 2002). The redox depends on temperature (glasses elaborated at higher temperature are more reduced) (Chopinet et al. 2002) and on the atmosphere during glass elaboration, though reaching the equilibrium between glass and atmosphere may be a very lengthy process, especially when a large quantity of glass is elaborated. But redox also varies with glass composition. For soda-lime silica matrices, the higher the content of modifying cations, such as Na^+ or Ca^{++}, of the glass, the stronger the oxidation tendency at constant oxygen partial pressure (Chopinet et al. 2002). When glasses contain lead, the influence of lead content is poorly documented. It is admitted that lead addition tends to lower the Fe^{2+}/Fe^{3+} ratio (Edwards et al. 1972), while increasing the Cu^+/Cu^{2+} ratio (Edwards et al. 1972; Freestone 1987), which certainly explains why many red glasses colored by cuprite Cu_2O crystals often have high lead content (see Chapter 6).

1.1.8. *Influence of lead oxide on glass durability*

It is generally admitted that lead glasses have better durability than soda-lime silica glasses. Nevertheless, since this durability depends on the set of present elements (and on their concentration), and on the testing conditions for assessing this durability (particularly the pH), the comparison between glass with lead and glass without lead is not very simple. It can nevertheless be stated that lead glasses have good durability for average pH. On the other hand, for acid pH, this durability is weakened (Bansal and Doremus 1986, p. 656; Lehman 2002, pp. 21–24) and may cause problems. This point will be developed further on (Chapters 13 and 14). Similarly, in alkaline medium, the durability of lead-rich glasses is lower than the one of classical soda-lime silica glasses (Bansal and Doremus 1986, p. 655).

1.1.9. *Influence of lead oxide on glass density*

Lead being a heavy element (compared to Si, Na, K or Ca), it is obvious that its introduction in a glass generates a net increase in density. Nevertheless, this increase is not very high for moderate contents; hence a lead crystal (25% by weight of PbO) has a density of 2.9, compared to 2.5 for a classical soda-lime silica glass, and it is not very easy to determine if an object is made from crystal or not just by weighing it. For much higher lead contents, the density difference then becomes very large compared to that of a classical soda-lime silica glass, which makes them easier to distinguish. Curves giving the variation of density of lead glasses depending on the lead content (see Figure 1.2), or more recently based on measurements on glasses of the PbO–SiO_2 system (Ben Kacem et al. 2017), are available.

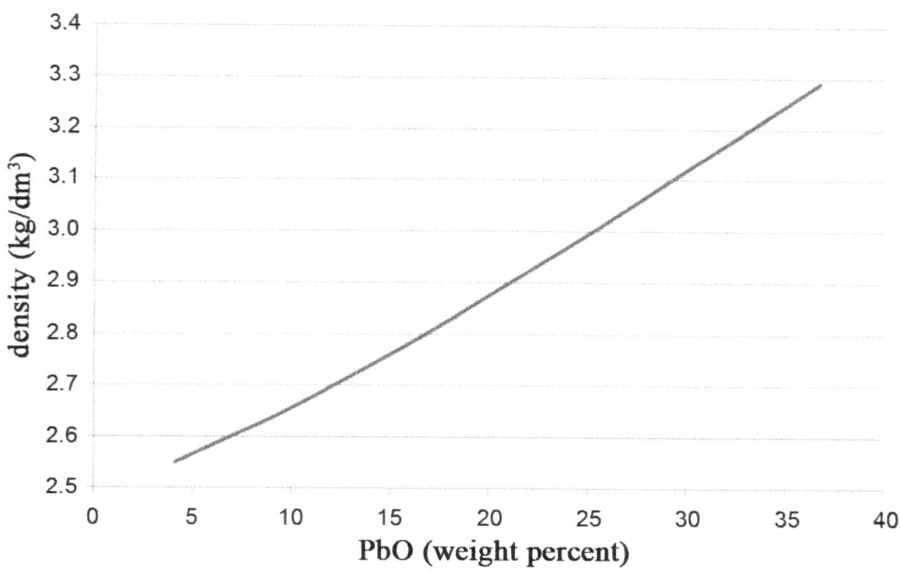

Figure 1.2. *Variation of the density of lead glasses as a function of PbO content according to Davison (2003, p. 10)*

1.1.10. *Influence of lead oxide on optical properties*

Just as the addition of lead to a glass increases its density, it also increases its refractive index, as well as its optical dispersion. This property is obviously widely used in optical glasses, as it is then possible to adjust the refractive index of a glass by varying its lead content. The composition-refractive index relation was the object of many works since the pioneering works of SCHOTT in the 1880s (Marker and Neuroth 1998), but this relation is however not simple, and various formulas were proposed (see the synthesis in Scholze (1974, pp. 123–133)). The lead content of a glass has a strong influence on the refractive index n_D (see, for example, Figure 1.3).

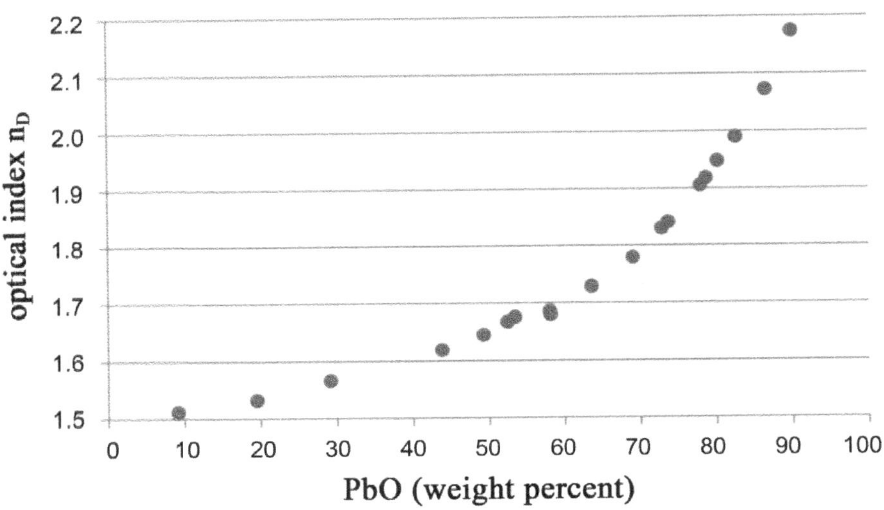

Figure 1.3. *Relation between the optical index n_D of glasses and their lead content according to Nordyke data (Nordyke 1984, p. 25)*

According to the classical distinction, there are crown glasses[3] with a low refractive index and a high Abbe number, and flint glasses[4], which by contrast, have a high refractive index and a low Abbe number. It is important to recall that dispersion corresponds to the variation of the refractive index with the wavelength, and that the relation between the Abbe number (ν_D) and this dispersion is:

3. These are generally alkaline silicates or alkaline and alkaline-earth silicates.

4. The high index of flint glasses is often, but not exclusively, due to the presence of lead in the matrix.

$$\nu_D = (n_D - 1)/(n_F - n_C)$$

where n_D is the refractive index of glass for $\lambda = 0.6563$ µm, n_F is the refractive index of glass for $\lambda = 0.4861$ µm and n_C is the refractive index of glass for $\lambda = 0.5893$ µm.

The higher the Abbe number, the weaker the glass dispersion of the various wavelengths. The Abbe number and the refractive index of a large number of glasses used in optics are often represented in a diagram. The example provided here refers to lead glasses (see Figure 1.4).

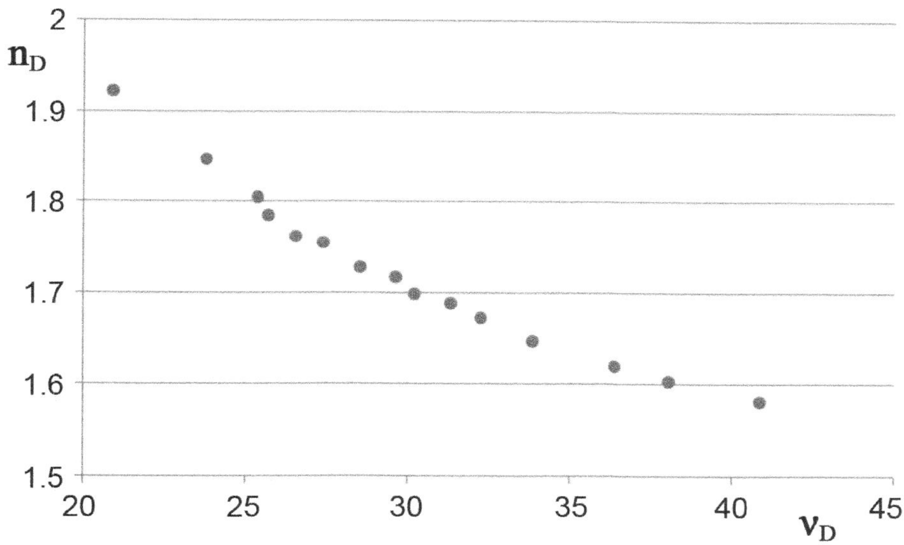

Figure 1.4. *Abbe diagram giving the refractive index n_D of various glasses depending on their Abbe number ν_D for various flint glasses containing lead (according to the values of the SCHOTT catalogue for optical glasses (SCHOTT 2020))*

Moreover, the increase in the refractive index of a glass makes it possible to increase the intensity reflected by the air–glass interface. For normal incidence, the reflected light intensity by such an interface is in fact $(n - 1)^2/(n + 1)^2$. But the higher the light intensity, the "brighter" the glass appears. Thus, when comparing a classical soda-lime glass (index n = 1.51) to a classical lead crystal (n = 1.55), the intensity reflected by the surface passes from 4.1% to 4.65%. This is a small increase, which is difficult to perceive by the eye, especially as the alteration of ancient objects tends to decrease the optical index of the surface and increase the rugosity, the object becoming less bright. On the other hand, for glazes whose PbO content may reach 70% by oxide weight, and whose index is practically 1.80, the effect is very clear: the intensity reflected by the surface is then 8.14%. It is interesting to note that rock crystal (quartz) has an index of 1.55; it has long been considered as a reference in terms of optical quality of the objects. The glass developed by the Venetians in the 15th century for manufacturing luxury objects is referred to as "cristallo" (Verità 1991, pp. 57–67) because it imitates rock crystal. This glass does not contain lead and therefore its index is of the order of 1.5.

1.1.11. *Influence of lead oxide on the mechanical properties*

Glass is an elastic material at ambient temperature. It is therefore characterized by a modulus of elasticity that connects the applied stress to the deformation. The modulus is all the higher as the elongation is weaker when the glass is subjected to tensile stress. The value of this modulus depends on the composition of the glass. The introduction of lead in the composition of a glass reduces the modulus of elasticity (by about 15–20% compared to a classical soda-lime silica glass) and also the hardness (Volf 1984, p. 458; Hynes and Jonson 1997). Consequently, the lead glass cutting and grinding is easier and the tool wear is lower than in the case of classical soda-lime silica glasses. Similarly, mechanical polishing is easier. It is however important to note that glass polishing is a complex phenomenon that involves both the physical properties of glass (its modulus of elasticity, its hardness) and chemical reactions at the glass/polishing paste interface (Cook 1990). It is therefore difficult to predict the behavior of a glass only based on its composition, but empirical measurements, following standardized tests, show that lead glasses tend to be more rapidly polished, which is widely taken advantage of in the industry of cut lead crystal glass.

It can also be noted that the resistance to cracking is much lower than in the case of lead glasses (Kato et al. 2010): consequently, if these glasses are submitted to an impact, they are more easily damaged with the development of cracks: they appear more fragile to scratch.

A crystal drinking glass can be easily distinguished from a soda-lime silica one by the specific sound it emits when tapped or when a wet finger is run on it. The vibration frequency of the sound waves depends on the geometry of the object, but also on the wave propagation velocity through matter. This velocity is lower in the lead crystal than in the soda-lime silica glass, because of its higher density and its lower Young's modulus. On the other hand, the sound wave is far less damped in the crystal glass: the sound "lasts" longer. The damping of sound waves is due to internal friction, which is weaker for crystal glass. It should be noted that the crystal sound is not so readily perceptible when the glass is thick. The sound properties of crystal were used by musicians (among them Mozart) to create pieces of music for instruments such as glass harmonica and the Cristal Baschet.

1.1.12. *Influence of lead content on the absorption of ionizing radiation*

It is a well-known fact that lead strongly absorbs ionizing radiation (X and gamma rays), which explains its use in the protection against this type of radiation. Lead-rich glasses can be used for making transparent windows while giving a radiation protection. These glasses may reach 71% by mass of PbO (Sovis 2020) and even 81% (Nordyke 1984, p. 9). The intensity of X-rays that is not absorbed by the material is given by:

$$I/I_0 = \exp\,[-(\mu/\rho)\cdot\rho\cdot x]$$

with:

- I = intensity of radiation that crosses the material;
- I_0 = initial intensity;
- μ/ρ = mass attenuation coefficient of the material for the given radiation;
- ρ = density of the material;
- x = thickness of the material through which the radiation passes.

But μ/ρ is much higher in the case of lead than in the case of lighter elements (typically silicon) for radiation of energy below 200 keV. As, on the other hand, the density of lead-rich glasses is also significantly higher than that of soda-lime silica, it is not surprising that lead glasses are used as protection against ionizing radiation.

These properties of lead concerning the absorption of ionizing radiation have also been used for manufacturing cathode ray tubes and television tubes: the protection against X-rays is obviously essential, and, on the other hand, the influence of lead on glass viscosity facilitates the fabrication of these objects of complex shape.

1.1.13. *Miscellanea*

Lead glasses are known for having very low electrical conductivity: Pb^{2+} ion is relatively large (0.132 nm) and prevents the movement of alkaline ions, commonly responsible for electrical conduction in glasses. This property is useful for applications in lighting, cathode-ray tubes and electronics (Schoenung 2008).

On the other hand, the fact of having available a low viscosity glass, which moreover has a good adhesion to metals that serve as electrode, renders the lead glass very useful for welding, thus opening the way for applications in the field of electronics and cathode-ray tubes (Nordyke 1984, p. 33; Pfaender 1996, p. 138).

1.2. Difficulties related to the use of lead oxide

1.2.1. *Elaboration difficulties*

Despite the advantages resulting from the addition of lead to glass compositions, it was not before the end of the17th century that the large-scale development of lead glasses started. It is worth considering why. It is most certainly largely due to the fact that mastering the melting of this type of glass raised problems. The use of lead glasses was already widespread before those times, but only for small-scale manufacturing (Mecking 2013, p. 2; Velde 2013, p. 75): small-size objects or small-series objects, glazes, which did not require the elaboration of large amounts of glass. Indeed, the lead oxide in this type of glass is susceptible to be reduced quite readily to metallic lead under the action of a reducing agent. It may be a matter of particles of metallic iron accidentally introduced in the composition, or of carbon particles resulting from incomplete combustion when using open crucibles or any other reducer or even the action of the flame itself (Wang et al. 2004). These particles of metallic lead fall at the bottom of the crucible (Volf 1984, p. 452) and attack the refractory (Neri et al. 1752) or the platinum of the crucible, which may lead to the puncture of the pot: the glass then spreads on the ground. The result is a loss all the greater as the pot size is larger, without counting the risks related to the sudden presence of melting glass on the ground. This phenomenon of refractory puncture by the metallic lead is well known to today's glassmakers using household cullet resulting from the collection of glass to be recycled: several furnaces were punctured this way, suddenly releasing dozens of tons of glass at 1,400°C (in this case the origin of lead is attributed to the recycling of champagne bottles whose neck was protected by a lead-rich metallic film). The solution found in the 17th century was the introduction in the composition of a powerful oxidizer (potassium nitrate or saltpeter) and perhaps the use of closed pots (Watts1990) (Chapters 6 and 7).

1.2.2. *Public health problems*

The risks associated with the use of lead (lead poisoning) have been known for a long time, as they were signaled by Hippocrates and then by Pliny (Hynes and Jonson 1997). Lead acts mainly by ingestion and by inhalation. It has powerful neurological effects. But it was only starting with the beginning of the 20th century that the risks related to its use have been thoroughly evaluated, and increasingly strict safety rules have been implemented in the glass and ceramic plants using lead (Singer and Germain 1960; Eppler 1983). These rules concern first of all the choice of raw materials in order to minimize dusts, especially those having a certain solubility in acid medium, and the volatile compounds of lead during the elaboration of glasses: a temperature of elaboration of lead glazes below 1,170°C is recommended in order to limit this volatilization (Lehman 2002). It is important to note that as soon as lead enters in the vitreous matrix, considered as inert, it is far less susceptible to be released, except for specific use cases, for example if the glass in which it is integrated comes into contact with an acid solution. It was not until the 1970s that this risk was taken into account and reflected by the adoption of very strict rules (Ben Kacem 2017) (Chapters 13 and 14).

Finally, besides the health problems related to lead as such, a further risk may occur. Indeed, lead glasses are generally refined using arsenic in the form of As_2O_3 or antimony in the form of Sb_2O_3, associated with an oxidizer, generally potassium nitrate KNO_3 (Chopinet et al. 2002), as the classical pair sulfate/reducer can no longer be used. But arsenic and antimony are themselves extremely toxic.

1.3. Conclusion

It can be noted that the introduction of lead in the composition of glass or glaze adds many interesting properties. However, the difficulties related to its use for significant productions were not properly addressed until the end of the 17th century, and the related public health problems have already practically stalled its use for glazes and limited it for lead glasses.

1.4. References

Bansal, N.P. and Doremus, R.H. (1986). *Handbook of Glass Properties*. Academic Press, New York.

Barber, D.J., Freestone, I.C., Moulding, K.M. (2009). Ancient copper red glasses: Investigation and analysis by microbeam techniques. In *From Mine to Microscope*, Shortland, A.J., Freestone, I.C., Rehren, T. (eds). Oxbow Books, Oxford.

Barton, J. and Guillemet, C. (2005). *Le Verre, science et technologie*. EDP Sciences, Les Ulis.

Ben Kacem, I. (2017). Du cristal au plomb jusqu'aux déchets domestiques : rôle du plomb dans les verres et les vitrocéramiques. Étude des systèmes $PbO\text{-}SiO_2$, $PbO\text{-}Al_2O_3\text{-}SiO_2$ et $PbO\text{-}CaO\text{-}SiO_2$. PhD Thesis, Université Paris-Est, Paris.

Ben Kacem, I., Gautron, L., Coillot, D., Neuville, D.R. (2017). Structure and properties of lead silicate glasses and melts. *Chem. Geol.*, 461, 104–114.

Charleston, R.J. (1960). Lead in glass. *Archaeometry*, 3, 1–4.

Chopinet, M.-H., Lizarazu, D., Rocanière, C. (2002). L'importance des phénomènes d'oxydo-réduction dans le verre. *C.R. Chimie*, 5(12), 939–949.

Cohen, B.M., Uhlmann, D.R., Shaw, R.R. (1973). Optical and electrical properties of lead silicate glasses. *J. Non-Cryst. Solids*, 12(2), 177–188.

Cook, L. (1990). Chemical processes in glass polishing. *J. Non-Cryst. Solids*, 120, 152–171.

Davison, S. (2003). *Conservation of Glass*. Butterworth and Heinemann, Oxford.

Edwards, R.J., Paul, A., Douglas, R.W. (1972). Spectroscopy and oxidation-reduction of iron and copper in Na_2O-PbO-SiO_2 glasses. *Phys. Chem. Glasses*, 13(5), 131–136.

Eppler, R.A. (1983). Glazes and enamels. In *Glass: Science and Technology, Volume 1: Glass-Forming Systems*, Uhlmann, D.R., Kreidl, N.J. (eds). Academic Press, Boston.

Eppler, R.A. and Eppler, D.R. (2000). *Glazes and Glass Coatings*. The American Ceramic Society, Westerville.

Fluegel, A. (2007). Glass viscosity calculation based on a global statistical modeling approach. *Glass Technol.: Europ. J. Glass Sci. Technol. A*, 48(1), 13–30.

Freestone, I.C. (1987). Composition and microstructure of early opaque red glass. In *Early Vitreous Materials*, Bimson, M., Freestone, I. (eds). British Museum Occasional, London.

Geller, R.F. and Bunting, E.N. (1936). The system K_2O-PbO-SiO_2. *J. Research NBS*, 17, 277–289.

Hynes, M.J. and Jonson, B. (1997). Lead, glass and the environment. *Chem. Soc. Rev.*, 26, 133–146.

Kaczmarczyk, A. and Hedges, R.E.M. (1983). *Ancient Egyptian Faience: An Analytical Survey of Egyptian Faience from Predynastic to Roman Times*. Aris & Phillips, Warminster.

Kato, Y., Yamazaki, H., Yoshida, S., Matsuoka, J. (2010). Effect of densification on crack initiation under Vickers indentation test. *J. Non-Cryst. Solids*, 356, 1768–1773.

La Delfa, S., Ciliberto, E., Pirri, L. (2008). Behavior of copper and lead as chromophore elements in sodium silicate glasses. *J. Cult. Herit.*, 9, e117–e122.

Lehman, R.L. (2002). *Lead Glazes for Ceramic Foodware*. The International Lead Management Center Research, Research Triangle Park.

Leiser, C.F. (1963). Importance of lead in glass. *The Glass Industry*, 509–576, 594, 630–632.

Marker, A.J. and Neuroth, N. (1998). Overview-optical glass: An engineered material. In *The Properties of Optical Glass*, Bach, H., Neuroth, N. (eds). Springer-Verlag, Berlin-Heidelberg.

Mecking, O. (2013). Medieval lead glass in Central Europe. *Archaeometry*, 55(4), 640–662.

Molina, G., Odin, G.P., Pradell, T., Shortland, A.J., Tite, M.S. (2014). Production technology and replication of lead antimonate yellow glass from New Kingdom Egypt and the Roman Empire. *J. Archaeol. Sci.*, 41, 171–184.

Neri, A., Merret, C., Kunckel, J. (1752). *Art de la Verrerie*. Durand et Pissot, Paris.

Nordyke, J.S. (1984). *Lead in the World of Ceramics*. American Ceramic Society, Columbus.

Paynter, S. and Jackson, C. (2018). Clarity and brilliance: Antimony in colourless natron glass explored using Roman glass found in Britain. *Archaeol. Anthropol. Sci.*, 11, 1533–1551.

Pfaender, H.G. (1996). *Schott Guide to Glass*. Chapman & Hall, London.

Robertshaw, P., Benco, N., Wood, M., Dussubieux, L., Melchiorre, E., Ettahiri, A. (2010). Chemical analysis of glass beads from medieval Al-Basra (Morocco). *Archaeometry*, 52(3), 355–379.

Schoenung, J.M. (2008). Lead compounds. In *Ceramic and Glass Materials: Structure, Properties and Processing*, Shackelford, J.F., Doremus, R.H. (eds). Springer, New York.

Scholze, H. (1974). *Le Verre, nature, structure et propriétés*. Le Dû, Institut du Verre, Paris.

SCHOTT (2020). Optical Glass 2020 [Online]. Available at: https://www.schott.com/d/advanced_optics/c36214d9-13c4-468c-bf40-8d438b89f532/1.33/schott-optical-glass-pocket-catalog-row.pdf [Accessed 14 November 2020].

Shelby, J.E. (2005). *Introduction to Glass Science and Technology*. The Royal Society of Chemistry, Cambridge.

Shortland, A.J. and Eremin, K. (2006). The analysis of second millennium glass from Egypt and Mesopotamia. Part 1: New WDS analyses. *Archaeometry*, 48(4), 581–603.

Singer, F. and Germain, W.L. (1960). *Les glaçures céramiques*. Borax Français, Saint-Germain-en-Laye, Cie Productrice des Borates, Brussels.

Sovis Glassolutions (2020). Caractéristiques du verre [Online]. Available at: http://www.sovis-optique.com/fr/sovis-optique-caracteristiques-caracteristiques-du-verre.html [Accessed 31 December 2020].

Tite, M.S., Freestone, I.C., Mason, R., Molera, J., Vendrell-Saz, M., Wood, N. (1998). Review article. Lead glazes in antiquity – Methods of production and reasons for use. *Archaeometry*, 40(2), 241–260.

Turner, W.E.S. (1956). Studies in ancient glasses and glassmaking processes. Part III: The chronology of the glassmaking constituents. *J. Soc. Glass Techn.*, 40, 39–52.

Velde, B. (2013). Compositions over several millennia in the western world. In *Modern Methods for Analysing Archaeological and Historical Glass*, Jannsen, K. (ed.). Wiley, Chichester.

Verità, M. (1991). Some technical aspects of ancient venetian glass. In *Technique et Science, Les arts du verre. Actes du colloque de Namur*, Namur, 20–21 October.

Verità, M., Bjornholt, B., James, L., Freestone, I., Henderson, J., Nenna, M.-D., Schibille, N. (2009). Glossary of mosaic glass terms [Online]. Available at: https://www.researchgate.net/publication/323907695_Glossary_of_Mosaic_Glass_Terms/references [Accessed 8 January 2021].

Volf, M.B. (1984). *Chemical Approach to Glass*. Elsevier, Amsterdam.

Wang, C., Krausch, G., Rädlein, E., Tratzky, S., Schramm, M., Weber, A. (2004). Studies of surface changes on industrial glasses with AFM, FE-SEM, EDX, SNMS and LM. Part 2: Surface changes by water drop etching, annealing at atmosphere and flame treatment. *Glass Sci. Technol.*, 77(6), 273–282.

Watts, D.C. (1990). Why George Ravenscroft introduced lead oxide into crystal glass. *Glass Technol.*, 31(5), 208–212.

Zarzycki, J. (1982). *Le verre et l'état vitreux*. Masson, Paris.

2

Lead Isotopes for the Study of Ancient Glass

Alicia VAN HAM-MEERT and Patrick DEGRYSE
KU Leuven, Belgium

2.1. Lead isotope chemistry

The element lead (Pb, from Latin: *plumbum*) has four common isotopes ^{204}Pb, ^{206}Pb, ^{207}Pb and ^{208}Pb, three of which (^{206}Pb, ^{207}Pb and ^{208}Pb) have a radiogenic origin. Lead-208 is the end product of a decay-chain starting from ^{232}Th, lead-206 from a chain starting with ^{238}U and lead-207 from a chain starting with ^{235}U.

In lead isotopic analysis, we are interested in the relative abundance of the different isotopes. Originally, lead ratios were only expressed as $^{207}Pb/^{206}Pb$ and $^{208}Pb/^{206}Pb$ (Nord and Billström 2018), as measurements of the low-abundance non-radiogenic lead-204 isotope were difficult. With analysis methods becoming increasingly sensitive, lead ratios are now measured and expressed with respect to ^{204}Pb, the only non-radiogenic isotope. Relative differences in isotopic composition between natural materials are caused by different ages of deposition and/or original uranium and thorium content.

2.2. The use of lead isotopes in archaeology

Pb is a heavy element with small relative mass differences between the isotopes. Because of this, no measurable fractionation of the isotopes of lead occurs during processing such as roasting, melting, casting and cupellation (Keisch 1970; Brill and

Lead in Glassy Materials in Cultural Heritage,
coordinated by Anne BOUQUILLON and Patrice LEHUÉDÉ.

Shields 1972, pp. 272–303; Stos-Gale and Gale 2009). Hence, the lead isotopic composition of an object containing Pb reflects the isotopic composition of the raw materials used during its manufacture (Stos-Gale and Gale 2009). This has been the major principle behind the use of lead isotopes in archaeology: the isotopic composition of an artifact is identical to the raw material from which it was derived, while the signatures of different raw materials may differ.

As early as 1970 variations in isotopic composition were used to determine the source of materials used by artists (Keisch 1970). According to this paper, isotopic analysis can be used for three different purposes: (1) looking at supply sources over time, (2) understanding exchange and relations between source, processing and distribution/cultural centers, and (3) discerning whether paintings are authentic and/or being able to identify them (Keisch 1970).

The main type of application and the most straightforward one has always been provenancing (Brill and Shields 1972). In the *Handbook of Archaeological Sciences*, Wilson and Pollard (2001, pp. 507–518) contributed a chapter entitled "The Provenance Hypothesis", tackling the chemical investigation of inorganic archaeological materials for provenancing purposes. In order to be able to provenance such materials, they note several requirements, some of which are no longer useful in the current paradigm (and which were later revised; Pollard et al. 2014). Here, we cite those still of importance today.

A chemical characteristic is used, which must be transmitted from raw materials to the finished product either unaltered or changed in a predictable way.

The difference in this chemical characteristic between potential raw material sources (inter-source variability) is larger than the variability within one single source (intra-source variability). This requires a good characterization of potential sources (more than one analysis per source and sufficient precision to discriminate between sources).

A match between an object and a mine, or a lead isotope province, does not necessarily entail that this is the origin of the raw material used. Indeed, raw material sources can have the same isotopic signature, such as the silver mines from Spain and Wales, or from Iran and Lavrion (Brill and Wampler 1967; Stos-Gale and Gale 2009). Provenance can, nevertheless, sometimes be achieved by combining isotopic information with other strands of evidence such as trace element composition or archaeological information. However, it is only possible to exclude a source with absolute certainty. Even if all analytical data match only one known source, there may always be another source, as of yet undiscovered, which has exactly the same properties. The common expression is therefore to say that a material is "consistent with a single source", rather than to assign it to this source with certainty. How "being consistent with a single source" is then interpreted and translated to human behavior depends on all available archaeological information, context and dating. Simply stating that X likely comes from Y is of little archaeological interest and even of little scientific interest. Only translation to human/societal behavior or choice adds to our understanding of ancient societies. Pollard et al. (2014) also note that when looking at provenance, we should take into account that between extraction/production at place Y and use/deposition at place X the object/raw material might have travelled and/or have been used in various places and over an extended period of time. This type of object biographical knowledge is often beyond our reach, and in those cases should remain a philosophical framework to keep in mind.

Lead isotopes have far more potential than provenancing alone. They can be used to distinguish groups or objects, to look at continuity in supply or to look at mixing and recycling within an assemblage. These questions have been at the heart of lead isotopic analysis in the last few years of research.

Interpreting lead isotope ratios necessitates knowing how lead entered the finished product. For example, in a copper object lead might enter with the copper source and hence be a tracer for it, but lead can also be added separately as an alloying element, which means the lead isotopic signature then is a tracer for the lead source.

As often in glass studies, Robert H. Brill and collaborators pioneered the use of lead isotopes, analyzing a variety of material such as Egyptian glass (Brill et al. 1974, pp. 9–27) or Japanese and Chinese glass (Brill et al. 1979). Another well-known geochemist, Karl Hans Wedepohl, used lead isotopes to study mainly medieval glass (Wedepohl and Baumann 2000).

In glass studies, lead isotopic analysis is the method which, compared to other isotope systems, most often yields results which cannot easily be interpreted in light of the archaeological context or other sample information (Huisman et al. 2017; Van Ham-Meert et al. 2019). Therefore, other isotopic systems used in vitreous materials research, together with the information they bring, are briefly discussed at the end of this chapter.

2.3. Lead isotopic analysis of glassy material

A number of instruments allow determining the lead isotopic composition of glass, and different sampling methods exist. These are summarized in Table 2.1.

Furthermore, other measurement methods exist but have not (yet) been applied in an archaeological context such as laser desorption/ionization mass spectrometry (LDI MS) and matrix-assisted laser desorption/ionization time-of-flight mass spectrometry (MALDI TOF MS) (Kamčeva et al. 2017) or laser ablation sector field inductively coupled plasma mass spectrometry (LA-SF-ICP-MS) (Ghazi and Millette 2004). More examples of lead isotopic analysis of glasses are discussed in Chapter 8 by Nadine Schibille.

The most common introduction systems (LA and nebulization of a solution of isolated Pb) and instrument (multi-collector–inductively coupled plasma–mass spectrometry [MC-ICP-MS]) are described in this section.

Sample preparation	Sample introduction	Instrumentation	Reference
1. Dissolution 2. Isolation of the target element	(Tl-spiked) solution	Multi-collector inductively coupled plasma mass spectrometry (MC-ICP-MS)	Van Ham-Meert et al. 2018; Huisman et al. 2017
Mounted pieces	Laser ablation	MC-ICP-MS	Van Ham-Meert et al. 2018
		ICP-dynamic reaction cell-MS	Resano et al. 2008
		ICP-quadrupole mass spectrometry	Marzo et al. 2007
• Mounted piece • Powder • Solution	Placed on a tip	Fast atom bombardment MS (FAB-MS)	Miller et al. 1986
Mounted piece covered with a gold layer	Sputtering	Ion microanalysis (IMA)–secondary ion mass spectrometry (SIMS)	Maeda et al. 1985; Medeghini et al. 2020

Table 2.1. *Summary of different measurement methods for determining lead isotopic composition in vitreous materials and their application*

2.3.1. *Sample introduction*

2.3.1.1. *Pneumatic nebulization*

Pneumatic nebulization is the most common introduction system for (MC)-ICP-MS. This requires the complete dissolution of the sample. As glass is resistant to acid and base attacks, dissolving it can be challenging. First, approximately 50–200 mg of glass is taken and finely crushed in an agate mortar. Often a combination of heat (130–220°C) and acid treatment such as aqua regia (1:3 $HCl:HNO_3$), hydrofluoric acid (HF) and/or perchloric acid ($HClO_4$) is used to achieve complete dissolution. After complete dissolution, lead must be isolated from the matrix. This is achieved through ion-exchange chromatography for which two types of resin are commonly used, either Sr spec or Pb spec resin. The isolated Pb is then taken up in a 3% HNO_3 solution and is ready for analysis.

2.3.1.2. *Laser ablation*

An alternative to solubilizing part of the sample is to use laser ablation. This can be seen as less invasive than taking a sample; however, it leads to a 10-fold decrease in precision. This reduction in precision is attributed to matrix effects in the plasma, introducing not just the target element, but all of the sample. Furthermore, sample and reference material (RM) are less alike than when a purified lead solution of known concentration is used. Finally, the sample holder for the laser is usually relatively small and a piece of glass has to be mounted in resin and polished down for obtaining an optimal signal. This mounted piece can then of course be used for subsequent analysis whether elemental or structural (such as Raman spectroscopy). Laser spot sizes of 100–200μm are used and long lines ablated to have enough measurement data.

2.3.2. *Ion detection MC-ICP-MS*

In MC-ICP-MS analysis, an aerosol (either the solution or from the laser) is introduced and ionized in the plasma. The ions (charged atoms) are accelerated through a strong potential toward the cups where they are counted. Along the road to the cups, a strong magnetic field bends the path of the different ions. The exact path depends on the mass to charge ratio (in short m/z), which is different for different isotopes. In static mode, the most common mode in MC-ICP-MS, all the isotopes are measured simultaneously to yield the best results.

2.3.3. *Data reduction and mass bias correction*

Two main factors have to be accounted for in this mathematical step. First, a blank subtraction is performed to remove the background signals or blank contributions on each peak. Next, potential interferences from other elements with the same m/z are corrected for. In the case of lead, there is only one interference ^{204}Hg on ^{204}Pb. In glass samples, the presence of mercury is rarely attested; however, traces can still be present from previous analysis on the same machine. To correct for any ^{204}Hg contribution, ^{202}Hg is monitored and the contribution of mercury on m/z 204 is estimated from this peak and the natural $^{204}Hg/^{202}Hg$ ratio. This correction is important as ^{204}Pb is the least abundant and only non-radiogenic lead isotope and any error on this can influence the final ratios significantly.

The main correction needed is fractionation. This term refers to the combination of instrumental and matrix effects, which contribute to the preferential ionization and/or transmission of heavier isotopes. This is especially important in the analysis of light elements where the relative mass difference is high. To correct for this mass discrimination, different methods can be applied and/or combined. Either a Tl-solution with a known $^{205}Tl/^{203}Tl$ is added or certain isotopes of lead are spiked. The first method is often preferred; this is the "internal standard". Furthermore, appropriate RMs are also included in the analysis: NIST SRM 981 for pneumatic nebulization, a glass standard such as Corning D or GSE-1G for laser ablation analysis; those are the "external standards".

Different mathematical models can be used for correction (i.e. mathematical laws); these are summarized in Table 2.2. R_{ij} is the corrected ratio, r_{ij} is the measured ratio, M_i and M_j are the mass of isotopes i and j, and g and β are coefficients. Russell's law and the power law can be used both for internal and external correction; the algebraic and geometric methods are only used for external correction in sample standard bracketing way where $R_{ij} = K_{ij}r_{ij}$ is the equation used for correction and the correction coefficient (K_{ij}) is determined using RMs as shown in Table 2.1.

Law	Mathematical formula
Russell's law	$R_{ij} = r_{ij}g^{M_i - M_j}$
Power law	$R_{ij} = r_{ij}\left(\frac{M_i}{M_j}\right)^{\beta}$
Algebraic	$K_{ij} = \frac{2 * R_{ij,RM}}{r_{ij,RM1} + r_{ij,RM2}}$
Geometric	$K_{ij} = \frac{R_{ij,RM}}{\sqrt{r_{ij,RM1} * r_{ij,RM2}}}$

Table 2.2. *Mathematical models used for mass bias correction*

2.3.3.1. *Internal correction*

Internal correction methods rely on monitoring a reference isotope ratio during each measurement run, either Tl or the lead spike. The value of the coefficient in the equations of Russell's law or the Power law is determined for those internal RMs and used to correct the isotopes of interest. This is thus based on the assumption that the internal reference and the target isotopes fractionate similarly.

2.3.3.2. *External correction*

External correction methods rely on measuring RMs alongside the samples. Usually, this is done in a sample standard bracketing fashion (i.e. alternation RMs and samples). The mathematical correction can either be based on the algebraic or geometric method, or using the values from the standard to determine the coefficients in Russell's law and the Power law.

2.3.3.3. *Mixed correction*

Finally, a combination of internal and external correction can be used where both an internal reference is added and external RMs are measured. Either first the internal and then the external correction is performed. Or, most commonly, the Woodhead method is applied. In this method, the "g"-coefficient of Russell's law is determined for both the internal RM and the target isotope ratio in the external standard. Those values are plotted, usually on a straight line, and an empirical mathematical equation is determined to express the g-coefficient of the target isotope as a function of the g-coefficient of the internal reference. For the samples, then the g-coefficient of the target isotope ratio is calculated based on the g-coefficient of the internal RM during that run. This method yields the best results in lead isotope analysis.

2.3.4. *Interpretation of results*

Once the lead isotopic composition is known, this needs to be interpreted in a meaningful way to address the archaeological question at hand.

2.3.4.1. *Lead as an ingredient of glass*

In order to be able to correctly interpret lead isotope data, it is important to know how lead entered the glass batch. Ultimately, lead isotopes only say something about the raw material it entered with, not about the other ingredients. This means elemental analysis is needed prior to lead isotope analysis. A combination of knowledge on the concentration of lead and the type of glass will allow users to know whether lead was added on purpose (e.g. in lead glaze, crystal or high-lead glass (such as the Islamic glass discussed in Chapter 8 by Nadine Schibille), lead-opacified or lead-colored glass) or incidentally (with the colorant such as Cu or Co, as part of the flux or silica source, etc.). In Figure 2.1, a summary is provided of the source of different elements for which different isotope systems are useful in glass analysis. It illustrates that lead can enter glass through many different pathways. Therefore, a good understanding of the type of glass being traced is needed.

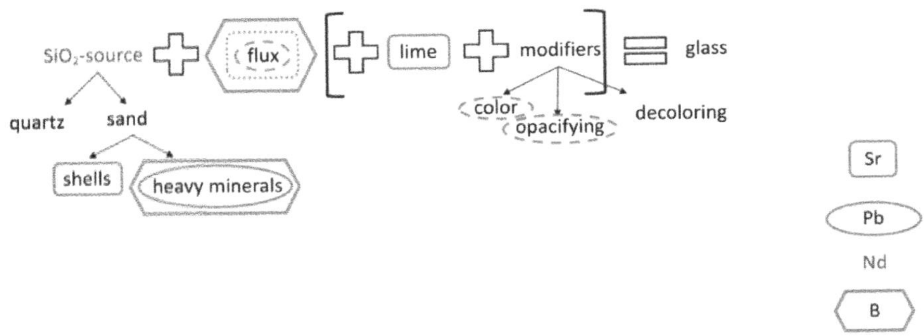

Figure 2.1. *Sources of different elements in glass. Full lines are main sources for an element, and dashed lines represent possible sources for an element (modified after Van Ham-Meert et al. 2018)*

2.3.4.2. *Mixing and recycling*

Another important factor in glass studies is the fact that glass is often recycled, remelted and mixed. Such mixing of different glass batches will affect the isotopic composition of the end product, which will be a weighted mean of the composition in the original components. Mixing and recycling cannot always be detected but elevated trace element concentrations, or mixed alkali or colorant/opacifier contents are strong indicators of recycling.

2.3.4.3. *Reference databases*

Ultimately, in order to provenance lead in glass, references are needed to compare the isotopic signature. For Sr isotope analysis, large isoscape maps and tools for modeling provenance exist, but this is not (yet) the case for lead isotope analysis, though few databases exist. For the Mediterranean and near east, the OXALID (Oxford Archaeological Lead Isotope Database) database (http://oxalid.arch.ox.ac.uk/, Stos-Gale and Gale 2009) provides a wealth of information. Blichert-Toft et al. (2016) mapped the lead isotopic composition of lead ores and K-feldspars in Europe and the Western Mediterranean. Hsu and

Sabatini (2020) recently published a lead isotope database for lead ores in China, especially for provenancing archaeological materials. Slowly more databases are becoming openly available, and as more archaeological objects are analyzed, the corpus against which isotopes can be compared increases, thus increasing the possibility to interpret the data.

2.3.4.4. *Lead isotope provinces*

As explained in the introduction when trying to provenance using geochemical data, we can never prove lead comes from a certain place; we can only exclude a number of places. The main challenge is that ores of geologically similar age have similar lead isotopic compositions. Based on this, Killick et al. (2020) argue that provenancing in Europe is hard because most ores, even far from each other, have the same isotopic signature, whereas places with many orogenesis events at different points in time (such as South Africa) allow more detailed provenancing.

Interpreting isotopic data in archaeology never happens in a vacuum; the historical and archaeological information around the object help to constrain the information obtained through isotopic analysis, hence allowing for more detailed information. Also, combining lead isotope data with information from other isotope systems (Sr, Nd, B, etc.) offers more options for obtaining useful information.

2.3.4.5. *Modeling*

In terms of dealing in a mathematically correct way with lead isotope ratios and the possible match between signatures of artifacts and possible raw materials, that is, "ore fields", the use of kernel statistics has been put forward (Pollard 2009, pp. 181–189). A method interpreting lead isotope ratios by calculating the definite integral under the kernel density estimate plot of different mining districts, using open-access software and legacy data, has recently been published (De Ceuster and Degryse 2020). A match with the reference data set may then indicate the true origin, while no match indicates an unknown origin, that is, not present in the data set of mineral resources. Moreover, mixing of metals and possible recycling can be inferred, and approached in more quantitative ways, as can the composition of materials with possible multiple lead sources in their raw material mixture. This approach also offers a graphical way in which to assess the correspondence of an artifact to several mining districts, such as with the use of biplots, but based on more sound mathematical grounds.

2.4. O, Sr, Nd and B isotopes for studying archaeological glass

2.4.1. *Sources of different elements in glass*

Other isotope systems have been used to investigate the provenance of raw materials and/or to look at recycling and mixing in glass making. Different systems allow insights into different raw materials (Degryse and Shortland 2020). Some enter the vitreous material together through the same raw material (Sr, B and Pb can all be present in the flux), and some are specific for one raw material (e.g. Nd as a tracer for the silica source).

2.4.2. *Sr isotopes in glass*

Apart from O and Pb, Sr isotopes have been used longest in glass isotope studies (Freestone et al. 2003; Henderson et al. 2005; Degryse et al. 2010; Brems et al. 2013a). Sr can substitute for calcium in most minerals and therefore is a tracer for the lime source in the glass. This can be either as a part of shell or limestone in sand, the lime in plant ash used for late Bronze Age glass making or in wood ash used for producing medieval stained glass windows, or even intentionally, separately added lime. The key to knowing what Sr is a tracer for is knowing the type of glass under study, that is, knowing the major elemental composition. Figure 2.2 provides a quick way to discriminate what Sr-isotopes are tracing for based on the chemical composition.

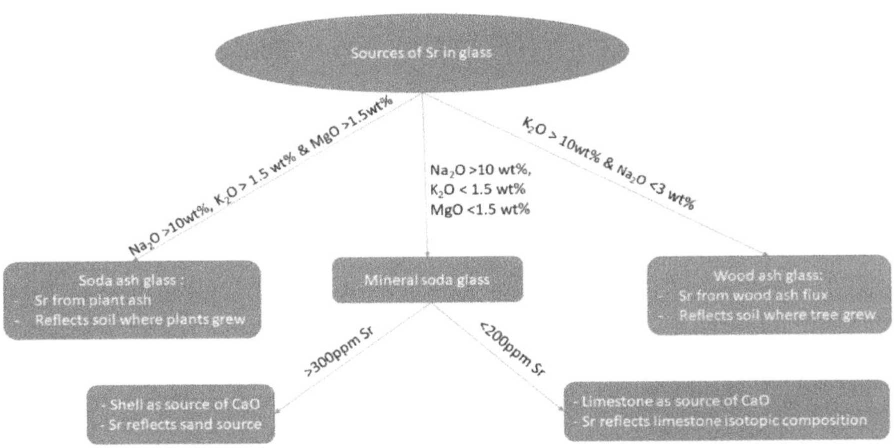

Figure 2.2. *Flowchart allowing to decide what raw material the Sr in glass is tracing for (after Van Ham-Meert 2019)*

Sr has four isotopes ^{84}Sr, ^{86}Sr, ^{87}Sr and ^{88}Sr. The isotopic composition of bedrock or a deposit depends on the original ^{87}Rb concentration (the parent for ^{87}Sr) and the age of the deposit. In archaeometry and geology, the isotope ratio used for tracing is $^{87}Sr/^{86}Sr$.

2.4.3. *Nd isotopes in glass*

Nd is present as a trace element in glass; usually only 4–10 ppm is detected. Nd enters the vitreous material with the silica source. Seven isotopes of Nd are known in nature: ^{142}Nd, ^{143}Nd, ^{144}Nd, ^{145}Nd, ^{146}Nd, ^{148}Nd and ^{150}Nd. The only radiogenic isotope, ^{143}Nd, is formed through alpha decay from ^{147}Sm. In geochemical studies, the isotopic ratio of interest is $^{143}Nd/^{144}Nd$. This ratio depends on the age of the rock and the initial Sm/Nd ratio. Because of the small difference in isotopic abundances between different deposits, Nd isotopes are reported as ε_{Nd} compared to the CHondritic Uniform Reservoir (CHUR) ($^{143}Nd/^{144}Nd$ = 0.512638) standard according to the following formula:

$$\varepsilon_{Nd} = \left(\frac{\left(\frac{^{143}Nd}{^{144}Nd}\right)_{sample}}{\left(\frac{^{143}Nd}{^{144}Nd}\right)_{CHUR}} - 1 \right) * 10{,}000$$

Nd isotopic analysis of glass is often performed in combination with Sr isotopic analysis; one providing information on the silica source, and the other on the flux source (Degryse et al. 2010; Brems et al. 2013b).

2.4.4. *B isotopes in glass*

Boron isotopes have been used in a limited number of studies (Devulder et al. 2014; Devulder et al. 2015; Van Ham-Meert et al. 2019). Boron is a light element and has two isotopes: ^{11}B and ^{12}B. In boron, isotopic studies results are expressed in permil (‰) relative to NIST SRM 951 whose signature is $^{11}B/^{10}B$ = 4.0436 (Catanzaro et al. 1970, p.70) using the following formula:

$$\delta^{11}B = \left(\frac{\left(\frac{^{11}B}{^{10}B}\right)_{sample}}{\left(\frac{^{11}B}{^{10}B}\right)_{NIST\,SRM951}} - 1 \right) * 1{,}000$$

Boron enters natron glass with the flux (i.e. the source of natron); there is also a contribution of the silica source, which lowers the isotopic signature (Devulder et al. 2014). The boron isotopic signature of different natron sources such as the Wadi Natrun in Egypt and lake Pikrolimni in Greece is different (29.8‰ and 10‰, respectively), making it an effective discriminator between both possible provenances. Devulder et al. (2014) also showed that despite the low mass and high relative mass difference between both isotopes, the high-temperature processing involved in glass making does not lead to fractionation.

2.4.5. *O isotopes in glass*

Another isotope system that has not received much attention in glass studies are oxygen isotopes (Henderson et al. 2005; Degryse et al. 2010; Blomme et al. 2016), though this is one of the earliest systems applied to vitreous materials studies (Brill 1970; Brill et al. 1999, pp. 303–322). Since most elements enter the glass as an oxide, the oxygen isotopic composition is a mixture signal of all the different raw materials. It is actually a weighed mean of the oxygen isotope contributions from all materials (Leslie et al. 2006; Blomme et al. 2017). It is most useful in combination with other isotopic systems such as Sr and Nd. Oxygen has three isotopes, ^{16}O, ^{17}O and ^{18}O, and the ratio of interest is $^{18}O/^{16}O$. Oxygen isotopic compositions are reported as $\delta^{18}O$, the per mil difference with respect to the Vienna Standard Mean Ocean Water (VSMOW). One of the main drawbacks of using oxygen isotopes for provenancing is the large number of sources with near-identical signatures.

2.5. Conclusion and future perspectives

The use of Pb isotopes for the study of glass is not as widespread as its use for the study of metallic archaeological materials. Sr and Nd isotopes are often the preferred isotopic systems for the study of ancient glass. However, lead is very useful as it provides insight on different raw materials and processes than Sr and Nd do. Therefore, it is believed its use can be increased in the years to come. Some caution, however, is needed as with any analytical technique. Lead isotopes are very reliable to exclude a possible source, but on their own they can never pinpoint a particular source. We also strongly believe that archaeometric studies and in particular lead isotopic analysis should strive to go beyond mere provenancing. Provenancing is not always possible, but looking at changes in supply, mixing, recycling, etc., yields precious information on technology and trade and can be fitted in the wider picture. This is only possible if different streams of information are brought together to answer archaeological questions.

Lead isotopic analysis is not always advised and can be costly. If elemental analysis is sufficient to answer the archaeological question asked, this is by far the preferred way, avoiding the production of data for its own sake. However, lead isotope ratios can be useful in distinguishing, for example, production processes (lead isotopes revealed the use of two batches of glass to produce bi-layered art nouveau glass (Schultheis et al. 2004) or to group glasses when the elemental composition does not allow this (e.g. Schultheis et al. 2004). Also, lead isotope analysis is still useful when lead is (part of) a deliberately added raw material and as an indicator for recycling. We suggest that, especially in the study of medieval stained glass windows, it might be interesting to compare the lead from the glass with the lead frame of the stained glass, and the lead contained in the grisaille (painting on the stained glass) or in contemporary paint. Lead isotopes may be useful when investigating the interrelation between metallurgy and glass working to see if colorants used in glass and metallurgical products have the same provenance. This is especially useful in the study of complete material assemblages from particular sites.

2.6. Acknowledgments

Our research into lead isotopes was financially supported by the Flemish Research Foundation (FWO Vlaanderen) under project no. G0C4315N and the KU Leuven Special Research Fund C14/19/060.

2.7. References

Blichert-Toft, J., Delile, H., Lee, C.-T., Stos-Gale, Z., Billström, K., Andersen, T., Hannu, H., Albarède, F. (2016). Large-scale tectonic cycles in Europe revealed by distinct Pb isotope provinces. *Geochem. Geophys.*, 17, 3854–3864.

Blomme, A., Elsen, J., Brems, D., Shortland, A., Dotsika, E., Degryse, P. (2016). Tracing the primary production location of core-formed glass vessels, Mediterranean Group I. *Journal of Archaeological Science*: *Reports*, 5, 1–9.

Blomme, A., Degryse, P., Dotsika, E., Ignatiadou, D., Longinelli, A., Silvestri, A. (2017). Provenance of polychrome and colourless 8th–4th century BC glass from Pieria, Greece: A chemical and isotopic approach. *J. Archaeol. Sci.*, 78, 134–146.

Brems, D., Ganio, M., Latruwe, K., Balcaen, L., Carremans, M., Gimeno, D., Silvestri, A., Vanhaecke, F., Muchez, P., Degryse, P. (2013a). Isotopes on the beach. Part 1: Strontium isotope ratios as a provenance indicator for lime raw materials used in Roman glass-making. *Archaeometry*, 55, 214–234.

Brems, D., Ganio, M., Latruwe, K., Balcaen, L., Carremans, M., Gimeno, D., Silvestri, A., Vanhaecke, F., Muchez, P., Degryse, P. (2013b). Isotopes on the beach. Part 2: Neodymium isotopic analysis for the provenancing of Roman glass-making. *Archaeometry*, 55, 449–464.

Brill, R.H. (1970). Lead and oxygen isotopes in ancient objects. *Philos. Trans. Roy. S. London*, 269, 143–164.

Brill, R.H. and Shields, W.R. (1972). Lead isotopes in ancient coins. In *Methods of Chemical and Metallurgical Investigation of Ancient Coinage*, Hall, E.T., Metcalf, D.M. (eds). Royal Numismatic Society, London.

Brill, R.H. and Wampler, J.M. (1967). Isotope studies of ancient lead. *American Journal of Archaeology*, 71(1), 63–77.

Brill, R.H., Barnes, I.L., Adams, B. (1974). Lead isotopes in some ancient Egyptian objects. In *Recent Advances in Science and Technology of Materials 3*, Bishay, A. (ed.). Plenum, New York.

Brill, R.H., Yamasaki, K., Lynus Barnes, I., Rosman, K.J.R., Diaz, M. (1979). Lead isotopes in some Japanese and Chinese glasses. *Ars Orientalis*, 11, 87–109.

Brill, R.H., Clayton, R.N., Mayeda, T.K., Stapleton, C.P. (1999). Oxygen isotope analyses of early glasses. In *Chemical Analyses of Early Glasses I*, Brill R.H. (ed.). Corning Museum of Glass, Corning.

Catanzaro, E.J., Champion, C.E., Garner, E.L., Marinenko, G., Sappenfield, K.M., Shields, W.R. (1970). *Boric Acid; Isotopic, and Assay Standard Reference Materials.* Institute for Materials Research National Bureau of Standards, Washington.

De Ceuster, S. and Degryse, P. (2020). A "match–no match" numerical and graphical kernel density approach to interpreting lead isotope signatures of ancient artefacts. *Archaeometry*, 62(S1), 107–116.

Degryse, P. and Shortland, A.J. (2020). Interpreting elements and isotopes in glass: A review. *Archaeometry*, 62(S1), 117–133.

Degryse, P., Boyce, A., Erb-Satullo, N., Eremin, K., Kirk, S., Scott, R., Shortland, A.J., Schneider, J., Walton, M. (2010). Isotopic discriminants between late Bronze Age glasses from Egypt and the Near East. *Archaeometry*, 52(3), 380–388.

Devulder, V., Vanhaecke, F., Shortland, A., Mattingly, D., Jackson, C., Degryse, P. (2014). Boron isotopic composition as a provenance indicator for the flux raw material in Roman natron glass. *Journal of Archaeological Science*, 46, 107–113.

Devulder, V., Gerdes, A., Vanhaecke, F., Degryse, P. (2015). Validation of the determination of the B isotopic composition in Roman glasses with laser ablation multi-collector inductively coupled plasma-mass spectrometry. *Spectrochimica Acta – Part B Atomic Spectroscopy*, 105, 116–120.

Freestone, I.C., Leslie, K.A., Thirlwall, M., Gorin-Rosen, Y. (2003). Strontium isotopes in the investigation of early glass production: Byzantine and early islamic glass from the near east. *Archaeometry*, 45(1), 19–32.

Ghazi, A.M. and Millette, J.R. (2004). Environmental forensic application of lead isotope ratio determination: A case study using laser ablation sector ICP-MS. *Environ. Forensics*, 5(2), 97–108.

Henderson, J., Evans, J.A., Sloane, H.J., Leng, M.J., Doherty, C. (2005). The use of oxygen, strontium and lead isotopes to provenance ancient glasses in the Middle East. *J. Archaeol. Sci.*, 32(5), 665–673.

Hsu, Y.-K. and Sabatini, B.J. (2019). A geochemical characterization of lead ores in China: An isotope database for provenancing archaeological materials. *PLoS One*, 14(4): e0215973.

Huisman, D.J., van der Laan, J., Davies, G.R., van Os, B.J.H., Roymans, N., Fermin, B., Karwowski, M. (2017). Purple haze: Combined geochemical and Pb-Sr isotope constraints on colourants in Celtic glass. *J. Archaeol. Sci.*, 81, 59–78.

Kamčeva, T.T., Nešic, M.D., Stoiljkovic, M.M., Popovic, I.A., Miletic, J.N., Rajcic, B.M., Petković, M.Ž., Veličković, S.R. (2017). Determination of isotopic distribution of lead by a matrix assisted laser desorption/ionization versus a laser desorption/ionization time of flight mass spectrometry. *Hemijska Industrija*, 71(1), 19–26.

Keisch, B. (1970). On the use of isotope mass spectrometry in the identification of artists' pigments. *Studies in Conservation*, 15(1), 1–5.

Killick, D.J., Stephens, J.A., Fenn, T.R. (2020). Geological constraints on the use of lead isotopes for provenance in archaeometallurgy. *Archaeometry*, 62(S1), 86–105.

Leslie, K.A., Freestone, I.C., Lowry, D., Thirlwall, M. (2006). The provenance and technology of Near Eastern glass: Oxygen isotopes by laser fluorination as a complement to strontium. *Archaeometry*, 48, 253–270.

Maeda, S., Takase, J., Sugitani, Y., Nagashima, K. (1985). Determination of lead isotope ratios in archeological products by ion microanalyzer. *Anal. Sci.*, 1(3), 227–230.

Marzo, P., Laborda, F., Pérez-Arantegui, J. (2007). A simple method for the determination of lead isotope ratios in ancient glazed ceramics using inductively coupled plasma – Quadrupole mass spectrometry. *At. Spectrosc.*, 28(6), 195–201.

Medeghini, L., Fayek, M., Mignardi, S., Coletti, F., Contino, A., De Vito, C. (2020). A provenance study of Roman lead-glazed ceramics using lead isotopes and secondary ion mass spectrometry (SIMS). *Microchem. J.*, 154, 104519.

Miller, J.M., Jones, T.R.B., Kenney, T., Rupp, D.W., Green, B.N., Bardoli, R.S. (1986). A fast atom bombardment study of the lead isotope ratios in early nineteenth century Niagara peninsula pottery glazes. *Can. J. Chem.*, 64(3), 488–491.

Nord, A.G. and Billström, K. (2018). Isotopes in cultural heritage: Present and future possibilities. *Herit. Sci.*, 6(25), 1–13.

Pollard, A.M. (2009). What a long strange trip it's been: Lead isotopes in archaeology. In *From Mine to Microscope*, Shortland, A.J., Freestone, I., Rehren, T., Tite, M.S. (eds). Oxbow Books, Oxford.

Pollard, A.M., Bray, P.J., Gosden, C. (2014). Is there something missing in scientific provenance studies of prehistoric artefacts? *Antiquity*, 88(340), 625–631.

Resano, M., Marzo, P., Pérez-Arantegui, J., Aramendía, M., Cloquet, C., Vanhaecke, F. (2008). Laser ablation-inductively coupled plasma-dynamic reaction cell-mass spectrometry for the determination of lead isotope ratios in ancient glazed ceramics for discriminating purposes. *J. Anal. At. Spectrom.*, 23(9), 1182–1191.

Schultheis, G., Prohaska, T., Stingeder, G., Dietrich, K., Jembrih-Simbürger, D., Schreiner, M. (2004). Characterisation of ancient and art nouveau glass samples by Pb isotopic analysis using laser ablation coupled to a magnetic sector field inductively coupled plasma mass spectrometer (LA-ICP-SF-MS). *J. Anal. At. Spectrom.*, 19(7), 838–843.

Stos-Gale, Z. (2004). Lead-isotope analyses of glass, glazes, and some metal artifacts. In *Serçe Limani: An Eleventh Century Shipwreck. Volume I: The Ship and Its Anchorage, Crew, and Passengers*, Bass, G.F., Matthews, S.D., Steffy, J.R., Van Doorninck, F.H. (eds). Texas University Press, College Station.

Stos-Gale, Z. and Gale, N.H. (2009). Metal provenancing using isotopes and the Oxford archaeological lead isotope database (OXALID). *Archaeol. Anthropol. Sci.*, 1(3), 195–213.

Van Ham-Meert, A., Chernonozhkin, S.M., Van Malderen, S.J.M., Van Acker, T., Vanhaecke, F., Degryse, P. (2018). Assessment of nanosecond laser ablation multi-collector inductively coupled plasma-mass spectrometry for Pb and Sr isotopic determination in archaeological glass: Mass bias correction strategies and results for corning glass reference materials. *Geostand. Geoanalytical Res.*, 42(2), 223–238.

Van Ham-Meert, A., Dillis, S., Blomme, A., Cahill, N., Claeys, P., Elsen, J., Eremin, K., Gerdes, A., Steuwe, C., Roeffaers, et al. (2019). A unique recipe for glass beads at iron age Sardis. *J. Archaeol. Sci.*, 108, 104974.

Wedepohl, K.H. and Baumann, A. (2000). The use of marine molluskan shells for Roman glass and local raw glass production in the Eifel area (Western Germany). *Naturwissenschaften*, 87(3), 129–132.

Wilson, L. and Pollard, A.M. (2001). The provenance hypothesis. In *Handbook of Archaeological Science*, Brothwell, D.R., Pollard, A.M. (eds). Wiley, New York.

PART 2

Structure of Lead Glasses: Influence on their Properties, Including Color

3

Structure and Properties of Lead Silicate Glasses

Daniel CAURANT[1], Gilles WALLEZ[2], Odile MAJÉRUS[1], Gauthier ROISINE[3] and Thibault CHARPENTIER[4]

[1]*IRCP, CNRS, Chimie ParisTech, France*

[2]*IRCP, Sorbonne University, Paris, France*

[3]*BUCT-Paris Curie Engineering School, Beijing, China*

[4]*NIMBE, CNRS, CEA, Paris-Saclay University, Gif-sur-Yvette, France*

3.1. Introduction

As already noted in Chapter 1, given their specific properties, glasses containing lead oxide have a very broad field of application. This extends from crystal glasses, enamels and glazes to optical glasses with a high refractive index, passing through glasses absorbing ionizing radiation and glasses for low-temperature welding of metals, ceramics or glasses (Frieser 1975; Rabinovich 1976; Takamori 1979; Volf 1984). For these various applications, glass composition must obviously be adapted in terms of its elaboration and forming (viscosity, expansion coefficient, glass transition temperature T_g) as well as in terms of its properties under the conditions of use (refractive index, transparency, color, chemical durability, electrical resistivity, etc.). These various properties, which concern both the liquid during its melting, forming and cooling (quenching), and the final glass, that is, the amorphous solid resulting from the quenching of the supercooled liquid below T_g, are closely related to its structure. In the case of oxide glasses that are of interest here, this corresponds to the way in which the polyhedrons associated with the various present cationic species surrounded by oxygen O^{2-} anions are arranged with respect to one another inside the network that constitutes the glass structure. Unlike crystals, this network

Lead in Glassy Materials in Cultural Heritage,
coordinated by Anne BOUQUILLON and Patrice LEHUÉDÉ.

has no long-range order (Figure 3.1). In the case of glasses containing lead, it is therefore particularly important to understand the way in which Pb^{2+} ions are inserted in the structure of this network depending on the lead oxide content and on the glass composition (nature of the forming oxide (SiO_2, B_2O_3, P_2O_5), effect of the presence of modifier oxides (alkaline and alkaline-earth oxides) and intermediate oxides (Al_2O_3))[1] and to evaluate their impact on the structure of this network, and on the environment, and the distribution of specific species that may be present, such as transition metal ions used for coloration (see Chapter 4).

1. A glass former oxide can be defined as an oxide that only by melting + more or less rapid cooling (quenching) readily yields a glass particularly due to the high viscosity of the corresponding liquid in the vicinity of the melting temperature T_f and in its supercooled state ($T < T_f$). A glass can be defined as a solid that has no long-range order (≥10Å) at the level of its atomic organization (unlike crystallized materials) but has a glass transition temperature T_g (range of temperature for which the liquid solidifies when cooled and the glass tends to soften when heated). A glass can be considered as a supercooled liquid that freezes during quenching. Compared to crystallized materials, glasses have therefore an excess of free energy and are in a non-equilibrium state from the perspective of thermodynamics (when heated above T_g they tend to crystallize). The main glass former oxides are SiO_2, B_2O_3, P_2O_5 and GeO_2 which readily lead to glasses and whose structure is a random continuous network of tetrahedrons (SiO_4, PO_4, GeO_4) or triangles (BO_3) connected only by the vertices (small surplus of free energy compared to the crystallized ordered state), by means of directional bonds of iono-covalent type (Zachariasen model (Zachariasen 1932)) (Figure 3.1(a)). As for a glass modifier oxide, it can be defined as an oxide that by itself it never leads to a glass by (even extremely rapid) melting + quenching of the melting liquid, particularly due to the very low viscosity of the liquid and to the structural organization of polyhedrons with respect to one another, which is difficult to maintain in a disordered state (excess of free energy between the disordered and ordered state much larger than for the glass former oxides). Higher or lower quantities of modifier oxides (particularly alkaline and alkaline-earth oxides) are however added to glass former oxides in the glasses to facilitate the melting of the mixture (role of fluxing agent), fining of the liquid (elimination of microbubbles and homogenization) and forming of objects (role of fluidizing agent), and to act on the properties of the final glass. At the structural level, modifier oxides (present in the form of cations) by means of non-directional ionic bonds play the role of compensator in the vicinity of negative charges present in the structure, such as those resulting from breaking the iono-covalent bonds such as Si-O-Si bonds (formation of non-bridging oxygen atoms, Si-O$^-$) (Figure 3.1(b)) or those due to the presence of negatively charged species, such as AlO_4^- or BO_4^-. Like modifier oxides, intermediate oxides – such as Al_2O_3 – never lead by themselves to glasses by melting + quenching, but at structural level they generally play a different role from that of modifiers and can enter the structure of the glass formers network. It is important to note however that in the absence of glass former oxide and in the presence of one or several modifier oxides, intermediate oxides can also lead in certain cases to glasses, as it is the case of aluminate glasses (Pye 1988).

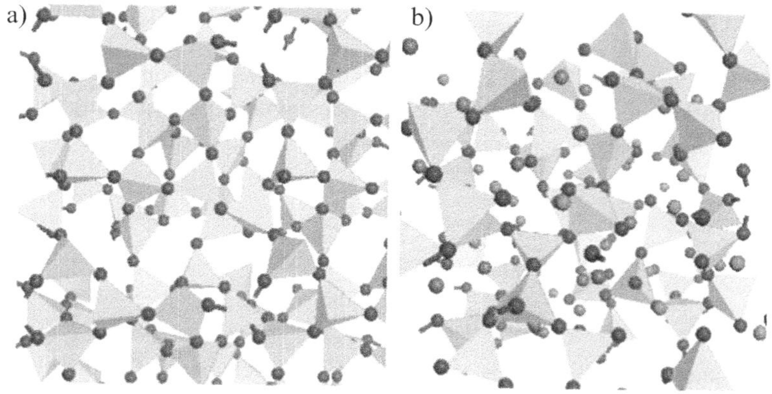

Figure 3.1. *Examples of three-dimensional representation of pure silica glass (a) and soda-lime silica glass (b) showing the SiO_4 tetrahedrons (in yellow), the bridging and non-bridging oxygen atoms (in red), and the modifier ions Na^+ (in purple) and Ca^{2+} (in green). Reproduced from Charpentier et al. (2013) with the permission of The Royal Society of Chemistry. For a color version of this figure, see www.iste.co.uk/bouquillon/lead.zip*

COMMENT ON FIGURE 3.1.– *Representation (a) of the pure silica glass showing the SiO_4 tetrahedrons all interconnected by bridging oxygen atoms and (b) a soda–lime silica glass (SiO_2-Na_2O-CaO) showing the breaking of the silicate network with the formation of non-bridging oxygen atoms connected to a single SiO_4 tetrahedron. The modifier ions Na^+ and Ca^{2+} compensate for the negative charge of the non-bridging oxygen atoms.*

This chapter focuses first on the chemistry of lead and more specifically on the properties and stereochemistry of the Pb^{2+} ion (species under which lead is present in oxide glasses), the nature of the Pb-O chemical bond and the crystalline structure of pure PbO oxide (stereochemistry and mode of connection of the polyhedrons). Then, since the lead-based oxide glasses of interest are overwhelmingly silicate glasses, the focus is on how the Pb^{2+} ions are inserted into the glassy network of silica and modify its structure, first of all in the case of the simple binary SiO_2-PbO system (for which the evolution of the structure of the binary crystalline phases will also be approached depending on their PbO content compared to glasses), then in the case of more complex systems such as the ternary SiO_2-PbO-R_2O (R: alkaline) and SiO_2-PbO-Al_2O_3 systems. The effect of these structural evolutions on certain properties of the glass or liquid will also be discussed, with the exception of optical properties (transmission, coloration) that are presented in Chapter 4.

3.2. Lead and lead oxides

This section focuses on the key elements facilitating the comprehension of lead chemistry in both crystallized and glassy oxides, with a particular highlight of the importance and consequences (physical and structural properties) of the existence of the lone $6s^2$ electron pair of the Pb^{2+} ion.

3.2.1. *Elementary lead and its specific properties*

3.2.1.1. *Lead oxidation states and* Pb^{2+} *ion*

Lead (atomic number 82) is the heaviest stable element of column 14 of the periodic table. Due to its electron configuration ([Xe] $4f^{14}$ $5d^{10}$ $6s^2$ $6p^2$), it has three common oxidation states, Pb^0, Pb^{2+} and Pb^{4+}, whose stability domains can be evaluated by the standard potentials of the ionization reactions in aqueous solution:

$$Pb^0 \rightleftharpoons Pb^{2+} + 2e^- \; E° = -0.13 \text{ V}$$

$$Pb^{IV}O_2 + 4H^+ + 2e^- \rightleftharpoons Pb^{2+} + 2H_2O \; E° = 1.47 \text{ V}$$

According to these two equations, bivalent lead is therefore stable within a large potential range, which includes particularly that of the O_2/O^{2-} pair (1–12 V). It is present in this oxidation state in the vast majority of lead-bearing minerals, such as litharge, massicot (PbO) and galena (PbS), which is the main ore, as well as in various carbonates, silicates, phosphates or sulfates and in oxide glasses (silicates, borates). Tetravalent lead, much rarer, can be found in minium (Pb_3O_4, or Pb^{IV} Pb^{II}_2 O_4) and plattnerite (PbO_2), especially known as the material of the positive electrode of lead-acid batteries. Even rarer is metallic (or native) lead. It is therefore logical that following the example of nature, the arts of fire practiced under an oxidizing potential close to that of the atmosphere can be used for the elaboration of solids containing exclusively Pb^{2+} ions, as in the case of oxide-based glasses.

With a coordination number of 8, the Pb^{2+} ion has a radius of 1.43 Å, comparable to that of strontium Sr^{2+} (1.40 Å) (Shannon 1976). Similar to the heavy alkaline-earth ions Ca^{2+}, Sr^{2+} and Ba^{2+} – large and lightly charged – their oxides have a basic behavior that renders them reactive in relation to acidic oxides (B_2O_3, CO_2, SiO_2, P_2O_5, SO_3, etc.) or amphoteric oxides (H_2O, Al_2O_3, Fe_2O_3, etc.) and leads to the formation of many mixed oxides. Lead oxide-based glasses, although they have no defined compositions, are part of this context. However, the outer electron configuration of the Pb^{2+} ion ([Xe] $4f^{14}$ $5d^{10}$ $6s^2$) is radically different from the saturated configuration of the ions in the second column (alkaline-earths). It can be graphically represented as a sphere comprising $4f$ and $5d$ orbitals, whose radius corresponds to that of the Pb^{4+} ion (1.08 Å) (Shannon 1976), and the external residual volume (nearly 60%) is filled by the lone $6s^2$ pair. Strongly screened by the inner orbitals, these two 6*s* electrons are both labile and very sensitive to the local electric field.

3.2.1.2. *Polarizability of Pb^{2+}ion*

Polarizability (α), which measures the capacity of a particle to polarize under the action of an electric field E ($P = \alpha \cdot E$, P being the induced polarization), is generally proportional to the volume for the same charge ions belonging to the same column. As shown in Figure 3.2, this relation strictly follows the alkaline-earth series, but the Pb^{2+} ion stands out very clearly due to the contribution of the $6s^2$ pair. This high polarizability proves to be a key parameter of the physical, chemical and structural properties of the Pb^{2+} ion and, more generally, of the ions of elements of block *p* having an ns^2np^0 outer electron configuration, such as Tl^+, Bi^{3+} and Te^{4+}, isoelectronic with Pb^{2+}.

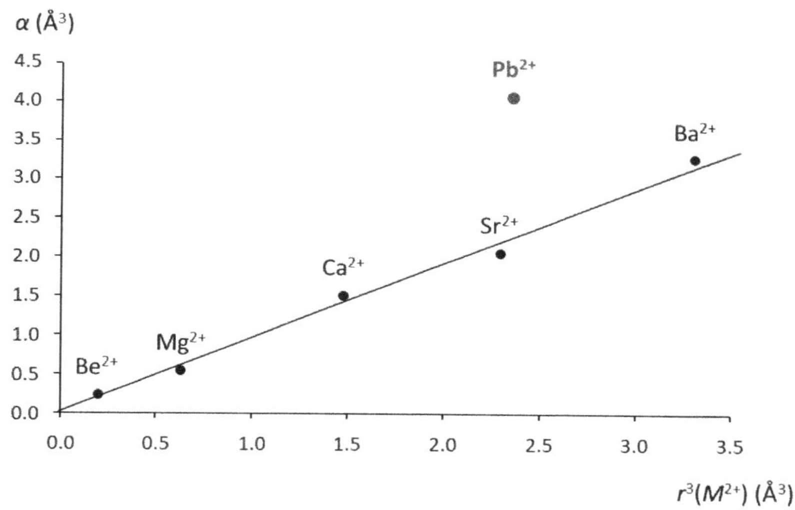

Figure 3.2. *Representation of polarizability α of Pb^{2+} ion and of alkaline-earth ions (Shannon and Fischer 2016) as a function of the cube of radius with a coordination number of 6*

Polarizability particularly determines the relative permittivity ε_r, as shown by the Clausius–Mosotti relation:

$$\frac{\varepsilon_r - 1}{\varepsilon_r + 2} = \frac{N\alpha}{3\varepsilon_0}$$

which yields:

$$\varepsilon_r = \frac{3\varepsilon_0 + 2N\alpha}{3\varepsilon_0 - N\alpha}$$

where N is the number of particles per unit volume, and ε_0 is the vacuum permittivity.

One of the main technological applications of lead oxide-based ceramics is logically found in the industry of dielectric, piezoelectric, pyroelectric and ferroelectric components, whose properties are directly linked to polarizability and where Pb^{2+} has gradually replaced Ba^{2+} in their composition. In the range of optical frequencies, polarizability is expressed through the refractive index n ($= \sqrt{\varepsilon_r}$). The predominantly positive effect of lead II can be illustrated by the series of crystal compounds $M^{II}CO_3$ with aragonite (orthorhombic, biaxial) structure, whose average refractive index, quasi-identical (1.62) for M = Ca, Sr and Ba, reaches 1.98 for cerussite $PbCO_3$. The shine and covering power of a pulverulent material being directly related to its refractive index, it can be readily understood why white lead (mixture of cerussite and hydrocerussite $Pb_3(CO_3)_2(OH)_2$) has been considered through the ages as top white pigment (Welcomme et al. 2007; Stols-Witlox 2011). Likewise, a specific effect of the introduction of lead in the composition of glasses and glazes is to increase the refractive index (Chapter 1, Figure 1.3).

3.2.2. *Lead stereochemistry in oxides*

The valence shell electron pair repulsion (VSEPR) theory is particularly helpful in understanding the distortions induced by non-bonding doublets of light elements of the p block and explaining the geometry of distorted molecules such as water and ammonia. A similar phenomenon can be observed in coordination complexes and crystalline structures containing heavy ions with the ns^2np^0 configuration, whose two common crystal arrangements of PbO (litharge and massicot) are typical examples.

In litharge (tetragonal symmetry, *P4/nmm*, a = 4.004 Å and c = 5.071 Å), a stable phase at ambient temperature, Pb^{2+} ions are surrounded by eight O^{2-} anions, as in a CsCl structure. However, here, instead of a cube, the polyhedron is shaped as a very elongated parallelepiped with a square base, in which lead occupies a highly off-centered position (Figure 3.3). Pb-O distances, which are extremely disparate (4 × 2.33 Å and 4 × 4.37 Å), reflect an imbalance in the distribution of bonding forces (Brown and Altermatt 1985), 99% of which are directed toward the closest four anions. Therefore, the actual coordination number is only 4, with a hemispherical distribution (known as "umbrella") specific to cations with the ns^2np^0 configuration, leading to the occupation of the apex of a square-based pyramid. Here, short bonds have the same length as a Pb^{4+}-O^{2-} bond (2.33 Å), which shows that the four close oxygen atoms are in contact with the sphere formed by the 4*f* and 5*d* orbitals of lead. In massicot (Figure 3.3), a PbO form that is stable above 490°C, metastable at ambient temperature, the hemispherical coordination number of lead is 4, although less regular due to the orthorhombic symmetry (*Pbcm*, a = 5.893 Å, b = 5.490 Å and c = 4.753 Å).

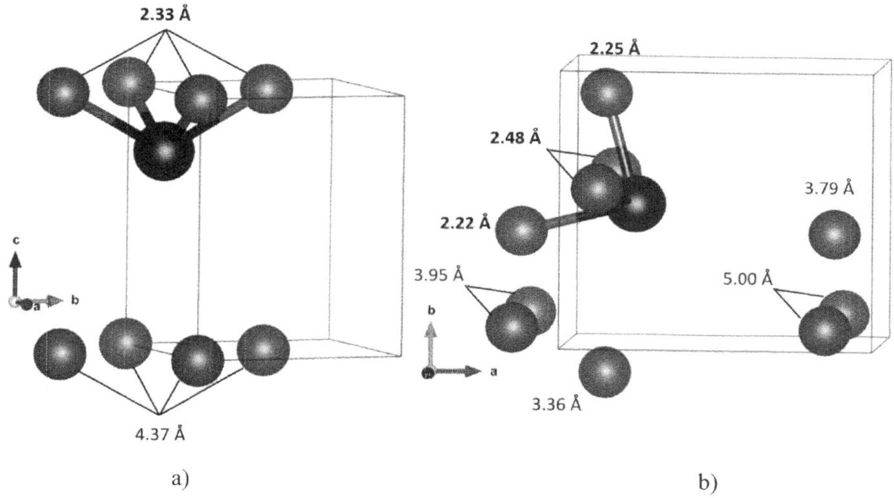

Figure 3.3. *Oxygenated environment of the Pb^{2+} ion and Pb-O distances in two varieties of PbO: litharge (a) and massicot (b). Pb (in black) and O (in red). For a color version of this figure, see www.iste.co.uk/bouquillon/lead.zip*

In both cases, the $6s^2$ pair is repelled to the opposite of short bonds due to the high polarizability of Pb^{2+} ions in the apparently unoccupied part of the oxygen polyhedron ("apparently" because these two electrons, which are a far less charged and coherent source of scattering than the 78 electrons of the lead core orbitals grouped in a small volume, are difficult to observe by X-ray diffraction, and even more difficult by neutron diffraction, for which only the nuclei are "visible"). From an electrostatic perspective, the $6s^2$ pair, however, plays an essential role, similar to that of a non-bonding pair in VSEPR, as its negative charge strongly repels the anions on the opposite side of strong bonds, hence the strong increase in Pb-O distance (4.37 Å) on this side of the polyhedron (Figure 3.3). The dipole moment of the Pb^{2+} ion thus influences the geometry of its environment. The representation of ions in actual proportions in the two PbO varieties (Figure 3.4) shows that it is off-centered by approximately 1 Å with respect to the nucleus and that it occupies roughly the same volume as an anion. Its presence in the apparently empty volumes of the structures makes it actually possible, with oxygen anions, to obtain pseudo-hexagonal layouts close to the ideal packing density contributing to the stability of these oxides.

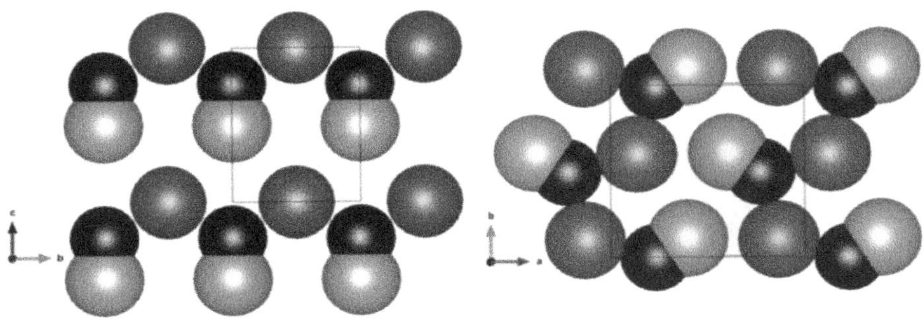

a) b)

Figure 3.4. *Representation in actual proportions of lead ions (in black) and of the* $6s^2$ *pair (in orange) in the atomic planes of litharge (a) and massicot (b). It is worth noting the quasi-hexagonal structure built by the* O^{2-} *anions (in red) and the* $6s^2$ *pairs. For a color version of this figure, see www.iste.co.uk/bouquillon/lead.zip*

Let us now focus on how the PbO_4 polyhedrons are connected in the two crystal varieties of PbO. In the case of litharge, these polyhedrons are interconnected by the edges and arranged head-to-tail to minimize the electron repulsions between non-bonding doublets of the lead atoms (Figure 3.5). In massicot, the pyramids are interconnected by the edges, and once more the arrangement makes it possible to minimize the electron repulsions between non-bonding doublets of Pb^{2+} ions. Litharge is the most stable form, as it is stabilized by the interactions between non-binding doublets, with an antiparallel orientation between near-neighbors of the same plane (001). This orientation is partly lost in massicot due to the distortion of polyhedrons, which leads globally to the weakening of this interaction at the crystal scale; massicot is therefore less stable.

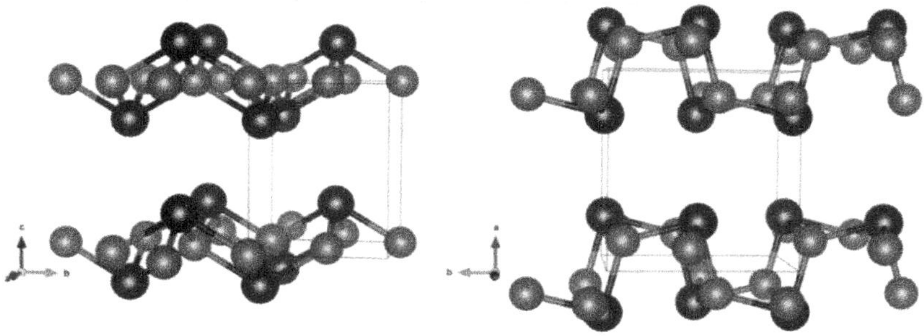

a) b)

Figure 3.5. *Crystal structures of two PbO polymorphs: litharge (a) and massicot (b). Pb (in black) and O (in red). For a color version of this figure, see www.iste.co.uk/bouquillon/lead.zip*

The origin of the shift of 6*s* orbital – normally spherical and nucleus centered – in ions with the ns^2np^0 configuration was the subject of many experimental and theoretical works following those of Gillespie and Nyholm in the 1950s (Gillespie and Nyholm 1957). The widespread model is that of a simple hybridization of 6*s* and 6*p* orbitals of Pb^{2+} allowing the lone pair to take the shape of a protuberant lobe; however, recent studies tend to prove the involvement of 2*p* levels of neighboring O^{2-} anions (Watson et al. 1999; Waghmare et al. 2003). Whatever the mechanism, the structural distortion of a crystalline solid by a stereochemical effect of "lone pair" ultimately follows a simple law: minimizing the network energy. In litharge, for example, the shift of the $6s^2$ pair "vacates" the opposite 4*f* and 5*d* orbitals, allowing the formation of four short Pb^{4+}-O^{2-} bonds, which are highly covalent and directional, globally more stable than the eight bonds with a high ionicity rate that would form in a symmetric structure of CsCl type. Adding that its electronegativity (1.87 on Pauling scale) is significantly higher than that of alkaline and alkaline-earth elements (χ_{Na} = 0.93; χ_K = 0.82; χ_{Mg} = 1.31; χ_{Ca} = 1.00), it is clear that the structural role of lead in glasses is much closer to that of glass former or intermediate elements (χ_{Si} = 1.9; χ_{Al} = 1.61), although closer to the alkaline and alkaline-earth elements due to its size. As will be noted further in this chapter (section 3.3.3), in oxide glasses such as silicates, the lone pair of Pb^{2+} ions also contributes to their incorporation in the form of polyhedrons of low coordination number (3 or 4) with the oxygen anions anisotropically distributed around lead (Fayon et al. 1999; Rybicki et al. 2001; Alderman et al. 2013).

The same energy contingency is, however, sometimes applicable to stereochemical activity when a stable structural form adapted to the composition makes it possible to host the cation in a steady environment. The already mentioned series of MCO_3 carbonates with aragonite structure offers an example. Metastable for Ca^{2+} that more often crystallizes as calcite, it is in fact perfectly adapted to larger cations, such as Sr^{2+} and Ba^{2+}, for which it is the stable form. However, the structure of cerussite $PbCO_3$ (Chevrier et al. 1992) is quasi-identical with that of $SrCO_3$ (Pannhorst and Lohn 1970), with an environment of Pb^{2+} whose regularity ($2.65 < \text{Pb-O} < 2.76$ Å) proves the absence of stereochemical activity of the $6s^2$ pair. However, the absence of permanent polarization of Pb^{2+} ions does not impede the previously mentioned strong polarizability. Moreover, it is important to note that the stereochemical activity of the lone pair of Pb^{2+} ions also depends on the nature of neighboring anions. Hence, unlike in PbO, the lone pair of Pb^{2+} ions is stereochemically inactive in PbS due to the anion–cation orbital interactions that differ for the two phases (Walsh and Watson 2005). The stereochemical activity of the $6s^2$ pair is therefore not systematic in the crystal structures containing lead II, but it is, however, frequent (it is the case of crystallized lead silicates, as will be seen in section 3.2), although sometimes less marked than in the two varieties of PbO. Pb^{2+} has, for this reason, an intermediate behavior between those of its isoelectronic neighbors, the stereochemical activity being rarer and less marked for Tl^+ and quasi-systematic and strong for Bi^{3+}, undoubtedly for reasons of size and charge.

In the biological domain, the effect of the lone pair widely contributes to lead toxicity, as it replaces cations that are essential to metabolic functioning, such as Ca^{2+}, Zn^{2+} or Fe^{2+} inside the structures of enzymes and complexes. This substitution is, on the one hand, easy and almost irreversible, as the stereochemical activity of the $6s^2$ pair allows the formation of strong bonds with the ligands, and, on the other hand, it is harmful, as it modifies the symmetry to the point of disrupting the functioning (Gourlaouen and Parisel 2007), which renders lead ingestion particularly toxic (lead poisoning, Chapters 13 and 14) (Patrick 2006).

3.3. Crystal phases and glasses of the SiO_2-PbO system

In lead oxide-based silicate glasses used both in industry and in crafts and art (massive glasses, glazes, enamels), SiO_2 and PbO are the majority of oxides forming the glass structure. Hence, in the crystal glass, the sum SiO_2 + PbO is generally approximately 85% of its mass (the rest being mainly alkaline oxides K_2O and Na_2O), while in transparent lead glazes deposited on the surface of ceramics, this sum commonly reaches over 90% in mass (the rest being then mainly composed of alumina most of the time (Tite et al. 1998) (Chapter 10)). Although the binary glasses of the SiO_2-PbO system actually have no practical use, as there is almost always an addition to their composition of other glass former, modifier or intermediate oxides (boron oxide, alkaline oxides, alumina, etc.) depending on the considered applications (Rabinovich 1976), it appears particularly important to properly understand the structure of these simple glasses depending on their PbO content and especially the manner in which the Pb^{2+} ion is inserted in the structure of the silicate glassy network for both low and highest PbO contents if the objective is to apprehend the structure of more complex lead silicate glasses. This explains the rich literature related to the structural study of glasses of the SiO_2-PbO system published for nearly 85 years (Bair 1936)! Despite the increasingly advanced and various methods of investigation and simulation implemented to explore the environment of lead, silicon and oxygen in these simple glasses, there are still many questions related to the way the Pb^{2+} ions are incorporated in these glasses and particularly for the highest PbO contents, as will be seen here. However, before focusing on the glassy system (the supercooled liquid frozen during quenching, which is in a non-equilibrium state), let us first examine the binary system at thermodynamic equilibrium depending on temperature and its composition and on the structure of mixed crystal phases existing in this system and more specifically on the manner in which the Pb^{2+} ions are incorporated into their crystal network.

3.3.1. *The binary SiO_2-PbO system at equilibrium*

Given the importance of this system, phase diagrams were proposed for the binary SiO_2-PbO system since the 1930s (Geller et al. 1934), which were subsequently completed and refined using new experimental results and calculations. To our knowledge, the most recent one belongs to Shevchenko and Jak (2018), combining calculations and experimental results (Figure 3.6).

As indicated in Chapter 1, this diagram shows the very clear fluxing effect of PbO when added to SiO_2 (strong decrease in the green curve corresponding to the liquidus of the system). However, it can be noted that for PbO contents ranging between 40 and 80 mol%, the liquidus temperature ($T_{liquidus}$) does not change much and ranges between 730 and 750°C, but it has a small increase when approaching the pure PbO phase (massicot, $T_{melting}$ = 886°C).

As long as the PbO content is below 50 mol%, the only stable crystal phases susceptible to forming inside the system are the various crystal varieties of silica (cristobalite, tridymite, quartz) and lead metasilicate $PbSiO_3$ (alamosite) with congruent melting. Therefore, there is no stable mixed crystal phase with less than 50 mol% PbO. It is important to note that contrary to the predictions of the calculations of Jak et al. (1997) (black curve, Figure 3.6), the experimental curve of liquidus exhibits no stable demixing dome, indicating the existence of phase separation phenomena inside the liquid ($T > T_{liquidus}$). However, the specific S-shape of the liquidus curve (green curve, Figure 3.6) suggests that metastable demixing may occur in the supercooled liquid ($T < T_{liquidus}$) (Calvert and Shaw 1970), as in the SiO_2-BaO system (Seward et al. 1968). The diagram shows that several mixed crystal phases may form in this system besides $PbSiO_3$ for PbO contents above 50 mol%:

– at high temperature, near the liquidus curve: Pb_2SiO_4 (lead cyclosilicate, with congruent melting), $Pb_{11}Si_3O_{17}$ (with congruent melting) and Pb_5SiO_7 (with incongruent melting). The Pb_4SiO_6 phase shown in older phase diagrams (Smart and Gasser 1974) would not be stable (Hirota and Hasegawa 1981);

– at lower temperatures in the solidus domain: $Pb_7Si_6O_{19}$ and $Pb_3Si_2O_7$ (these two phases decompose before melting).

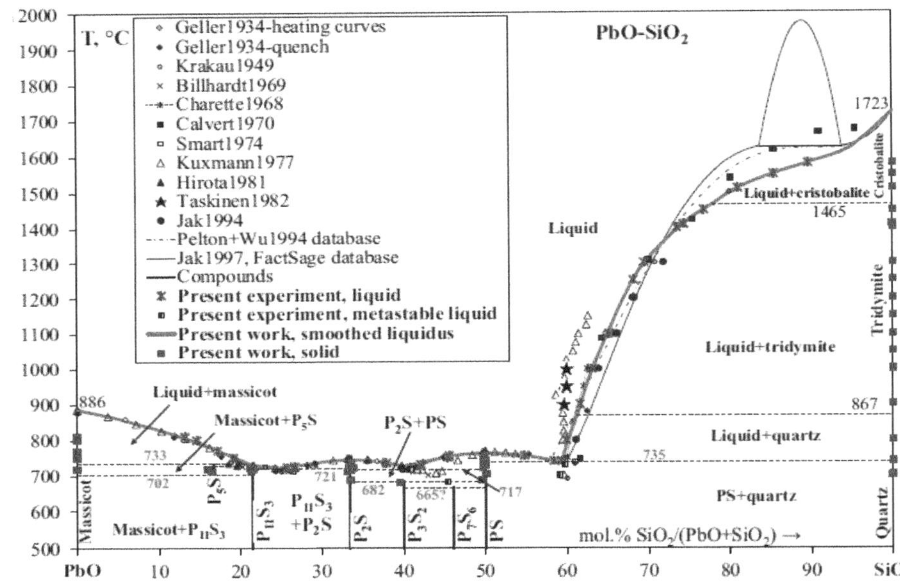

Figure 3.6. *Phase diagram of the PbO-SiO$_2$ system (mol%) above 500°C (Shevchenko and Jak 2018). Massicot: PbO phase stable at high temperature, for mixed crystal phases: P = PbO and S = SiO$_2$. For a color version of this figure, see www.iste.co.uk/bouquillon/lead.zip*

COMMENT ON FIGURE 3.6.– *Diagram combining several experimental and calculated results from sources in the literature with recent experimental results (red and blue dots correspond to compositions and temperatures explored by the authors of the study and the green liquidus curve is a smoothing result achieved in this same study). This diagram also contains the phase diagram calculated by the same team in a previous study (Jak et al. 1997).*

3.3.2. *Crystal phases of the binary SiO_2-PbO system*

This section presents the evolution of the structure of mixed crystal phases mentioned in the previous section depending on the lead oxide content according to the data in the literature[2] (mol% PbO): *alamosite* $PbSiO_3$ (50%) (Boucher and Peacor 1968), $Pb_7Si_6O_{19}$ (53.8%) (Siidra et al. 2014), $Pb_3Si_2O_7$ (60%) (Petter and Harnik 1971), Pb_2SiO_4 (66.6%) (Dent Glasser et al. 1981), and $Pb_{11}Si_3O_{17}$ (78.6%) (Hirota and Hasegawa 1981; Kato 1982). Three of these structures are represented in Figure 3.7. Pb_5SiO_7 (83.3%), which is stable only at high temperature (Hirota and Hasegawa 1981), is the only one whose structure remains unsolved. The focus is here on the effect of PbO addition to the silicate network and on the manner in which Pb^{2+} ions are incorporated in this structure (type of occupied sites, existence of connections between them and with the silicate units). The absence of mixed crystallized oxides with a PbO content below 50 mol% (or 79 mass%) is a tendency found among the known complex silicates (exception: margarosanite $PbCa_2Si_3O_9$ has a 43 mol% PbO rate). Therefore, the estimation of the structural behavior of crystal glasses and lead glazes (with PbO contents on the order of 10 and 30 mol%, respectively) based on the crystal phases will only be possible by extrapolation or interpolation with the pure SiO_2 pole. However, as will be seen in section 3.3.3.1, binary glasses of the SiO_2-PbO system with up to 83 mol% PbO content can be obtained; therefore, for PbO contents that are as high as those of crystal phases (Figure 3.7). An examination of the crystal structures shows that the $6s^2$ pair of Pb^{2+} is systematically active from a stereochemical perspective; thus, the average number of short bonds ($d_{Pb\text{-}O} < 2.5$ Å, with a bond valence above 0.35 v.u.[3]) varies only from 2.3 to 3.0 per Pb^{2+} ion, with no correlation with the PbO rate being noted.

As it disperses SiO_4 tetrahedrons, PbO addition is expected to mechanically reduce their cross-linking. Hence, the silicate network, formed of infinite *meta* chains in $PbSiO_3$, evolves toward disilicate units in $Pb_3Si_2O_7$, following the same tendency as for the addition of CaO. For these two families, the rate of bridging oxygen atoms (involved in the Si-O-Si bonds) evolves in line with the theoretical model following a law in $(2 - 3x)/(2 - x)$, with $x = M^{II}/(M^{II} + Si)$, which explains the depolymerization of the silicate network (Figure 3.8). The same is applicable to $Pb_7Si_6O_{19}$, whose more complex structure expands in $Pb_{21}(Si_7O_{22})_2(Si_4O_{13})$. The

2. Although most of the structural studies conducted on the binary crystal phases of SiO_2-PbO system involved X-ray diffraction, NMR studies conducted on ^{29}Si and ^{207}Pb nuclei allowed the correlation of the chemical shift changes to the evolution of the environment of silicon and lead in the structure of crystals, thus providing very useful data for understanding the evolution of the environment of these same species in glasses (Bessada et al. 1994; Fayon et al. 1997).

3. v.u.: unit of bond valence.

analysis of crystal structures suggests that in lead glasses and glazes ($0.1 < x < 0.3$), lead oxide has a depolymerizing effect similar to that of a standard modifier. For $x = 2/3$, the theoretical threshold of total depolymerization (Figure 3.8), silicate tetrahedrons become effectively independent in Ca_2SiO_4 (calcium orthosilicate), but its lead homolog has a radically different structure of oxycyclosilicate type (Pb_2OSiO_3, Figure 3.7) due to which Si-O-Si cross-linkages (cycles of four tetrahedrons) persist. The same is valid for $Pb_{11}Si_3O_{17}$ ($Pb_{11}O_6(Si_2O_7)(SiO_4)$), which has an even higher lead rate and contains silicate dimers (Figure 3.7). This singularity is very likely explained by the stereochemistry of the $6s^2$ pair, which allows the establishment of strong bonds in the opposite direction of silicate units; certain anions are thus grabbed by three or four Pb^{2+} cations and isolated from the silicate network (Figure 3.7); the latter can therefore maintain a certain degree of polymerization, for example in the form of cycles or dimers (Figure 3.7).

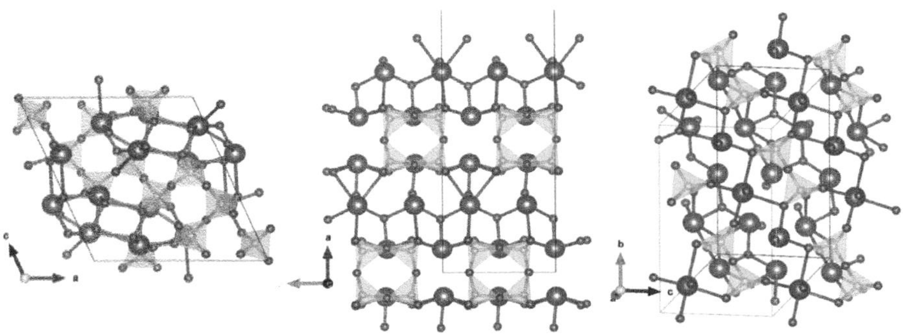

Figure 3.7. *Crystal structures of three lead silicates (in-depth representation of 1/2 unit cell) in ascending order of PbO content from left to right (50, 66.6 and 78.6 mol%) showing the various modes of connection between SiO_4 tetrahedrons (in yellow) as well as the asymmetry of the coordination polyhedron for PbO_n. Pb (in black) and O (in red). For a color version of this figure, see www.iste.co.uk/bouquillon/lead.zip*

The strong Pb-O bonds (with O^{2-} ions that are not bound to silicon) are at the origin of a second form of cross-linking, which does not exist with alkaline or alkaline-earth modifiers. It is, however, more difficult to define than that of the silicate network due to the irregularity of the lead environment. Taking into account only the Pb-O bonds whose length is below 2.5 Å, the formation of Pb-O-Si and Pb-O-Pb cross-linkages can be noted, and their evolution is represented in Figure 3.9. The separation of lead, which occurs beyond $x = 0.66$, generates a significant decrease in the former to the benefit of the latter. In the domain of glasses and glazes

(0.1 < *x* < 0.3), the interpolation between SiO_2 and the nearest crystallized lead silicates suggests the existence of a significant primary cross-linking of approximately 0.2–0.7 "strong" Pb-O-Si cross-linkages per oxygen. On the other hand, the Pb-O-Pb cross-linkages are approximately three to four times rarer (Figure 3.9). Within the lead-rich crystal phases of the SiO_2-PbO system (≥50 mol% PbO, Figure 3.6), the connection between the polyhedrons occupied by Pb^{2+} ions is generally performed by the edges (Figure 3.7).

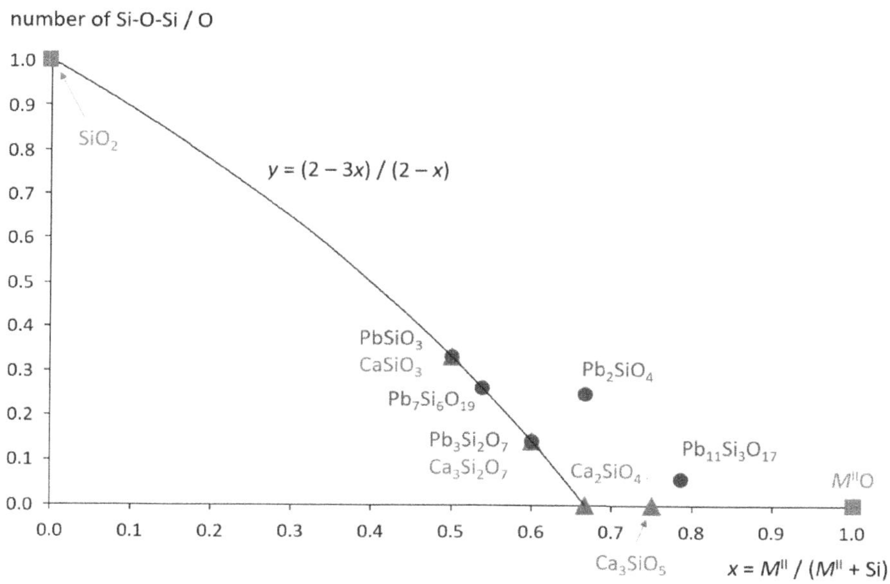

Figure 3.8. *Evolution of the Si-O-Si cross-linkage rate per oxygen atom depending on the molar ratio $M^{II}/(M^{II} + Si)$ for M = Pb and Ca in crystallized lead silicates (●) or calcium silicates (▲). For a color version of this figure, see www.iste.co.uk/bouquillon/lead.zip*

This approach based on the comparison with known crystal structures, although weakened by their high PbO rates (≥50 mol%) for the crystal glasses and lead glazes (≤30 mol% PbO), is more realistic for PbO-rich glasses having contents within the same range as crystal phases (section 3.3.3.1), therefore suggesting that lead contributes to the depolymerization of the silicate network similarly to calcium. However, the stereochemical activity of the $6s^2$ pair allows the formation of Pb-O-Si cross-linkages based on strong Pb-O bonds. The strong Pb-O-Pb cross-linkages would be marginal for crystal glasses and lead glazes but would have a significant and increasing contribution for glasses with much higher PbO contents (Figure 3.9).

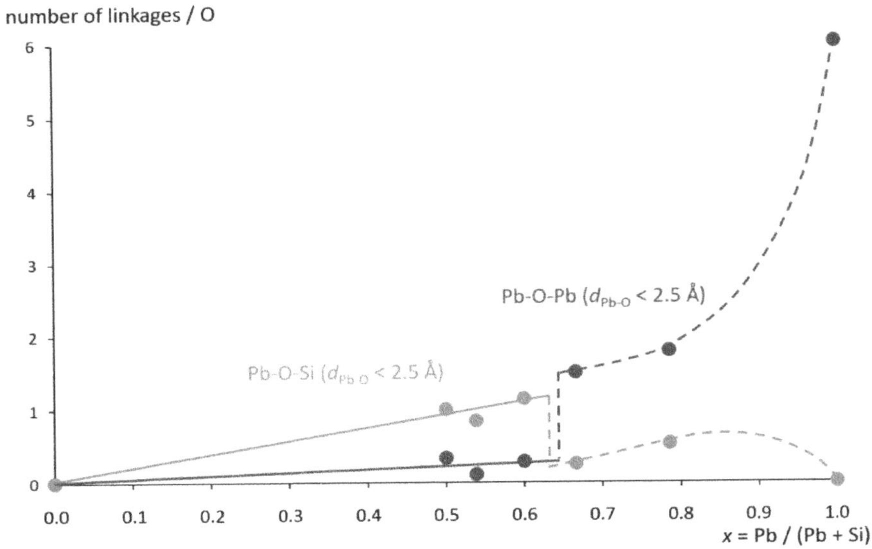

Figure 3.9. *Evolution of the rate of Pb-O-Si and Pb-O-Pb cross-linkage per oxygen atom depending on the molar ratio of Pb/(Pb + Si) in the crystal phases of lead silicates. For a color version of this figure, see www.iste.co.uk/bouquillon/lead.zip*

3.3.3. *Glasses of the binary SiO_2-PbO system*

3.3.3.1. *Overview of glasses of the binary SiO_2-PbO system*

After the crystal phases of lead silicates, let us now focus on the impact of adding increasing quantities of PbO to the well-known glass former oxide vitreous silica SiO_2 and more particularly on its structure and the manner in which Pb^{2+} ions are incorporated into it[4] and on the evolution of certain physical properties. The study of the structure of these glasses and the structure-property relations is particularly important from both fundamental and applied perspectives, as SiO_2 and PbO are the two predominant oxides in most of the lead oxide-based silicate glasses (crystal glass industry, glazes and enamels, optics, etc.). Moreover, the very wide domain of vitrifiable composition in this system compared to other binary systems, such as alkaline or alkaline-earth silicates, is still the subject of many questions in

4. As mentioned above (section 3.2.1.1), lead is overwhelmingly present in the form of Pb^{2+} ions in oxide glasses. The presence of Pb^{4+} ions would only step in for highly lead-rich glass compositions melted under highly oxidizing atmosphere. This is in line with the evolution of the stability of lead oxides with temperature ($PbO_2 < Pb_3O_4 < PbO$).

terms of their structure and the role of lead, particularly for the PbO-richest compositions. Indeed, it is possible to obtain binary SiO_2-PbO glasses by melting-quenching up to PbO contents of over 83 mol%[5] (95 mass%) (Feller et al. 2010), while for SiO_2-R_2O systems (R: alkaline), it is increasingly difficult to obtain glasses above 55 mol% R_2O (Imaoka and Yamazaki 1963) due to the very strong tendency to crystallize of the corresponding supercooled liquids during cooling, associated both with a decrease in their viscosity and with the very strong depolymerization of the silicate network[6]. In the lead system, for compositions very rich in PbO, the amorphous structure of the liquid can be stabilized by quenching, even though SiO_2 is in a minority. This implies a different structural behavior of Pb^{2+} ions compared to that of alkaline and alkaline-earth ions in these glasses and particularly for high PbO content (which is in agreement with the comparison of crystal phases of alkaline-earth and lead silicates, Figure 3.8), although, like them, its gradual addition to SiO_2 induces both a strong decrease in the liquidus temperature of the system (Figure 3.6) and the viscosity of the liquid (Figure 3.10).

It is important to note that even if the liquid is still homogeneous for $T > T_{liquidus}$ (Figure 3.6), the compositions rich in SiO_2 (>65 mol%) have a demixing tendency (metastable, therefore sub-liquidus) during the cooling of the liquid (Calvert and Shaw 1970; Vogel 1994). However, the tendency toward phase separation in these binary glasses is suppressed when K_2O is added (as is the case in the crystal glass), which can be explained by the quasi-absence of demixing tendency in the binary SiO_2-K_2O (Charles 1966) due to the low-field strength[7] of the K^+ cation (F ~ 0.14 Å^{-2}) compared to the Pb^{2+} cation (F ~ 0.36 Å^{-2}) (Vogel 1994). The substantial tendency of binary glasses SiO_2-PbO with low PbO content to demixing coupled with their

5. Binary glasses with PbO contents above 90 mol% were sometimes mentioned in the literature (Ellis et al. 1979), but due to the lacking results of chemical analysis, such data must be carefully considered given the strong evaporation tendency of PbO for lead-rich compositions (>70–80 mass% PbO) (Andersen 1919; Volf 1984). Moreover, melting liquids that are very rich in PbO (>90 mass%) have a very strong crystallization tendency (Faivre and Weiss 1963, p. 587).

6. For binary SiO_2-R_2O glasses, assuming the gradual formation of SiO_4 Q^n entities with a decreasing number n of bridging oxygen atoms (Figure 3.1(b)) when R_2O is added (Q^n represents the SiO_4 tetrahedrons counting n bridging oxygen atoms), it appears that between 50 and 60 mol% R_2O there are only highly depolymerized entities Q^2 and Q^1 (binary model). Beyond 66.6 mol% R_2O, the glass would only be composed of completely depolymerized Q^0 (SiO_4^{4-}) entities. It is therefore possible to obtain binary SiO_2-PbO glasses quite above this last value. For further details, see section 3.3.3.2.1.

7. The field strength F of a cation M^{z+} can be defined as the ratio $F = z/d(M\text{-}O)^2$, where d(M-O) is the average distance between M and the first neighbor O^{2-} anions. The values of the field strength for K^+ and Pb^{2+} ions were calculated from the average values of the distance d(M-O) in binary silicate glasses (d(K-O) ~ 2.64Å and d(Pb-O) ~ 2.40Å).

very high liquidus temperature (Figure 3.6) requiring melting temperatures of approximately 1,600–1,700°C explains the small number of studies in the literature on the compositions of lead silicates very rich in silica (Bair 1936; Calvert and Shaw 1970). A certain number of studies have also been conducted on the crystallization of binary SiO_2-PbO glasses with or without nucleating agents (such as P_2O_5) but will not be developed here (Carr and Subramanian 1982a, 1982b; Lippmaa et al. 1982; Neiman et al. 1982; Kaur et al. 2013; Pena et al. 2020). Hence, according to Kaur et al. (2013), as long as the PbO content < 50 mol%, glasses resist well to crystallization at heating, while when the PbO content ≥ 50 mol%, the majority crystallization of phases such as $PbSiO_3$ and Pb_2SiO_4 was evidenced. Before going further into the structural details and in particular on the manner in which Pb^{2+} ions are inserted in the vitreous structure and how they react with the silicate network (section 3.3.3.2), it is interesting to present the effect of the gradual addition of PbO to silica on certain physical properties of the glass and of the liquid (Kaur et al. 2013).

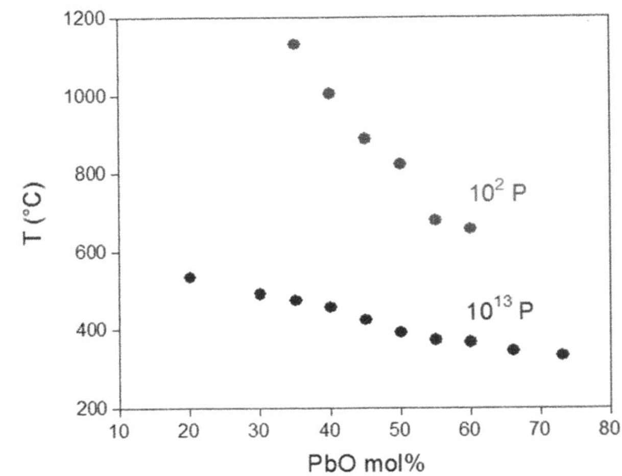

Figure 3.10. *Evolution as a function of PbO content of the temperature corresponding to viscosities of 10^2 and 10^{13} Poise for liquids of the SiO_2-PbO system (according to Martlew data (Martlew 2005)). It can be noted that the temperature corresponding to a given viscosity decreases with the PbO content, which shows its fluxing effect*

In addition to the above-mentioned effect on the viscosity of the melting liquid (strong decrease with the PbO content, Figure 3.10), highlighting the fluxing effect of the lead oxide, as is the case for alkaline oxides and boron oxide, the incorporation of increasing PbO contents into the silica leads to a decrease in the glass transition temperature, electrical resistivity of glasses and surface tension of associated liquids. Moreover, the molar volume tends also to decrease until 60 mol% PbO and then slightly increases (Shelby 1983; Kaur et al. 2013; Ben Kacem 2017). The effect of the lead content on the mechanical properties of binary SiO_2-PbO glasses has also been studied, showing, for example, a tendency of their hardness and Young's modulus to decrease when the PbO concentration increases (Yoshihoto and Soga 1987; Kaur et al. 2013). The increase in density and in the refractive index presented in Chapter 1 is simply explained by the high atomic mass of lead compared to silicon and oxygen (density) and by the strong polarizability of the Pb^{2+} ion and its high atomic weight (refractive index). Furthermore, it is important to note that the coefficient of thermal expansion of binary SiO_2-PbO glasses increases linearly with their molar content in PbO (Greenough et al. 1981; Shelby 1983) but remains below that of binary SiO_2-Na_2O glasses (Scholze 1980); this property is important, particularly for glazes (crazing problem). The chemical durability in aqueous medium of the glasses containing lead oxide was briefly addressed in Chapter 1 (section 1.1.8) and is presented in detail in Chapter 13 for industrial glasses containing both lead and alkaline ions (crystal glass); therefore, it will not be covered here. However, related to the binary SiO_2-PbO glasses without alkaline ions, it is important to note the study conducted by El-Shamy and Taki-Eldin (1974) on the chemical durability of glasses depending on pH, which shows a non-monotonic evolution of their leaching tendency depending on their PbO content. It is also important to mention the more recent works of Mizuno et al. (2005) on the durability of binary SiO_2-PbO glasses conducted in an acid medium, which the authors tried to link to the distribution of lead-based structural units PbO_n in the vitreous structure as a function of lead concentration.

3.3.3.2. *Structure of the glasses of the binary SiO_2-PbO system*

According to certain works in the literature (Sorokina et al. 1996), the structural role of lead in binary silicate glasses evolves with the PbO content, passing from modifier to glass former oxide, which might explain the larger expansion of the vitreous field of lead silicates compared to alkaline silicates, as mentioned above (section 3.3.3.1). Before presenting more precisely these works and other studies stating that the structural role of lead would not evolve fundamentally as a function of its concentration (Smets and Lommen 1982; Mizuno et al. 2005; Ben Kacem et al. 2017) and proposing various structural models, let us focus only on lead oxide PbO and its capacities, or lack thereof, to lead to a glass by melting-quenching.

It is important to note first that PbO is not a glass former oxide, since unlike silica, for example, it does not lead to the formation of a glass by melting + quenching. This may be due to several reasons. First, the very low viscosity of liquid PbO in the vicinity of its melting temperature ($\eta \sim 0.01$ Pa.s (0.1 P) (Sinn 1979)), which is higher only by a factor of 10 than that of water at ambient temperature ($\eta \sim$

0.001 Pa.s (0.01 P) (Korson et al. 1969)) and significantly lower than that of melted silica in the vicinity of its melting temperature ($\eta \sim 10^6$ Pa.s (10^7 P) (Martlew 2005)) explains the ease of PbO crystallization from the liquid by the strong mobility of species and therefore the difficulty of maintaining the latter in a supercooled state and consequently obtaining glass by quenching. Moreover, from a structural point of view, according to Zachariasen rules related to glass former oxides M_xO_y (Zachariasen 1932), these must be such that, on the one hand, the MO_n polyhedrons are loosely coordinated (3 or 4) – which corresponds to small highly charged cations $M^{+2y/x}$ such as Si^{4+}, B^{3+} and P^{5+} – and connected with one another only by the vertices (not by the edges or the faces), allowing a significant flexibility in the structural arrangement, without getting a too high increase in the free energy of the system. According to the same rules and for the same reasons, it is stipulated that the oxygen anions should not connect more than 2 MO_n polyhedrons. The coordination number of Pb^{2+} ions is 4 in crystal PbO, as already seen above (Figures 3.3 and 3.5). The coordination number of O^{2-} ions should also be 4 to have electroneutrality. This explains why the PbO_4 units are connected by the edges and not by the vertices, which is contrary to Zachariasen's rules. Being relatively constrained locally, this structure may seem unfavorable to accepting significant disorder at the level of polyhedral layout without this leading to a significant increase in its free energy[8]. From a thermodynamic point of view, this should be in favor of the crystallized state compared to the vitreous state and could structurally explain why PbO is not a glass former oxide. Likewise, according to Dietzel's theory (Dietzel 1948) connecting the cationic field strength F and the structural role of oxides in glasses, it appears that Pb^{2+} ion has a field strength of approximately 0.36 Å^{-2}[9], which would put PbO in the class of modifier oxides for which F ranges between 0.1 and 0.4 Å^{-2} (F ranges between 1.5 and 2.0 Å^{-2} for glass former oxides). However, it is important to recall here that due to its strong polarizability (section 3.2.1.2), from the point of view of its bonds with the nearest neighbor oxygen ions, the Pb^{2+} ion can be considered a Pb^{4+} ion with a higher field strength F = 0.48 Å^{-2}. This would put PbO in the class of intermediate oxides (F ranging between 0.5 and 1.0 Å^{-2}) with quite directional Pb-O bonds, which are ionocovalent and asymmetrically arranged in its environment (Figures 3.3 and 3.5). This type of Pb-O bond is therefore very different from the one existing between alkaline and alkaline-earth cations and O^{2-} anions in the crystallized oxides (ionic bonds distributed quite symmetrically around cations).

8. It is important to note however that due to the two-dimensional character of the crystalline structure of PbO (litharge and massicot, Figure 3.6), it may probably be easier to accept local constraints due to disorder than three-dimensional crystalline structures such as those of alkaline and alkaline-earth oxides, whose role as modifier oxides is well established.

9. Calculation made according to the crystalline PbO structure and Shannon's ionic radii (Shannon 1976) (rPb^{2+} = 0.98Å, rO^{2-} = 1.38Å).

It can then be understood that the mode of incorporation of Pb^{2+} ions in a silicate glass may be very different[10] from that of traditional modifier cations (Na^+, K^+, Ca^{2+}, etc.), which are far less polarizable[11] and for which the bonds with the oxygen anions are weaker, ionic and therefore non-directional. Relying on the crystal structure of PbO and assuming that the environment of Pb^{2+} ions in SiO_2-PbO glasses was also very asymmetrical with quite strong directional bonds due to their very strong polarization by O^{2-} anions, Fajans and Kreidl (1948) proposed in the 1940s that this particular type of lead environment was at the origin of the largest stability of the liquids of SiO_2-PbO system in terms of crystallization (and therefore of their easier vitrification) compared to alkaline silicate compositions for which the crystallization processes are facilitated by the easier reorganization of species in the corresponding liquids (weaker and non-asymmetric bonds between alkaline cations and O^{2-} anions)[12]. However, the study of the structure of binary SiO_2-PbO glasses started approximately 10 years earlier, in 1936, by X-ray scattering (Bair 1936), only 1 year after the well-known works of Warren and Loring on the X-ray scattering study of the structure of soda silicate glasses (Warren and Loring 1935) and 4 years after the famous Zachariasen's publication on the structure of oxide glasses (Zachariasen 1932). However, the model proposed by Bair for lead silica glasses (Bair 1936), based on his processing of X-ray scattering data, was very close to the structural model of soda silicate glasses of Warren and Loring (random continuous silicate network with homogeneous distribution of Pb^{2+} ions near non-bridging oxygen atoms of SiO_4 tetrahedrons). Since these first works aimed at clarifying the structure of binary SiO_2-PbO glasses, many studies have been conducted using various structural investigation methods, such as X-ray and neutron scattering, X-ray absorption (EXAFS[13] and XANES[14]), XPS[15], multinuclear NMR[16]

10. Indeed, as noted in section 3.3.2, for crystallized lead silicates the environment of Pb^{2+} ions is also strongly asymmetric.

11. Indeed, according to Kordes, the values of polarizability (in Å^3) are the following: Na^+ (0.175) < Ca^{2+} (0.469) < K^+ (0.821) < Pb^{2+} (3.623) (Dimitrov and Komatsu 2010).

12. It is interesting to signal that according to Vogel (1994), the fact that the vitrifiable domain of the P_2O_5-PbO system is less extended than that of SiO_2-PbO system (66 mol% compared to 83 mol%) could be explained by the very strong polarization exerted by the P^{5+} ions on the O^{2-} anions of the phosphate network, which would have a weaker polarizing effect on the Pb^{2+} ions than in the case of silicate glasses (Si^{4+} ion having a weaker field strength than P^{5+} ion), thus leading to weaker Pb-O bonding forces and a higher coordination number for Pb^{2+} ions with the oxygen atoms associated with phosphorus (Hoppe et al. 2003). For further information on the structure and properties of the glasses of the binary P_2O_5-PbO system, please see some previous studies (Dayanand et al. 1996; El-Egili et al. 2003; Matori et al. 2013).

13. EXAFS: Extended X-ray Absorption Fine Structure.

14. XANES: X-ray Absorption Near Edge Structure.

15. XPS: X-ray Photoelectron Spectroscopy.

(^{207}Pb, ^{29}Si, ^{17}O), Raman and IR spectroscopies, as well as molecular dynamics. These studies confirmed the hypothesis of Fajans and Kreidl of a very asymmetric environment of Pb^{2+} ions in silicate glasses, such as the study of Mydlar et al. in 1970 (Mydlar et al. 1970). Currently, all the authors agree that in lead silicate glasses, the majority of Pb^{2+} ions are found in very asymmetric environments, such as in PbO and lead silicate crystals (Figures 3.3, 3.5 and 3.7), which also have a sterochemically active $6s^2$ pair (sections 3.2.2 and 3.3.2). However, the authors still diverge on the structural role of lead, its type of environment and its evolution as a function of PbO content in glass. A part of these structural works on binary SiO_2-PbO glasses was the object of past (Rabinovich 1976) or more recent (Alderman 2013) bibliographic syntheses. Based on some of these studies, and particularly on the more recent ones, the following sections of this chapter focus on the manner in which the silicate vitreous network is affected by the addition of increasing quantities of lead oxide. Then, the structural role of Pb^{2+} ions and the evolution of their environment and their distribution in the vitreous network will be examined.

3.3.3.2.1. Vitreous silicate network

Prior to any lead oxide addition, the structure of pure silica glass is exclusively formed of SiO_4 tetrahedrons connected to one another by their vertices forming a continuous network, as seen at the beginning of this chapter (Figure 3.1(a)). In this case, each SiO_4 tetrahedron has four bridging oxygen atoms. To introduce the classification of various types of SiO_4 tetrahedrons present in glasses composed of silica and one or more other oxides, such as PbO, let us consider the well-known case of binary SiO_2-Na_2O glasses. When Na_2O is added to the silica glass, Si-O-Si bonds are broken (Na_2O has the role of modifier oxide): the vitreous silicate network becomes partially depolymerized, and non-bridging oxygen atoms – whose negative charge is locally compensated by Na^+ ions – are formed (Figure 3.1(b)). To account for the gradual depolymerization of SiO_4 tetrahedrons depending on the Na_2O content, the notation Q^n is introduced, representing the SiO_4 tetrahedrons counting n bridging oxygen atoms. A pure silica glass thus contains only Q^4 units, and for binary glasses, a Q^3 unit corresponds to a SiO_4^- tetrahedron, a Q^2 unit to a SiO_4^{2-} tetrahedron, a Q^1 unit to a SiO_4^{3-} tetrahedron and a Q^0 unit to a SiO_4^{4-} tetrahedron that is completely isolated from the rest of the silicate network. The more Q^n units with low n, the more difficult it becomes to vitrify the liquid, as it is increasingly fluid and is readily organized in the form of crystals. Using this notation, the chemical equation corresponding to the introduction of Na_2O in pure silica glass is $2Q^4 + Na_2O \rightarrow 2Q^3$. It is in fact an acid–base reaction (in the sense of oxo-acidity (Flood and Förland 1947)) between the basic oxide Na_2O giving O^{2-} anions and the

16. NMR: Nuclear Magnetic Resonance (MAS NMR = NMR with magic angle spinning; MQMAS NMR = multi-quantum NMR with magic angle spinning).

acidic oxide SiO_2 that accepts them. Moreover, in the melting liquid, there is an equilibrium (disproportionation reaction) between Q^n units ($2Q^n \leftrightarrow Q^{n-1} + Q^{n+1}$) that can be found in the glass obtained after quenching (Maekawa et al. 1991). The description of the silicate network particularly involves the study of its degree of polymerization, hence following up the evolution of the proportions of various units Q^n that compose it. In the case of simple silicate compositions, this can be done, for example, using ^{29}Si NMR and Raman (Maekawa et al. 1991; O'Shaughnessy et al. 2020) spectroscopies, while the follow-up of the evolution of the proportions of the various types of oxygen atoms (bridging, non-bridging or not connected to the silicate network) can be achieved by ^{17}O NMR (after glass enrichment in ^{17}O)[17] and XPS spectroscopies, for example (Dalby et al. 2007; Lee and Kim 2015).

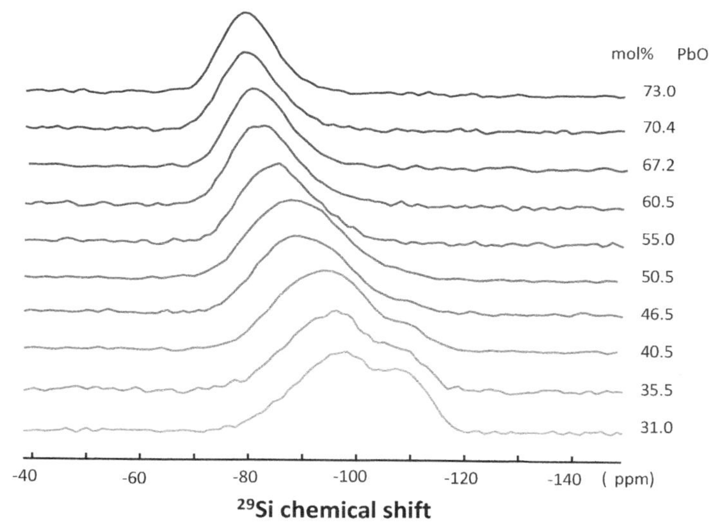

Figure 3.11. *Evolution of ^{29}Si MAS NMR spectra of lead silicate glasses depending on their PbO content (Fayon 1998). 7 T magnetic field, 10 kHz rotational frequency (with the kind permission of F. Fayon)*

In the case of interest here, several studies focused on the evolution of the structure of the silicate network of lead silicate binary glasses and clearly

17. Given the very low natural abundance of ^{17}O isotope (0.038%) the only one detectable by NMR due to its non-zero nuclear spin I (I = 5/2) compared to the ^{16}O (99.75%) isotope, of zero nuclear spin, it is necessary to enrich the glasses in ^{17}O. In the case of silicate glasses, this is classically done by preparing ^{17}O enriched silicon oxide SiO_2, for example, from a mixture of $SiCl_4$ and water enriched in ^{17}O.

highlighted the fact that the introduction of PbO leads to an increasing depolymerization of the silicate network. For example, Bessada et al. (1994), Fayon et al. (1998) and Fayon (1998) have used ^{29}Si NMR to compare the spectra of glasses to those of crystalline phases and to quantify the Q^n units present in the glasses containing 31–73 mol% PbO (Figures 3.11 and 3.12).

It has thus been shown that from 31 mol% to approximately 40 mol% PbO, the Q^3 units were predominant and the proportion of Q^2 units increased mainly to the detriment of Q^4 units (Fayon 1998; Fayon et al. 1998) (Figure 3.12). Then, Q^1 units appear, and their proportion increases with the PbO content mainly to the detriment of Q^3 units and become predominant beyond 60 mol% PbO. Finally, the Q^0 units appear and are the predominant units with Q^1 units beyond 72 mol% PbO (Figure 3.12). This structural evolution was more recently confirmed over a more extended range of PbO contents (33–83 mol%) by Feller et al. (2010), who showed that the glass prepared with 83 mol% PbO contained almost exclusively Q^1 species (24%), entities composed of only two SiO_4 tetrahedrons connected by a vertex, and Q^0 species (72%), namely, SiO_4 tetrahedrons with no connection to another SiO_4 tetrahedron, thus confirming the possibility of obtaining a vitreous state in the absence of an extended silicate vitreous network for binary SiO_2-PbO compositions. The comparison of ^{29}Si chemical shifts of Q^n units between binary glasses and crystalline phases of the same composition has, on the other hand, evidenced a number of Pb^{2+} ions present as second neighbors of silicon atoms that are more significant for the glasses (Bessada et al. 1994). Furthermore, it is interesting to note that the equilibrium constants of $2Q^n \leftrightarrow Q^{n-1} + Q^{n+1}$ reactions (disproportionation) in the supercooled liquid and frozen in the glass during quenching, which were determined by ^{29}Si NMR from proportions of various Q^n units in the glasses (Figure 3.12), have an order of magnitude above that of alkaline silicate glasses (Fayon et al. 1998), which tends to show a more significant clustering tendency of Pb^{2+} ions compared to alkaline ions. The persistence of Q^2 and Q^1 units far beyond 66 mol% PbO (threshold beyond which only Q^0 units should remain if PbO fully transferred its O^{2-} anions to the silicate network by acidic-basic reaction) suggests that an increasing fraction of oxygen atoms in the structure are only connected to Pb^{2+} ions, the rest being present in the form of bridging oxygen atoms (Si-O-Si) and non-bridging oxygen atoms (Si-O-Pb). This has been directly confirmed by ^{17}O NMR for PbO-rich (60–71 mol%) glasses enriched in ^{17}O (Figure 3.13) (Lee and Kim 2015).

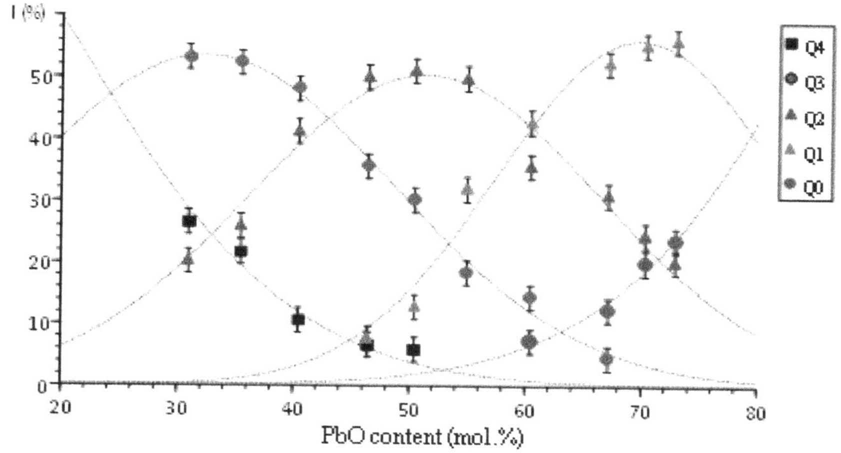

Figure 3.12. *Evolution of the relative proportions of Q^n units determined from ^{29}Si MAS NMR spectra of lead silicate glasses (Figure 3.11) as a function of their PbO content (the dotted curves serve only as a guide) (Fayon 1998) (with the kind permission of F. Fayon). For a color version of this figure, see www.iste.co.uk/bouquillon/lead.zip*

Other authors have used Raman spectroscopy to follow up the evolution of the structure of lead silicate glasses and try to quantify the Q^n units by sometimes drawing comparisons with the crystalline phases (Furukawa et al. 1978; Feller et al. 2010; Kaur et al. 2013; Ben Kacem et al. 2017). Similar to ^{29}Si NMR, a very strong evolution of Raman spectra with the PbO content was observed in conjunction with an increasing depolymerization of the silicate network. Moreover, in line with the ^{17}O NMR results (Figure 3.13), the joint formation of Pb-O-Pb bonds was evidenced, reflecting an increasing tendency of Pb^{2+} ions to bind to one another (clusterization). The studies conducted by XPS (O 1s) on binary SiO_2-PbO glasses have also evidenced a strong decrease in the proportion of bridging oxygen atoms with the PbO content; however, the resolution of spectra is insufficient to separate the contribution of non-bridging oxygen atoms from that of oxygen atoms that are not bound to silicon (Smets and Lommen 1982; Sorokina et al. 1996; Dalby et al. 2007). However, some of these studies highlight the persistence of bridging oxygen atoms, that is, Si-O-Si bonds in the most PbO-rich compositions (Smets and Lommen 1982; Dalby et al. 2007), in agreement with the ^{29}Si NMR studies (Figure 3.12). Finally, it is important to mention the studies conducted by X-ray and neutron scattering combined with structural modeling using the reverse Monte Carlo method (Suzuya et al. 1999; Kohara et al. 2010), which led to a distribution of Q^n units as a function of composition (Figure 3.14) consistent with the ^{29}Si NMR results

(Figure 3.12). On the other hand, the structural models obtained by molecular dynamics and which do not take into account the polarizability of Pb^{2+} ions that leads to anisotropic local environments yield average results; these work using symmetric interaction potentials (Ribycka et al. 1999).

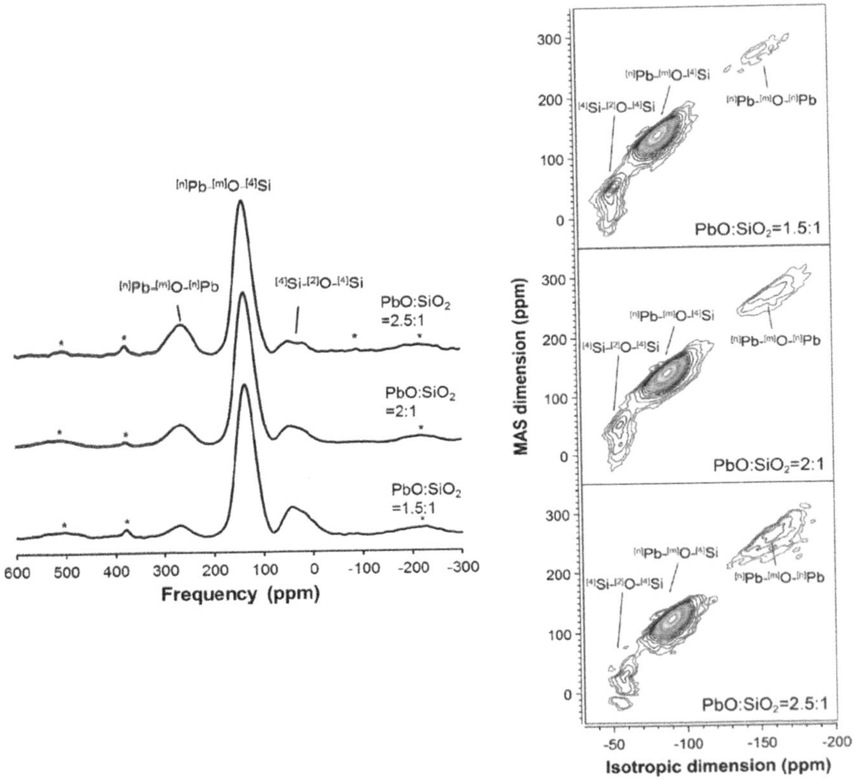

a) b)

Figure 3.13. *(a) ^{17}O MAS NMR spectra of lead silicate glasses enriched in ^{17}O and (b) 3QMAS NMR spectra of the same lead silicate glasses enriched in ^{17}O showing the various existing types of oxygen atoms. Reproduced with permission from Lee and Kim (2015). Copyright (2020) American Chemical Society. For a color version of this figure, see www.iste.co.uk/bouquillon/lead.zip*

COMMENT ON FIGURE 3.13.– *(a) The peaks marked by asterisks are rotational bands. (b) The PbO/SiO_2 molar content ratios are indicated in the figures and correspond to 60 (1.5:1), 66 (2:1) and 71 (2,5:1) mol% PbO. The various types of* *oxygen atoms evidenced are bridging oxygen atoms (*$^{[4]}Si$*-*$^{[2]}O$*-*$^{[4]}Si$*), non-bridging oxygen atoms (*$^{[n]}Pb$*-*$^{[m]}O$*-*$^{[4]}Si$*) and oxygen atoms bound only to lead (*$^{[n]}Pb$*-*$^{[m]}O$*-*$^{[n]}Pb$*).*

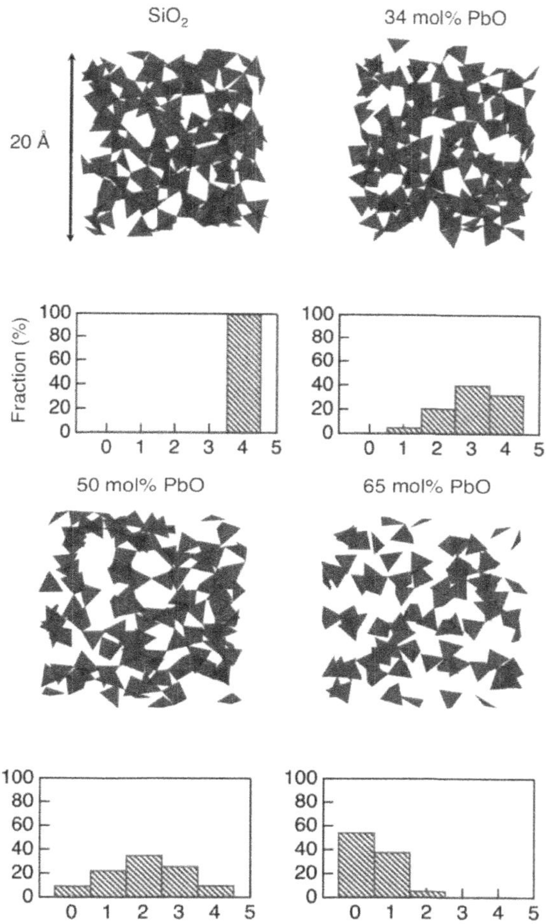

Figure 3.14. *Evolution of the silicate vitreous network (only* SiO_4 *tetrahedrons are represented) of lead silicate and pure silica glasses depending on their PbO content. Reproduced with permission from Kohara et al. (2010). Copyright (2020) by the American Physical Society*

COMMENT ON FIGURE 3.14.– *The structural models have been obtained by the reverse Monte Carlo method from X-ray and neutron scattering results. The* *distribution of Q^n (n = 0, 1, 2, 3, 4) units determined for various glasses from these models is also presented in the form of histograms (Kohara* et al. *2010).*

All these results related to the evolution of the silicate network of SiO_2-PbO glasses are therefore qualitatively close to at least the evolution of the structure of crystalline phases of lead silicates presented in section 3.2 (Figures 3.7–3.9). Indeed, both cases show a strong depolymerization of the silicate (vitreous and crystalline) network with the persistence of SiO_4 tetrahedrons bound to one another for the highest PbO contents (Figures 3.7, 3.8 and 3.12): Q^1 and Q^2 units for glasses and Si_4O_{12} rings and Q^1 units for Pb_2SiO_4 (66 mol% PbO) and $Pb_{11}Si_3O_{17}$ (78 mol% PbO) crystalline phases, respectively. In both cases, there is a significant proportion of oxygen atoms exclusively bound to lead (Pb-O-Pb bonds) (Figures 3.9 and 3.13).

3.3.3.2.2. Environment, distribution and structural role of lead

Various techniques can be used to extract information on the lead environment in glasses. Some of these techniques make it possible to specifically probe lead, such as NMR[18], EXAFS and XANES and XPS, while other techniques, such as X-ray and neutron scattering, provide very useful information on the environment of Pb^{2+} ions (coordination number, Pb-O distance) and probe other elements that are present in the vitreous network. The studies published in the literature on lead environment and incorporation using these investigation techniques are numerous and sometimes contradictory, so the structural role of lead in silicate glasses is still under debate. However, the majority of authors agree currently that Pb^{2+} ions are inserted in the structure of binary silicate glasses with their lone pair $6s^2$ stereochemically active in the form of very asymmetric PbO_n polyhedrons with low coordination number (n = 3 – 4) of pyramidal type with short Pb-O bonds with strong covalent character: lead would therefore have a local environment close to the one occupied in the crystalline phases of lead silicates and in the pure lead oxide (sections 3.2.2 and 3.3.2). Moreover, for high PbO contents, PbO_n units tend to bind to one another to form increasingly larger percolating domains until building the main frame of the vitreous network, including increasingly depolymerized silicate units (Q^1, Q^0), similar to crystallized lead silicates (Figure 3.7). The existence of such domains is in agreement with the ^{17}O NMR-evidenced increasing proportion of oxygen atoms exclusively bound to lead (Pb-O-Pb bonds, Figure 3.13). These conclusions rely, among others, on X-ray absorption (EXAFS, XANES) and X-ray and neutron scattering studies.

18. ^{207}Pb is the only stable isotope of lead (natural abundance 22.1%) having a non-zero nuclear spin (I = ½); it can therefore be studied by NMR. However, given its high polarizability, the NMR signals obtained are large and poorly resolved.

The X-ray and neutron scattering studies allow the extraction of radial distribution functions concerning the various pairs of atoms present in the structure and from which it is possible to find the interatomic distances and coordination numbers (Figure 3.15). Hence, Hoppe et al. (2003) and Takaishi et al. (2005) found Pb-O distances on the order of 2.30 Å for the nearest oxygen atoms with a coordination number of 3, and these distances and coordination numbers varied relatively little with the PbO content in glasses (Figure 3.15).

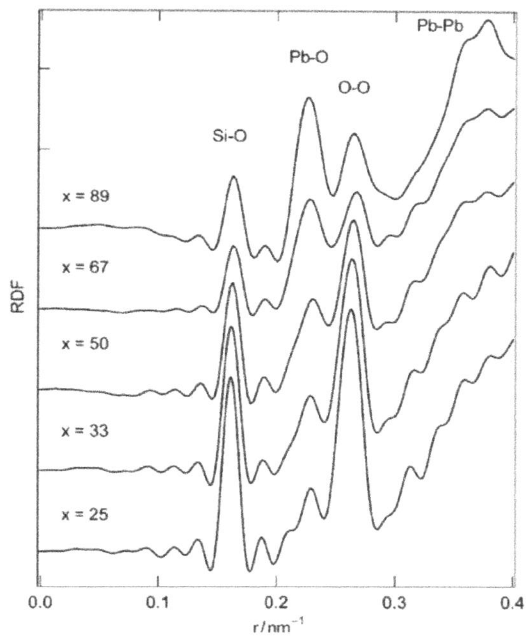

Figure 3.15. *Radial distribution functions calculated from the neutron scattering data for glasses of composition (1-x)SiO_2 – xPbO (x in mol%) (Takaishi et al. 2005)*

COMMENT ON FIGURE 3.15.– *For neutron scattering (shown here), all correlations can be seen, particularly those associated with Si-O and O-O, as in this case lead has a behavior similar to the other atoms. On the other hand, in the case of X-ray scattering (not shown here), which is largely dominated by lead due to its significantly higher atomic number compared to that of other species present (Si and O), the resulting information is essentially related to Pb-O and Pb-Pb correlations.*

Based on their X-ray and neutron scattering results, and particularly on the Pb-Pb distance that remains quasi-constant (3.83–3.86 Å) for all the studied glasses

(25–89 mol%), Takaishi et al. proposed a structural model of SiO_2-PbO glasses involving Pb_2O_4 units composed of two trigonal pyramidal PbO_3 connected by an edge and oriented in opposite directions to minimize the repulsion between the free pairs $6s^2$ of the two Pb^{2+} ions, these Pb_2O_4 units having a zero global charge (Figure 3.16) (Mizuno et al. 2005; Takaishi et al. 2005). These entities would be present starting with the lowest PbO contents studied by the authors (25 mol%). According to this model, lead behaves as a forming element starting with the lowest lead concentrations.

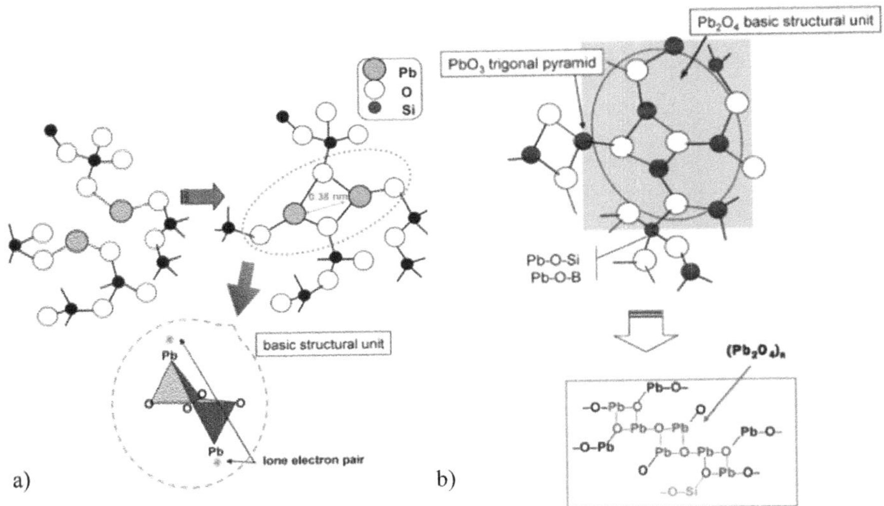

Figure 3.16. *Structural model of lead silicate glasses proposed by Takaishi et al. (2005) based on the X-ray and neutron scattering results*

COMMENT ON FIGURE 3.16.– *The Pb_2O_4 basic unit consisting of two PbO_3 pyramids with triangular base connected by an edge and with an opposed layout of the free $6s^2$ pairs of lead is presented. (a) For low PbO contents, Pb_2O_4 units are incorporated into the silicate network; (b) for high PbO contents, there is a connection between the Pb_2O_4 units. The 3.8 Å distance between the Pb^{2+} ions is indicated.*

Based on their scattering results and on a structural model obtained with the reverse Monte Carlo simulation method for glasses with PbO contents between 30 and 65 mol%, Kohara et al. (2010) did not confirm the existence of Pb_2O_4 dimeric units in lead silicate glasses but a lead environment predominantly in the form of distorted PbO_4 units of square-based pyramidal type (Figure 3.17). Such an environment is close to that of lead in the crystalline forms of PbO (Figures 3.3 and 3.5) and in the crystalline phases of lead silicates (Figure 3.7). A number of PbO_n units connected to one another strongly increasing with the lead content were evidenced in the structure of these glasses. The same authors have also presented a comparison of the evolution of the silicate network and the lead network depending on the lead oxide percentage, which clearly shows the predominance of the lead network starting from at least 50 mol% PbO (Figure 3.17). An equivalent study conducted by Suzuya et al. (1999) on the 50 mol% PbO composition also led these authors to propose a lead environment in the form of irregular PbO_4 pyramids; these units are assumed to form rings constituting the lead network. These models involving an increasing connectivity of PbO_4 units to the detriment of the silicate network are close to the one proposed by several other authors (Bessada et al. 1994; Wang and Zhang 1996).

Alderman et al. (2013) and Alderman (2013) have more recently proposed a more convincing structural description of lead silicate glasses (0–80 mol% PbO) also based on X-ray and neutron scattering results, but using a structural refinement of their experimental data employing empirical interatomic interaction potentials[19] (Figures 3.18 and 3.19). Their modeling was optimized particularly by considering Pb^{2+} ions as dipoles to better account for the specific stereochemistry of the sites occupied by these ions, induced by the non-bonding $6s^2$ lone pair of lead. These authors note first, as for crystals, an increase with the PbO content of the average Si-O distance ascribed to the evolution of the silicon coordination sphere and, more particularly, an increase in the average number of Pb^{2+} ions around the non-bridging oxygen atoms. The authors also note a slight evolution of the lead environment between 35 and 80 mol% PbO: the short Pb-O distances (≤2.7 Å) tend to become shorter and more numerous around the Pb^{2+} ions until reaching four bonds. Conversely, the number of long Pb-O distances (between 2.7 and 3.27 Å) tends to decrease. The environment of Pb^{2+} ions tends therefore to evolve toward that of litharge (Figure 3.5) when the lead oxide rate increases and therefore when the silicon number in the second sphere of coordination of lead decreases. Consequently, the oxygen environment evolves strongly with the lead oxide rate passing from an OSi_2 environment to an increasingly lead-rich environment as lead oxide is added, tending to an OPb_4 environment on average, as noted in the crystals. This is reflected by the presence, in all glasses, of PbO_n polyhedrons connected by edges, which makes it possible to effectively have several Pb^{2+} ions compensating

19. This technique referred to as EPSR (Empirical Potential Structure Refinement) was implemented for the study of the local environment around Pb^{2+} ions, the way in which PbO_n polyhedrons are connected and the layout of free $6s^2$ pairs. This method, which takes into account the strong polarizability of Pb^{2+} ions, is therefore more suitable than the simple reverse Monte Carlo method implemented in other studies on lead silicate glasses.

for a given non-bridging oxygen atom and to have more than two non-bridging oxygen atoms around a Pb^{2+} ion, as is the case in crystals. However, the presence of the Pb_2O_4 unit proposed by Takaishi et al. (2005) as a basic unit of the lead network independent of the PbO content (Figure 3.16) was not observed in this study. Finally, an important conclusion of these works is the emphasis on the formation of lacunar areas in lead oxide-rich glasses due to the organization of the non-bonding doublets of Pb^{2+} ions (Figure 3.19). This organization of non-bonding doublets of lead is found in lead-rich crystallized lead silicates such as $Pb_{11}Si_3O_{17}$ (Figure 3.7), litharge and massicot (Figure 3.5), which form channels in the crystalline structure. These areas of deficiency (voids) are not present in glasses with low lead oxide content (Figure 3.18), whose structure is closer to that of alamosite $PbSiO_3$ (Figure 3.7).

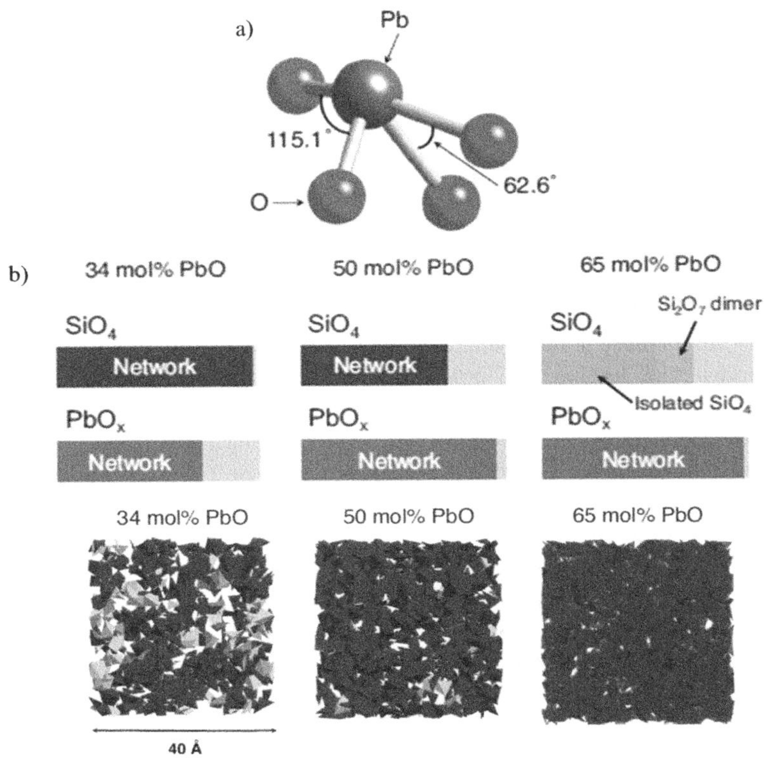

Figure 3.17. *Local and extended structure of lead silicate glasses (30–65 mol% PbO) according to X-ray and neutron scattering results and the reverse Monte Carlo simulation method (RMC) (Kohara et al. 2010). For a color version of this figure, see www.iste.co.uk/bouquillon/lead.zip*

COMMENT ON FIGURE 3.17.– *(a) Distorted PbO_4 pyramidal unit resulting from RMC simulations. (b) (Up) Comparison of evolution depending on the lead content of silicate networks (blue) and lead networks (red) composed of SiO_4 and PbO_x interconnected units. The proportion of SiO_4 and PbO_x units that do not participate in these networks is represented in yellow, green and pink. (Down) Evolution of the lead network with lead content (the interconnected PbO_x units are represented in red, and the rest are represented in yellow; the SiO_4 tetrahedrons are not represented). Reproduced with permission from Kohara* et al. *(2010). Copyright (2020) by the American Physical Society.*

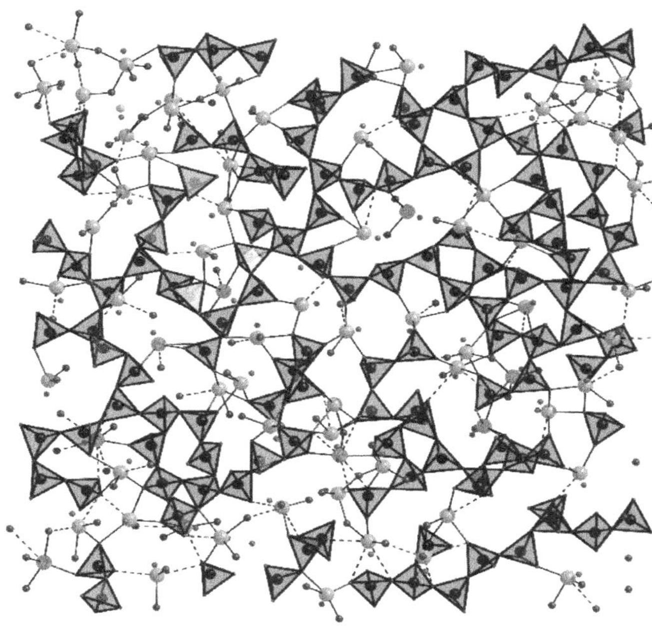

Figure 3.18. *Structural model proposed by Alderman for the 35PbO–65SiO_2 (mol%) glass. Reproduced with the permission of O. Alderman from Alderman (2013). For a color version of this figure, see www.iste.co.uk/bouquillon/lead.zip*

COMMENT ON FIGURE 3.18.– *SiO_4 tetrahedrons are represented in blue, Pb^{2+} ions by the large gray spheres, and the oxygen atoms by small red spheres. The Pb-O bonds whose length is below 2.7 Å are in the solid line, and those ranging between 2.7 and 3.27 Å are in the dotted line. The non-bonding doublets of Pb^{2+} ions are represented by small pink spheres.*

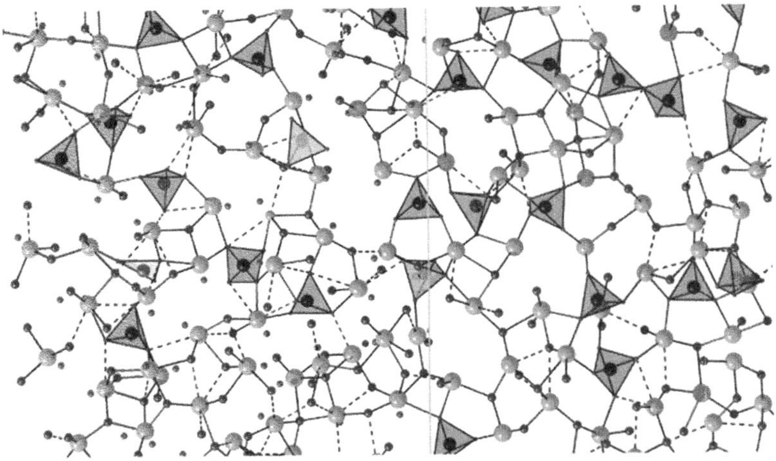

Figure 3.19. *Structural model proposed by Alderman for 80PbO–20SiO$_2$ glass. To the right of the vertical separation line (yellow), the non-bonding doublets of Pb^{2+} ions are omitted, while to the left, they are represented (pink spheres). The green triangle is a SiO$_3$ defect resulting from the calculation. Reproduced with the permission of O. Alderman from Alderman (2013). For a color version of this figure, see www.iste.co.uk/bouquillon/lead.zip*

3.3.3.3. *Summary of the structural evolution of binary SiO$_2$-PbO glasses and the role of lead*

All the structural results published on the binary SiO$_2$-PbO glasses presented in this chapter tend to highlight that the structural role of lead does not fundamentally change with the PbO content and that it is closer to the role of a glass former oxide. Its short-distance environment is very asymmetric, of deformed pyramidal type with triangular or square base (PbO$_3$, PbO$_4$) with lead at the vertex and in all cases the lone 6s^2 pair of Pb^{2+} ions stereochemically active, similar to the PbO crystalline phases (massicot and litharge) and lead silicates (sections 3.2.2 and 3.3.2). The lead environment would therefore slightly evolve as a function of PbO content, with only a slight reinforcement of short-distance PbO bonds, more tetracoordinated (PbO$_4$) environments and the increasing presence of Pb-O-Pb connections with bonds by vertices and edges between the PbO$_n$ polyhedrons forming an increasingly significant lead network, as in the case of crystalline phases (Figures 3.7 and 3.9). Similar to the latter (Figures 3.7 and 3.8), a substantial evolution of the silicate network with the PbO content is observed, with gradual depolymerization (Si-O-Si bonds disappear to the benefit of Si-O-Pb bonds, Figures 3.12 and 3.14), silicon atoms having an increasing number of Pb^{2+} ions as second neighbors. However, for

PbO contents above that of lead oxycyclosilicate (Pb_2SiO_4, 60 mol% PbO) from which complete depolymerization of the silicate network was expected – as is the case for alkaline silicate systems, for example – with only isolated SiO_4 units (Q^0), small-size units having Si-O-Si bonds are still present even for the lead-richest compositions (Figures 3.12–3.14). The presence of these units in the lead network for lead-rich compositions as well as the existence of an environment distribution for Pb^{2+} ions and Pb-O-Pb and Pb-O-Si connections in the structure of glasses (and thus of corresponding liquids during melting and before quenching) must therefore contribute to slowing down the crystallization of the supercooled liquid during quenching and therefore make it easier to obtain a glass. On the other hand, in the case of alkaline and alkaline-earth cations that are bound to O^{2-} anions only by ionic and non-directional bonds and that have quite spherical environments in the silicate glasses, it is clear that the reorganization of the liquid leading to crystals is facilitated, and therefore, it is not possible to vitrify silicate liquids for high alkaline and alkaline-earth oxide contents. All these structural results can be used to explain to a large extent the variation in the physical properties of the binary SiO_2-PbO glasses as a function of PbO content, as mentioned in section 3.3.3.1. Hence, the depolymerization of the silicate network, the fact that the average strength of the bonds between oxygen and its Si and Pb neighbors decreases with the lead content (Pb-O bonds being weaker than Si-O bonds) and the strong polarizability of Pb^{2+} ions (and therefore their strong deformability) explain the decrease with the PbO content of the liquid viscosity, glass transition temperature and electrical resistivity (Figure 3.10). Likewise, the increase in the thermal expansion coefficient of SiO_2-PbO glasses with the PbO content can be explained by a decrease in the average strength of bonds, as for alkaline silicate glasses; however, Pb-O bonds have an iono-covalent character, unlike the bonds between alkaline R^+ ions and O^{2-} anions, which are purely ionic, and the expansion coefficient of alkaline silicate glasses remains higher than that of lead silicate glasses for the same R_2O and PbO molar content (Scholze 1980). Finally, the decrease in the molar volume of SiO_2-PbO glasses when the lead concentration increases for the lowest PbO contents (Shelby 1983; Kaur et al. 2013; Ben Kacem et al. 2017) might be explained by the decrease in the free volume existing in the silicate glasses rich in SiO_2 following the gradual breaking of the Si-O-Si bonds. The stabilization of the free volume, and even its reincrease from 50 mol% PbO, could be explained by the formation of an increasingly extended lead network and by the presence of an increasing number of free $6s^2$ pairs associated with lead creating deficient area (voids) in the structure (Figure 3.19). It is important to note that the strong polarizability of Pb^{2+} ion and the existence of the lone $6s^2$ pair stereochemically active may be at the origin of the significantly higher solubility of metallic oxides such as Cr_2O_3 in silicate glasses containing lead oxide compared to classical silicate glasses containing alkaline and alkaline-earth oxides and in which the chromium oxide is little soluble (Ashour and Shehata 1972; Volf 1984).

3.4. Glasses of the SiO_2-PbO-R_2O system (R= Na, K)

3.4.1. *Overview of ternary SiO_2-PbO-R_2O glasses*

After the binary SiO_2-PbO glasses, let us focus on the structure and properties of ternary SiO_2-PbO-R_2O glasses, where R represents an alkaline element such as Na or K. These ternary glasses have many applications in fields such as luxury glassmaking (crystal glass making), glazes and enamels and glasses absorbing ionizing radiation, as mentioned in Chapter 1. Hence, the crystal glass belongs to the ternary SiO_2-PbO-K_2O system with molar contents of approximately 77% SiO_2, 11% K_2O and 10% PbO (with, according to the production, a small amount of Na_2O, Al_2O_3 and traces of fining agents such as Sb_2O_3), which leads to PbO mass contents of approximately 28% in this case. While, for example, tin-opacified lead alkaline mixed heritage glazes have mass contents of approximately 20–40% PbO and 5–12% Na_2O + K_2O, compared to contents of approximately 45–60% PbO and less than 2% Na_2O + K_2O for lead transparent glazes (Tite et al. 1998, Chapter 10). In the case of crystal glass, there is a multiple interest of adding lead to the binary glass of potassium silicates (melting, forming, machining, optical properties, napping of glazes, see Chapter 1). In parallel, the addition of K_2O to the binary SiO_2-PbO glass makes it possible to reduce the demixing tendency of lead silicate glasses having low or moderate PbO contents (<35 mol%) (section 3.3.3.1) (Calvert and Shaw 1970; Vogel 1994) and facilitates the melting of the vitrifiable mixture. Given the importance both at industrial and heritage levels of ternary SiO_2-PbO-R_2O glasses, many studies related to their hydrolytic resistance in submerged media or in humid atmospheres have been conducted with two purposes: sanitation (Angeli et al. 2016) and conservation (Palomar et al. 2020). The alteration of this type of glass is widely addressed in Chapter 13.

3.4.2. *Structure of ternary SiO_2-PbO-R_2O glasses*

The structure of binary SiO_2-PbO glasses as a function of their PbO content is largely presented in section 3.3.2. This section focuses on the effect of the simultaneous presence of PbO and an alkaline oxide (such as Na_2O or K_2O) on the structure of the glass and the lead environment. Alkaline oxides are well known for their modifier role in binary SiO_2-R_2O silicate glasses, leading to a more or less marked depolymerization of the silicate network as a function of the R_2O content (Maekawa et al. 1991). It is interesting to consider here the way in which the structure of the binary lead silicate glasses is affected when a modifier oxide (here R_2O) is added to their composition and particularly on the way in which the environment and the distribution of Pb^{2+} ions can be modified. However, unlike the structural studies conducted on binary lead silicate glasses, which are particularly abundant in the literature, those related to ternary SiO_2-PbO-R_2O glasses are less numerous, despite the important industrial applications of these glasses. Several spectroscopy methods were implemented for the study of these glasses, such as Raman, IR and XPS spectroscopies.

At the same time, while the crystalline phases of the PbO-SiO_2 system were the object of a quasi-systematic structural study (section 3.3.2), very little is known about more complex crystalline compounds, such as those of ternary SiO_2-PbO-R_2O systems. Thus, despite the considerable number of possible compositions, crystallized lead silicates containing alkaline or alkaline-earth cations were the object of structural studies only as part of both rare and ancient works related to $K_2Pb_2Si_2O_7$ disilicate (McMurdie 1941) and its isotype $K_2Pb_2Ge_2O_7$ (Bassi et al. 1965), cyclosilicate $PbCa_2Si_3O_9$ or *margarosanite* (Freed et al. 1969) and oxy-monosilicate $Li_{10}PbO_2(SiO_4)_2$ (Brandes et al. 1994). Despite different lead contents, these structures show that the depolymerizing effect of alkaline and alkaline-earth oxides – at various degrees depending on their content – helps to promote, similar to binary lead silicates, asymmetric environments for Pb^{2+} ions (lone $6s^2$ pair stereochemically active) and strong Pb-O-Pb cross-linkages. This effect is particularly prominent for the crystalline phase $Li_{10}PbO_2(SiO_4)_2$, which has infinite -Pb-O- chains thanks to its exceptionally high rate of alkaline elements that fully depolymerize the silicate network.

3.4.2.1. *The silicate vitreous network*

There are several studies on ternary SiO_2-PbO-R_2O glasses in the literature varying the relative proportions of the three oxides. Thus, by adding increasing Na_2O contents to a glass of composition $50SiO_2$–50PbO (mol%) or increasing PbO contents to a glass of composition $75SiO_2$–$25Na_2O$ (mol%), studies conducted by ^{29}Si NMR (Shrikhande et al. 2001, 2007) have clearly highlighted a depolymerization of the silicate network, that is, a decrease in the proportion of bridging oxygen atoms connecting the silicon atoms, as is the case when increasing PbO contents are added to SiO_2 (Figure 3.12). A similar structural evolution has been observed during the gradual substitution of SiO_2 by PbO (up to 50 mol%) in an industrial crystal glass whose composition is close to $77.9SiO_2$–10.7PbO–$11.4K_2O$ (mol%), as presented in Figure 13.1. For this latter glass, ^{29}Si NMR has shown that it was essentially composed of Q^3 units (59%) and Q^4 units (41%), which corresponds to 25.7% of non-bridging oxygen atoms and is in perfect agreement with what is obtained by calculation from its composition assuming that K_2O and PbO fully transfer their O^{2-} anions to the silicate network (Angeli et al. 2016) (complete acid–base reaction). This may suggest that these two oxides have similar effects on the silicate network and a similar structural role. However, for higher PbO contents (>30 mol%), there is an increasing disagreement between the result of the calculation of the proportion of non-bridging oxygen atoms based on the glass composition and the NMR results (Chapter 13). This strongly suggests that for quite high lead contents, a part of the oxygen brought by PbO is in the form of Pb-O-Pb bonds, as is the case of lead-rich binary SiO_2-PbO glasses (section 3.3.3.2). A study conducted by XPS (O 1s) on ternary glasses by gradually substituting PbO with Na_2O also confirms this (Gee et al. 2001).

3.4.2.2. *Environment and distribution of lead and alkaline elements in the vitreous network*

To our knowledge, unlike the binary SiO_2-PbO glasses, there are no EXAFS, XANES, X-ray or neutron scattering or ^{207}Pb NMR data available in the literature for the ternary SiO_2-PbO-R_2O glasses that may provide direct information on the lead environment in these glasses. The situation is therefore different from the binary SiO_2-PbO glasses for which it was highlighted by these techniques that lead in the form of Pb^{2+} ions (with its stereochemically active lone $6s^2$ pair) was localized in very asymmetric sites connected to the silicate network by quite strong Si-O-Pb bonds, with the possibility that these sites are also connected to one another for high PbO contents. However, by analogy with the above-mentioned ternary crystalline phases of alkaline or alkaline-earth lead silicate, it is highly probable that in the ternary glasses of the same systems, the Pb^{2+} ions are also located in asymmetric environments with their stereochemically active lone $6s^2$ pair. Moreover, a recent study conducted by NMR on ternary SiO_2-PbO-K_2O glasses enriched in ^{17}O brought abundant information on the way in which Pb^{2+} ions are distributed with respect to K^+ ions in the vitreous network structure depending on the PbO content (substituted for SiO_2) (Figure 3.20) (Angeli et al. 2016). This study comes close to Lee et al.'s (2015) study conducted on binary SiO_2-PbO glasses (Figure 3.13) and is presented in detail in Chapter 13 (Figure 13.2). It shows the evolution of the proportion of various types of connections between Si, Pb, O and K in the glass structure depending on its lead content: Si-O-Si, Si-O-K, Si-O-Pb, Si-O-(Pb,K) and Pb-O-Pb. It appears that for the lowest PbO contents, such as that of crystal glass (~11 mol%), Pb^{2+} ions share with a part of the K^+ ions the non-bridging oxygen atoms of the structure in the form of mixed Si-O-(Pb,K) bonds; however, a significant part of K^+ ions remains connected only to non-bridging oxygen atoms in the form of Si-O-K bonds (Figure 3.20).

It is important to note that for these PbO contents, the presence of oxygen atoms isolated from the silicate network in the form of Pb-O-Pb bonds is not observed, which shows that PbO oxide fully reacts with SiO_2, in agreement with the above-mentioned ^{29}Si NMR results. On the other hand, when the PbO content increases up to at least 30 mol%, all the K^+ ions share non-bridging oxygen atoms with the Pb^{2+} ions. It is only for higher PbO contents (>30 mol%) that Pb-O-Pb connections are highlighted, as suggested by the ^{29}Si NMR and (O 1s) XPS studies mentioned in section 3.4.2.1. This is consistent with the increasing clusterization of PbO_n entities and the gradual formation of a lead network for the binary lead silicate glasses and is in agreement with the structure of the above-mentioned ternary crystalline phases of lead-rich alkaline or alkaline-earth silicates. This was confirmed by a molecular dynamics study and adjustment of the NMR results by the reverse Monte Carlo method (Chapter 13, Figure 13.4). Moreover, Si-O-Pb connections without K^+ ions in their nearest neighborhood should be increasingly abundant. The significant change in the environment of potassium ions with the lead content highlighted for these glasses is in agreement with the clear evolution of the (Na 1s) XPS spectrum observed with the lead content for $70SiO_2$–xPbO–(30-x)Na_2O (mol%) glasses (Gee et al. 2001). These structural evolutions, particularly the formation of mixed Si-O-(Pb,K) bonds to the detriment of simple Si-O-K bonds, explain the effect of the partial substitution of SiO_2 by PbO on the electrical resistivity (increase) of alkaline silicate glasses (Scholze 1980), with the large Pb^{2+} ions impeding the mobility of alkaline ions.

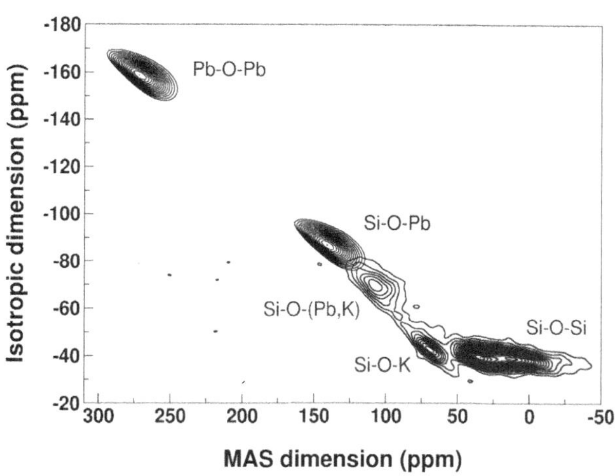

Figure 3.20. *^{17}O MQ-MAS spectrum of a crystal glass ($77.1SiO_2$–10.6PbO–$11.3K_2O$–$0.8Na_2O$–$0.2Sb_2O_3$ mol%) in blue shows the contribution of oxygen atoms of Si-O-Si, Si-O-K and Si-O-(Pb,K) bonds (Angeli et al. 2016). The position of the contribution of Si-O-Pb and Pb-O-Pb bonds obtained by Lee et al. (2014) for PbO-rich binary SiO_2-PbO glasses is represented in black. For a color version of this figure, see www.iste.co.uk/bouquillon/lead.zip*

3.5. Glasses of the SiO_2-PbO-Al_2O_3 system

3.5.1. *Overview of ternary SiO_2-PbO-Al_2O_3 glasses*

Although the glasses of the ternary SiO_2-PbO-Al_2O_3 system have ancient applications, such as glazes deposited on the surface of ceramic objects[20] (Tite et al. 1998) and much more recently as materials containing rare earth ions for optics and dosimetry (Bhargavi et al. 2012, 2014, 2015; Rao et al. 2013, 2014), as well as for protection against ionizing radiation (Singh et al. 2014), only a limited number of works have been published on the glasses of this system. This is particularly the case for studies conducted on their structure and on the impact of the presence of alumina on the environment of Pb^{2+} ions in the vitreous network compared to the extent of the studies conducted on binary SiO_2-PbO glasses (section 3.3.3). Some examples are essentially the recent studies conducted by Bhargavi et al. (2012, 2014, 2015), Rao et al. (2013, 2014) and Ben Kacem (2017) using vibrational or optical spectroscopies, by Roisine (2018) using NMR and Raman spectroscopies and X-ray and neutron scattering, and the more ancient ones of Prabakar and Rao (1991) using NMR spectroscopy and of Roy (1990) using infrared spectroscopy. It is important to note, however, that other structural studies have been conducted on more complex glasses of the SiO_2-PbO-Al_2O_3-B_2O_3 system containing boron in addition to aluminum, but they are not covered here (Sawvel et al. 2005; Saini et al. 2009; Monika et al. 2012). Although the extension of the vitrifiable domain in the ternary SiO_2-PbO-Al_2O_3 system has not been reported in the literature, it can be stated that the Al_2O_3 content of the studied glasses never exceeds 30 mol% (Ben Kacem 2017) and that glasses with only 10 mol% SiO_2 (10SiO_2–80PbO–10Al_2O_3) could be obtained by classical melting-quenching (Prabakar and Rao 1991). It is interesting to note that hyperquenching can be used to prepare binary Al_2O_3-PbO glasses without silicon and with PbO contents ranging between 50 and 95 mol% (Kantor et al. 1973; Morikawa et al. 1981).

20. In the case of highly lead-rich lead glazes, the origin of the alumina in their composition (< 10% Al_2O_3 mass) (Tite et al. 1998; Roisine 2018) may be either the raw materials deposited at the surface of ceramic objects before firing, or the partial dissolution in the thin layer of melting liquid rich in SiO_2 and in PbO of a fraction of alumina initially present in the paste constituting the objects.

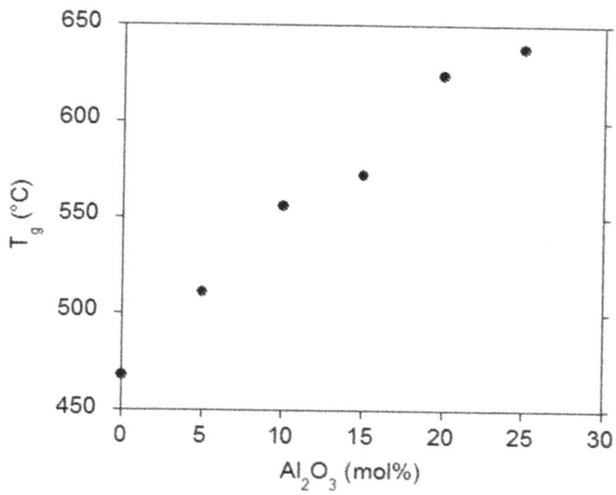

Figure 3.21. *Evolution of the glass transition temperature T_g with increasing Al_2O_3 contents to the $40SiO_2$–$60PbO$ (mol%) composition (according to the data of Roisine 2018)*

There is a multifaceted interest in adding aluminum to the composition of lead silicate glasses due to the significant effect of this element on their properties. Hence, Al_2O_3 is assumed to increase the mechanical resistance of binary glasses of the SiO_2-PbO system and decrease their crystallization tendency (Mylyanych et al. 1999). Moreover, according to the phase diagram of the SiO_2-PbO-Al_2O_3 system (Chen et al. 2001), the introduction of alumina in the binary SiO_2-PbO system would reduce the demixing tendency of the liquid. However, the liquidus temperature tends to significantly increase, particularly for the lead-richest compositions corresponding to the domain of lead glazes, which may pose melting problems and therefore limit the amount of alumina that can be added. Moreover, for this type of PbO-rich composition, the glass transition temperature and the viscosity of the liquid increase rapidly with the alumina rate (Figures 3.21 and 3.22).

It is also important to note the existence of a large number of ternary crystalline phases (11 phases according to Chen et al. 2001) in this domain of composition that may contribute, in addition to the strong increase in viscosity of the melting liquid, to facilitate its vitrification during cooling according to the "frustration" principle (Mysen and Richet 2005). Finally, the positive effect of the presence of aluminum in lead silicate glasses containing rare earth ions is worth mentioning, as it leads to a better dispersion of the latter in the vitreous structure (destruction of clusters) (Rao et al. 2013).

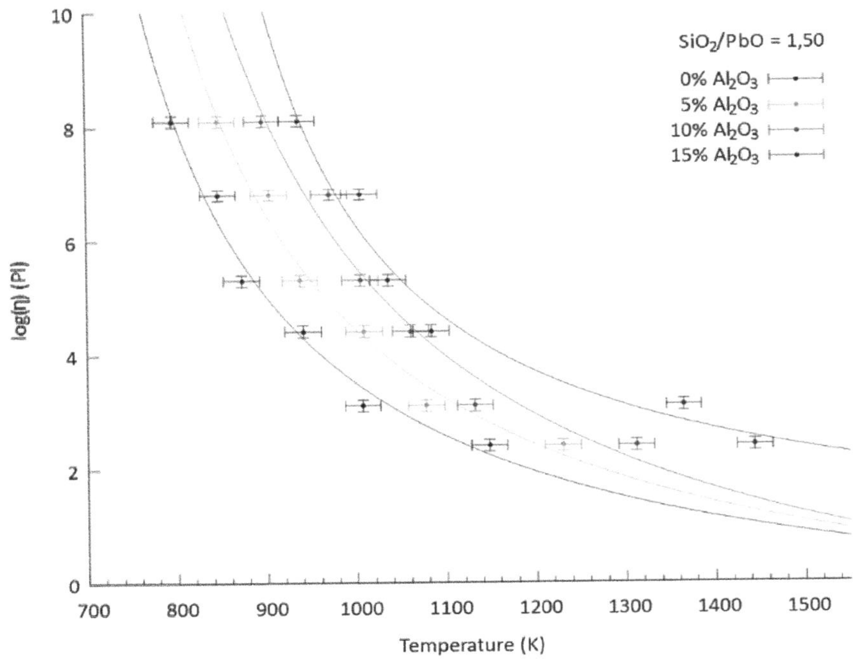

Figure 3.22. *Evolution of viscosity η (in poiseuilles Pl) as a function of temperature after adding increasing Al_2O_3 contents to the $40SiO_2$–60PbO (mol%) composition (Roisine 2018). For a color version of this figure, see www.iste.co.uk/bouquillon/lead.zip*

3.5.2. *Structure of ternary SiO_2-Al_2O_3-PbO glasses*

3.5.2.1. *Crystalline phases of the SiO_2-Al_2O_3-PbO system*

Similar to the above-mentioned ternary systems of alkaline or alkaline-earth lead silicates (section 3.4.2), the known structures of ternary Pb-Si-Al-O oxides are very few and come down to lead feldspar $PbAl_2Si_2O_8$ (Benna et al. 1996) and other zeolite-type structures (Ronay and Seff 1985; Yeom et al. 1997). However, unlike the crystalline phases of SiO_2-PbO, SiO_2-PbO-R_2O and SiO_2-PbO-R'O (R: alkaline and R': alkaline-earth) systems, in these structures, the lengths of Pb-O bonds are relatively regular. Their environment in $PbAl_2Si_2O_8$ is thus very similar to that of Sr^{2+} ions in the equivalent feldspar, highlighting the absence of stereochemical activity of the lead $6s^2$ pair in this very aluminum-rich phase. This is likely explained by the strength of the aluminosilicate skeleton and its maximal cross-linking. Bridging two SiO_4 and AlO_4 tetrahedrons, an oxygen anion is available at only approximately 0.25 u.v. to share with the neighboring Pb^{2+} ions, which prevents the formation of strong bonds and hence the highly structuring Pb-O-Pb cross-linkages observed in the depolymerized silicates of SiO_2-PbO, SiO_2-PbO-R_2O and SiO_2-PbO-R'O systems. In this type of aluminum-rich crystalline structure, Pb^{2+} ions compensate for the charge of the AlO_4^- tetrahedral units, unlike the previous systems considered in this chapter.

3.5.2.2. *Aluminosilicate vitreous network and the lead environment*

Since the earliest structural studies conducted on ternary SiO_2-Al_2O_3-PbO glasses (Prabakar and Rao 1991), ^{27}Al NMR revealed that aluminum was incorporated in their structure essentially in the form of AlO_4^- tetrahedral entities and that a part of the Pb^{2+} ions present in the structure acted as charge compensators of these entities for Al_2O_3 contents of 10 mol% and PbO contents ranging between 30 and 80 mol%. Concerning the silicate network, the studies conducted more recently on various series of ternary glasses by Raman and ^{29}Si NMR have shown that the addition of alumina or its partial substitution to PbO in the binary SiO_2-PbO glasses led in all cases to a strong evolution of their structure (Ben Kacem 2017; Roisine 2018) (Figures 3.23 and 3.24).

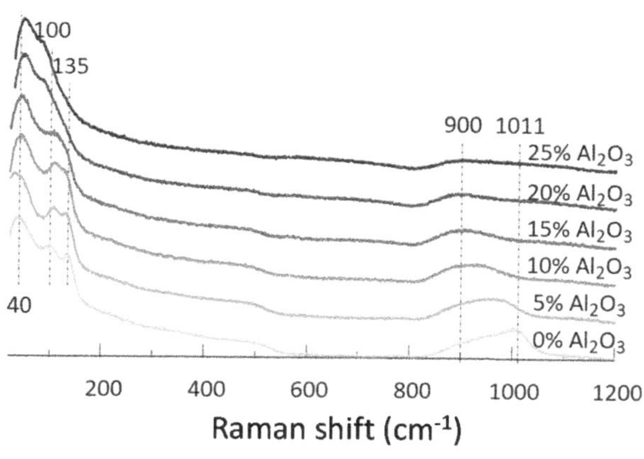

Figure 3.23. *Evolution of Raman spectra of ternary SiO_2-Al_2O_3-PbO glasses obtained by adding increasing Al_2O_3 contents (0–25 mol%) to the binary $60SiO_2$–40PbO (mol%) glass (Roisine 2018)*

It can be observed that by adding increasing quantities of alumina to lead silicate glasses, there is a shift toward low energies and a change in the shape of the Raman band between 800 and 1,200 cm^{-1} associated with elongation vibrations of T-O bonds of tetrahedral entities TO_4 (T = Si, Al) (Figure 3.23). The evolution of this band can only be explained by the gradual incorporation of aluminum in the form of AlO_4^- units in the silicate network and not by a depolymerization of the latter according to ^{17}O NMR results (Figures 3.25 and 3.26) and the evolution of physical properties (strong increase of T_g and viscosity, Figures 3.21 and 3.22).

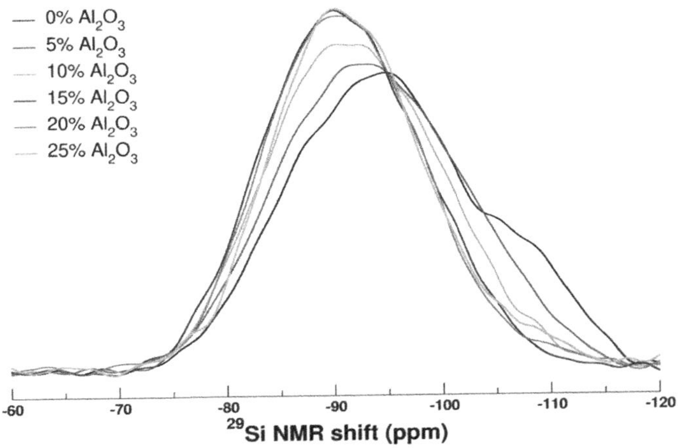

Figure 3.24. *Evolution of ^{29}Si MAS NMR spectra of ternary SiO_2-Al_2O_3-PbO glasses obtained by adding increasing Al_2O_3 contents (0–25 mol%) to the binary $60SiO_2$–40PbO (mol%) glass. For the glass without Al_2O_3, the contributions toward – 110 and – 95 ppm correspond to the Q^4 and Q^3 entities, respectively (Roisine 2018). For a color version of this figure, see www.iste.co.uk/bouquillon/lead.zip*

Indeed, it is known for other silicate glasses without lead that the presence of fourfold coordinated aluminum in the silicon environment (as a second neighbor) leads to a shift of the elongation vibration bands of Q^n units toward low energies (Rossano and Mysen 2012). The evolution of the ^{29}Si NMR signal toward the higher values of the chemical shift with the Al_2O_3 content (Figure 3.24) can be similarly explained; the chemical shift of Q^n units increases when a growing number of AlO_4^- tetrahedrons are present as second neighbors of SiO_4 units (Kirkpatrick et al. 1985). The low energy region (100–200 cm^{-1}) of Raman spectra – corresponding to Pb-O vibrations (Ben Kacem 2017) – also has a strong evolution with the content of alumina added to the binary SiO_2-PbO glasses (Figure 3.23). A strong evolution in this region of the spectrum has also been observed if PbO is gradually replaced by Al_2O_3 with a constant SiO_2 content (Ben Kacem 2017). This shows a significant change in the local environment of Pb^{2+} ions, which may be explained by the gradual disappearance of quite strong Pb-O-(Pb,Si) bonds to the benefit of weaker Pb-O-Al bonds associated with the role of the charge compensator of AlO_4^- units played by an increasing number of Pb^{2+} ions. Indeed, the direct study of the aluminum environment by ^{27}Al NMR in these glasses (Figure 3.27) shows that it is still predominantly present in the form of AlO_4^- (and therefore necessarily locally compensated by Pb^{2+} ions, which are the only cations susceptible to play this role in ternary glasses), with the increasing but sill minority contribution of AlO_5 units highlighted by ^{27}Al MQMAS NMR for the highest alumina contents (>15 mol% Al_2O_3) (Roisine 2018).

Concerning the oxygen atoms in the structure, the ^{17}O NMR study of ^{17}O-enriched ternary glasses also confirms a strong evolution of the lead environment following the addition of alumina in the binary SiO_2-PbO glasses (Figures 3.25 and 3.26). Indeed, a strong decrease in the contribution of Si-O-Pb bonds to the benefit of Si-O-Al bonds can be noted, thus showing a strong repolymerization of the vitreous network (strong decrease in the proportion of non-bridging oxygen atoms). Moreover, the small number of Pb-O-Pb bonds present in the binary glass seems to disappear in the ternary glass (Figure 3.25), which may suggest a better dispersion of lead in the aluminosilicate network, in agreement with the results obtained by Raman spectroscopy (region of low-frequency spectra).

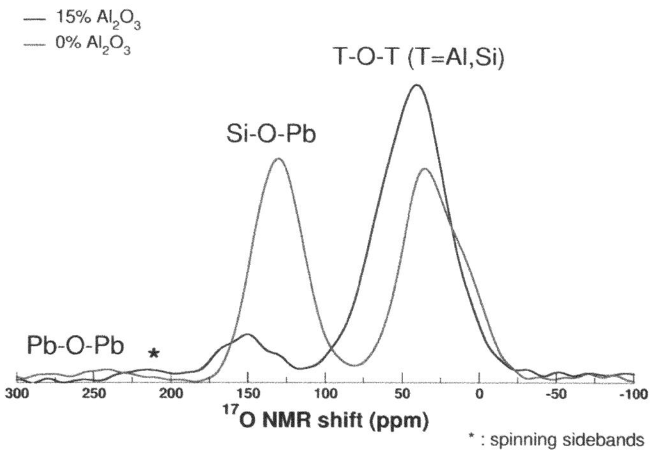

Figure 3.25. *Comparison of ^{17}O MAS NMR spectra of $60SiO_2$–40PbO and $51SiO_2$–34PbO–$15Al_2O_3$ (mol%) glasses enriched in ^{17}O (*rotational band) (Roisine 2018). For a color version of this figure, see www.iste.co.uk/bouquillon/lead.zip*

However, additional studies conducted on the same glasses by X-ray scattering did not reveal a significant evolution of the average Pb-Pb distance during the addition of alumina to the binary SiO_2-PbO glasses (Roisine 2018). According to the structural results concerning the stereochemistry of Pb^{2+} ions in Al_2O_3-rich crystalline phases such as $PbAl_2Si_2O_8$ (section 3.5.2.1), for which the free $6s^2$ pairs are stereochemically inactive, it may be thought that the same is valid for ternary SiO_2-Al_2O_3-PbO glasses, at least for Pb^{2+} ions playing the role of charge compensator of AlO_4^- units, and excess Pb^{2+} ions are likely to preserve their $6s^2$ pair active, similar to the binary SiO_2-PbO glasses.

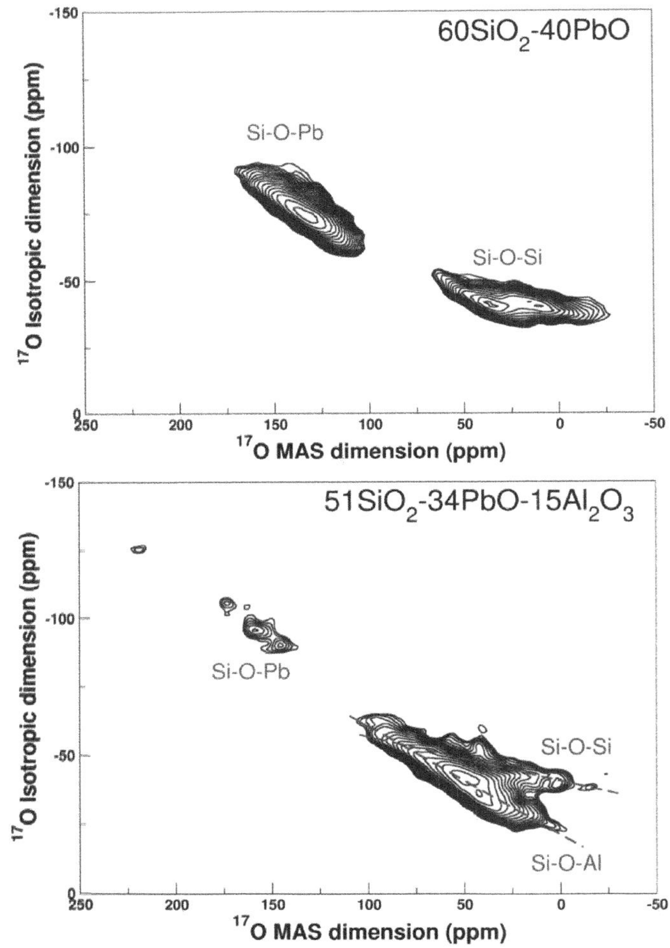

Figure 3.26. *^{17}O MQMAS NMR spectra of 60SiO2–40PbO and 51SiO2–34PbO–15Al2O3 glasses in mol% enriched in ^{17}O (Roisine 2018). For a color version of this figure, see www.iste.co.uk/bouquillon/lead.zip*

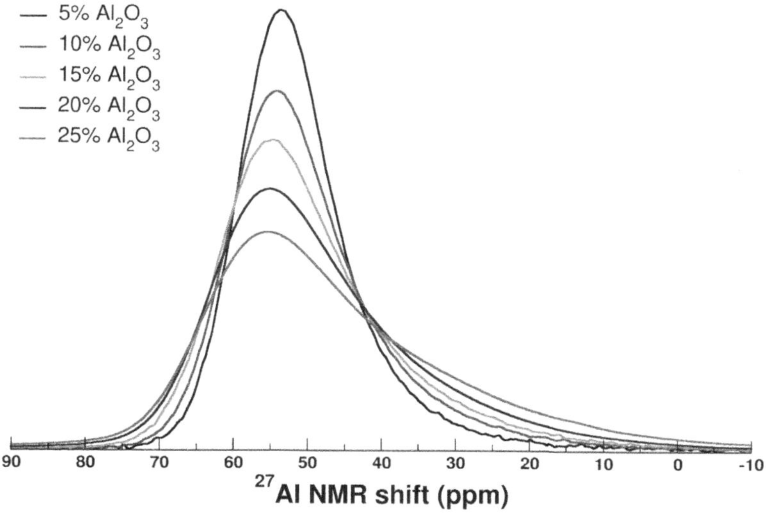

Figure 3.27. *Evolution of ^{27}Al MAS NMR spectra of ternary SiO_2-Al_2O_3-PbO glasses obtained by adding increasing Al_2O_3 contents (0–25 mol%) to the binary $60SiO_2$–40PbO (mol%) glass (Roisine 2018). For a color version of this figure, see www.iste.co.uk/bouquillon/lead.zip*

COMMENT ON FIGURE 3.27.– *The signal broadening and shift toward higher values of the chemical shift are due to an increase in the quadrupolar coupling constant C_Q and an increase in the isotropic chemical shift δ_{iso} of ^{27}Al, reflecting a strong evolution of the distortion of AlO_4^- sites when Al_2O_3 content increases.*

3.6. Conclusion

Given its high polarizability and its specific electron configuration (lone $6s^2$ pair), Pb^{2+} ion has a unique stereochemistry, which leads to obtaining binary SiO_2-PbO silicate glasses that may contain very high PbO contents, significantly higher in particular than the quantity of alkaline R_2O oxides that can be introduced in the binary glasses of alkaline silicates (SiO_2-R_2O). Although the first structural studies conducted on binary glasses of lead silicates date back over 80 years, there are still unanswered questions on the structural role of PbO in silicate glasses and on the evolution of the distribution of Pb^{2+} ions as a function of the lead concentration. However, the implementation of many investigation methods coupled with numerical simulation methods has led to many publications and to increasingly precise knowledge of the structure of these glasses. Hence, for the binary glasses of the SiO_2-PbO system, it appears that – as for the PbO crystalline phases and lead silicates – Pb^{2+} ions are located in very asymmetrical sites of low coordination (3,4) with their stereochemically active lone $6s^2$ pair. Although the addition of PbO leads to a depolymerization of the silicate network, the structural role of Pb^{2+} ions is very different from that of bivalent ions, such as alkaline-earth ions. Indeed, on the one hand, significantly stronger bonds are established starting with the low PbO contents between the Pb^{2+} ions and the non-bridging oxygen atoms of the silicate network, and, on the other hand, for high PbO contents, an increasing proportion of O^{2-} ions are found connected only to Pb^{2+} ions, gradually forming an increasingly extended lead network. Nevertheless, even for the strongest PbO contents, a significant proportion of SiO_4 units remain connected to one another, thus contributing to the stabilization of the lead network in relation to crystallization. On the other hand, there are significantly fewer structural studies conducted on other lead-rich systems, but they are very instructive. Hence, in alkaline silicate glasses, for low lead contents, Pb^{2+} ions tend to be incorporated in areas containing alkaline ions, where they associate with non-bridging oxygen atoms, while for high PbO contents, Pb-O-Pb connections appear, as a signature of the formation of an increasingly extended lead network. In these glasses, Pb^{2+} ions preserve their stereochemically active lone $6s^2$ pair and are located in very asymmetric sites by analogy with crystalline phases. On the other hand, in the case of lead aluminosilicate glasses, Pb^{2+} ions tend to play the role of charge compensators of the AlO_4^- units and likely lose the stereochemical activity of their $6s^2$ pair. However, questions still remain regarding the evolution of the lead network as a function of Al_2O_3 content in these glasses.

3.7. References

Alderman, O.L.G. (2013). The structure of vitreous binary oxides: Silicate, germanate and plumbite networks. PhD thesis, University of Warwick, Coventry.

Alderman, O.L.G., Hannon, A.C., Holland, D., Feller, S., Lehr, G., Vitale, A.J., Hoppe, U., Zimmerman, M.V., Watenphule, A. (2013). Lone-pair distribution and plumbite network formation in high lead silicate glass, $80PbO.20SiO_2$. *Phys. Chem. Chem. Phys.*, 15, 8506–8519.

Andersen, O. (1919). The volatilization of lead oxide from lead silicate melts. *J. Am. Ceram. Soc.*, 2, 784–789.

Angeli, F., Jollivet, P., Charpentier, T., Fournier, M., Gin, S. (2016). Structure and chemical durability of lead crystal glass. *Environ. Sci. Technol.*, 50, 11549–11558.

Ashour, G.M. and Shehata, A.M. (1972). Study of crystallization of chromium oxide in glass (chrome aventurine). *Trans. Indian Ceram. Soc.*, 31, 136–143.

Bair, C.J. (1936). The constitution of lead oxide-silica glasses I: atomic arrangement. *J. Am. Ceram. Soc.*, 19, 339–347.

Bassi, G. and Lajzerowicz-Bonneteau, J. (1965). Structure du pyrogermanate de plomb potassium $K_2Pb_2Ge_2O_7$. *Bull. Soc. Fr. Minéral. Crist.*, 88, 342–344.

Ben Kacem, I. (2017). Du cristal au plomb jusqu'aux déchets domestiques : rôle du plomb dans les verres et les vitrocéramiques : étude des systèmes PbO-SiO_2, PbO-CaO-SiO_2 et PbO-Al_2O_3-SiO_2. PhD Thesis, Université Paris-Est, Paris.

Ben Kacem, I., Gautron, L., Coillot, D., Neuville, D.R. (2017). Structure and properties of lead silicate glasses and melts. *Chem. Geol.*, 461, 104–114.

Benna, P., Tribaudino, M., Bruno, E. (1996). The structure of ordered and disordered lead feldspar ($PbAl_2Si_2O_8$). *Am. Mineral.*, 81, 1337–1343.

Bessada, C., Massiot, D., Coutures, J., Douy, A., Coutures, J.-P., Taulelle, F. (1994). ^{29}Si MAS-NMR in lead silicates. *J. Non-Cryst. Solids*, 168, 76–85.

Bhargavi, K., Reddy, M.S., Rao, P.R., Rao, N.N., Rao, M.S., Kumar, V.R., Veeraiah, N. (2012). The structural influence of aluminium ions on emission characteristics of Sm^{3+} ions in lead aluminium silicate glass system. *Mater. Res. Bull.*, 47, 267–273.

Bhargavi, K., Rao, M.S., Veeraiah, N., Sanyal, B., Gandhi, Y., Baskaran, G.S. (2014). Thermal luminescence of Ho_2O_3-PbO-Al_2O_3-SiO_2 glasses exposed to gamma radiation. *Int. J. Appl. Glass Sci.*, 6, 128–136.

Bhargavi, K., Sanyal, B., Rao, M.S., Kumar, V., Gandhi, Y., Baskaran, G.S., Veeraiah, N. (2015). γ-Ray induced thermoluminescence characteristics of the PbO-Al_2O_3-SiO_2:Pr^{3+} glass system. *J. Lumin.*, 161, 417–421.

Boucher, M.L. and Peacor, D.R. (1968). The crystal structure of alamosite, $PbSiO_3$. *Z. Kristallogr.*, 126, 98–111.

Brandes, R. and Hoppe, R. (1994). $Li_{10}Si_2Pb^{II}O_{10} = Li_{20}[(SiO_4)_4(OPbO_2PbO)]$ – daserste "gemischte" silicat-plumbat(II). *Z. Anorg. Allg. Chem.*, 620, 2026–2032.

Brown, I.D. and Altermatt, D. (1985). Bond-valence parameters obtained from a systematic analysis of the inorganic crystal structure database. *Acta Cryst.*, B41, 244–247.

Calvert, P.D. and Shaw, R.R. (1970). Liquidus behavior in the silica-rich region of the system PbO-SiO_2. *J. Am. Ceram. Soc.*, 53, 350–352.

Carr, S.M. and Subramanian, K.N. (1982a). Spherulitic crystal growth in P_2O_5-nucleated lead silicate glasses. *J. Non-Cryst. Solids*, 60, 307–312.

Carr, S.M. and Subramanian, K.N. (1982b). Effect of P_2O_5 on the crystallization of lead silicate glasses. *J. Am. Ceram. Soc.*, 65, 346–349.

Charles, R.J. (1966). Metastable liquid immiscibility in alkali metal oxide-silica systems. *J. Am. Ceram. Soc.*, 49, 55–62.

Charpentier, T., Menziani, M.C., Pedone, A. (2013). Computational simulations of solid state NMR spectra: A new era in structure determination of oxide glasses. *RCS Adv.*, 3, 10550–10578.

Chen, S., Zhao, B., Hayes, P.C., Jak, E. (2001). Experimental study of phase equilibria in the $PbO-Al_2O_3-SiO_2$ system. *Metal. Mater. Trans. B*, 32B, 997–1005.

Chevrier, G., Giester, G., Heger, G., Jarosch, D., Wildner, M., Zemann, J. (1992). Neutron single-crystal refinement of cerussite, $PbCO_3$, and comparison with other aragonite-type carbonates. *Z. Kristallogr.*, 199, 67–74.

Dalby, K.N., Wayne Nesbitt, H., Zakaznova-Herzog, V.P., King, P.L. (2007). Resolution of bridging signals from O 1s spectra of silicate glasses using XPS: Implications for O and Si speciation. *Geochem. Cosmochim. Acta*, 71, 4297–4313.

Dayanand, C., Bhikshamaiah, G., Jaya Tyagaraju, V., Salagram, M., Krishna Murthy, A.S.R. (1996). Structural investigations of phosphate glasses: A detailed infrared study of the $x(PbO)-(1-x) P_2O_5$ vitreous system. *J. Mater. Sci.*, 31, 1945–1967.

Dent Glasser, L.S., Howie, R.A., Smart, R.M. (1981). The structure of lead orthosilicate, $2PbO.SiO_2$. *Acta Cryst. B*, 37, 303–306.

Dietzel, A. (1948). Structure and properties of glass. *Glastechn. Ber.*, 22, 41–50.

Dimitrov, V. and Komatsu, T. (2010). An interpretation of optical properties of oxides and oxide glasses in terms of the electronic polarizability and average single bond strength (review). *J. Univ. Chem. Technol. Metall.*, 45, 219–250.

El Shamy, T.M. and Taki-Eldin, H.D. (1974). The chemical durability of $PbO-SiO_2$ glasses. *Glass Technol.*, 15, 48–52.

El-Egili, K., Doweidar, H., Moustafa, Y.M., Abbas, I. (2003). Structure and some physical properties of $PbO-P_2O_5$ glasses. *Physica B*, 339, 237–245.

Ellis, E., Johnson, D.W., Breeze, A., Magee, P.M., Perkins, P.G. (1979). The electronic band structure and optical properties of oxide glasses. *Philos. Mag. B*, 40, 125–137.

Faivre, R. and Weiss, R. (1963). Le plomb. In *Nouveau traité de chimie minérale*, volume 8, Pascal, P. (ed.). Masson, Paris.

Fajans, K. and Kreidl, N.J. (1948). Stability of lead glasses and polarization of ions. *J. Am. Ceram. Soc.*, 31, 105–114.

Fayon, F. (1998). Caractérisation du rôle structural du plomb dans des matrices vitreuses : approche par RMN haute résolution solide mono et bidimensionnelle. PhD Thesis, University of Orléans, Orléans.

Fayon, F., Farnan, I., Bessada, C., Coutures, J., Massiot, D., Coutures, J.-P. (1997). Empirical correlations between ^{207}Pb NMR chemical shifts and structure in solids. *J. Am. Chem. Soc.*, 119, 6837–6843.

Fayon, F., Bessada, C., Massiot, D., Farnan, I., Coutures, J.-P. (1998). ^{29}Si and ^{207}Pb NMR study of local order in lead silicate glasses. *J. Non-Cryst. Solids*, 232–234, 403–408.

Fayon, F., Landron, C., Sakurai, K., Bessada, C., Massiot, D. (1999). Pb^{2+} environment in lead silicate glasses probed by Pb-L_{III} edge XAFS and ^{207}Pb NMR. *J. Non-Cryst. Solids*, 243, 39–44.

Feller, S., Riley, A., Edwards, T., Croskrey, J., Schue, A., Liss, D., Stentz, D., Blair, S., Kelley, M., Smith, G. et al. (2010). A multispectroscopic structural study of lead silicate glasses over an extended range of compositions. *J. Non-Cryst. Solids*, 356, 304–313.

Flood, H. and Förland, T. (1947). The acidic and basic properties of oxides. *Acta Chem. Scand.*, 1, 592–604.

Freed, R.L. and Peacor, D.R. (1969). Determination and refinement of the crystal structure of margaro-sanite $PbCa_2Si_3O_9$. *Z. Kristallogr.*, 128, 213–228.

Frieser, R.J. (1975). A review of solder glasses. *Electro. Comp. Sci. Tech.*, 2, 163–199.

Furukawa, T., Brawer, S.A., White, W.B. (1978). The structure of lead silicate glasses determined by vibrational spectroscopy. *J. Mater. Sci.*, 14, 268–282.

Gee, I.A., Holland, D., McConville, C.F. (2001). Atomic environments in binary lead silicate and ternary alkali lead silicate glasses. *Phys. Chem. Glasses*, 42, 339–348.

Geller, R.F., Creamer, A.S., Bunting, E.N. (1934). The system PbO-SiO_2. *J. Res. Natl. Bur. Stand.*, 13, 237–244.

Gillespie, R.J. and Nyholm, R.S.Q. (1957). Inorganic stereochemistry. *Rev. Chem. Soc.*, 11, 339–380.

Gourlaouen, C. and Parisel, O. (2007). Is an electronic shield at the molecular origin of lead poisoning? A computational modeling experiment. *Angew. Chem.*, 119, 559–562.

Greenough, D.R., Dentschuk, P., Palmer, S.B. (1981). Thermal expansion of lead silicate glasses. *J. Mater. Sci.*, 19, 599–603.

Hirota, K. and Hasegawa, Y.T. (1981). Phase relations in the system PbO-$PbSiO_3$. *Bull. Chem. Soc. Jpn.*, 54, 754–756.

Hoppe, U., Kranold, R., Ghosh, A., Landron, C., Neuefeind, J., Jovari, P. (2003). Environments of lead cations in oxide glasses probed by X-ray diffraction. *J. Non-Cryst. Solids*, 328, 146–156.

Imaoka, M. and Yamazaki, T. (1963). Studies of the glass-formation range silicate systems. Investigations on the glass-formation range 2. *J. Ceram. Assoc. Jpn.*, 71, 215–223.

Jak, E., Hayes, P.C., Degterov, S., Pelton, A.D., Wu, P. (1997). Thermodynamic optimization of the systems PbO-SiO_2, PbO-ZnO, ZnO-SiO_2 and PbO-ZnO-SiO_2. *Metall. Mater. Trans. B.*, 28, 1011–1018.

Kantor, P., Revcolevschi, A., Collongues, R. (1973). Preparation of iron sesquioxide glasses by ultra-fast quenching from the melt ("splat-cooling"). *J. Mater. Sci.*, 8, 1359–1361.

Kato, K. (1982). Die Kristallstruktur des Bleisilicats $Pb_{11}Si_3O_{17}$. *Acta Cryst. B*, 38, 57–62.

Kaur, A., Khanna, A., Singla, S., Dixi, A., Kothiyal, G.P., Krishnan, K., Aggarwal, S.K., Sath, V., González, F., González-Barriuso, M. (2013). Structure–property correlations in lead silicate glasses and crystalline phases. *Phase Transit.*, 86, 759–777.

Kirkpatrick, R.J., Smith, K.A., Schramm, S., Turner, G., Yang, W.-H. (1985). Solid-state nuclear magnetic resonance spectroscopy of minerals. *Ann. Rev. Earth Planet. Sci.*, 13, 29–47.

Kohara, S., Ohno, H., Takata, M., Usuki, T., Morita, H., Suzuya, K., Akola, J., Pusztai, L. (2010). Lead silicate glasses: Binary network-former glasses with large amounts of free volume. *Phys. Rev. B*, 82, 134209.

Korson, L., Drost-Hansen, W., Millero, F.J. (1969). Viscosity of water at various temperatures. *J. Phys. Chem.*, 73, 34–39.

Lee, S.K. and Kim, E.J. (2015). Probing metal-bridging oxygen and configurational disorder in amorphous lead silicates: Insights from ^{17}O solid-state nuclear magnetic resonance. *J. Phys. Chem. C*, 119, 748–756.

Lippmaa, E., Samoson, A., Mägi, M., Teeäär, R., Schraml, J., Götz, J. (1982). High resolution ^{29}Si NMR study of the structure and devitrification of lead-silicate glasses. *J. Non-Cryst. Solids*, 50, 215–218.

Maekawa, H., Maekawa, T., Kawamura, K., Yokokawa, T. (1991). The structural groups of alkali silicate glasses determined from ^{29}Si MAS-NMR. *J. Non-Cryst. Solids*, 127, 53–64.

Martlew, D. (2005). Viscosity of molten glasses. In *Properties of Glass-Forming Melts*, Pye, L.D., Montenero, A., Joseph, I. (eds). CRC Press, Boca Raton.

McMurdie, H.F. (1941). $K_2Pb_2Si_2O_7$. *J. Res. Natl. Bur. Stand.*, 26, 489.

Mizuno, M., Takahashi, M., Takaishi, T., Yoko, T. (2005). Leaching of lead and connectivity of plumbate networks in lead silicate glasses. *J. Am. Ceram. Soc.*, 88, 2908–2912.

Monika, Falconieri, M., Baccaro, S., Sharma, G., Thind, K.S., Singh, D.P. (2012). Role of aluminum oxide in the structure of heavy metal oxide borosilicate glasses. *Phys. Status Solidi A*, 209, 1438–1444.

Morikawa, H., Jegoudez, J., Mazières, C., Revcolevschi, A. (1981). Glass forming ability and crystallization in the PbO-Al_2O_3 system. *J. Non-Cryst. Solids*, 44, 107–112.

Mydlar, M.F., Kreidl, N.J., Hendren, J.K., Clayton, G.T. (1970). X-ray diffraction study of lead silicate glasses. *Phys. Chem. Glasses*, 11, 196–204.

Mylyanych, L., Sheredko, M.A., Melnyk, S.K. (1999). Study of glass structures and crystalline phases in the PbO-Al_2O_3-SiO_2 system. *J. Anal. At. Spectrom.*, 14, 513–521.

Mysen, B. and Richet, P. (2005). *Silicate Glasses and Melts*. Elsevier Sciences, Amsterdam.

Neiman, T.S., Yinnon, H., Uhlmann, D.R. (1982). Crystallization kinetics of lead metasilicate. *J. Non-Cryst. Solids*, 48, 393–403.

O'Shaughnessy, C., Henderson, G.S., Wayne Nesbitt, H., Bancroft, G.M., Neuville, D.R. (2020). The influence of modifier cations on the Raman stretching modes of Q_n species in alkali silicate glasses. *J. Am. Ceram. Soc.*, 103, 3991–4001.

Palomar, T., Mosa, J., Aparicio, M. (2020). Hydrolytic resistance of K_2O-PbO-SiO_2 glasses in aqueous and high-humidity environments. *J. Am. Ceram. Soc.*, 103, 5248–5258.

Pannhorst, W. and Lohn, J. (1970). ZurKristallstruktur von Strontianit, $SrCO_3$. *Z. Kristallogr.*, 131, 455–459.

Patrick, L. (2006). Lead toxicity, a review of the literature. Part 1: Exposure, evaluation, and treatment. *Altern. Med. Rev.*, 11, 2–22.

Pena, R.B., Sampaio, D.V., Lancelotti, R.F., Cunha, T.R., Zanotto, E.D., Pizani, P.S. (2020). In-situ Raman spectroscopy unveils metastable crystallization in lead metasilicate glass. *J. Non-Cryst. Solids*, 546, 120254.

Petter, W., Harnik, A.B., Keppler, U. (1971). Die Kristallstruktur von Blei-Barysilit, $Pb_3Si_2O_7$. *Z. Kristallogr.*, 133, 445–458.

Prabakar, S. and Rao, K.J. (1991). MAS NMR studies of ternary silicate glasses. *Phil. Mag. B*, 64, 401–411.

Pye, L.D. (1988). Aluminate glasses – A review. *Proc. SPIE 0929, Infrared Optical Materials IV*, 149–156.

Rabinovich, E.M. (1976). Lead in glasses. *J. Mater. Sci.*, 11, 925–948.

Rao, M.S., Sudarsan, V., Brik, M.G., Bhargavi, K., Rao, C.S., Gandhi, Y., Veeraiah, N. (2013). The de-clustering influence of aluminum ions on the emission features of Nd^{3+} ions in PbO–SiO_2 glasses. *Opt. Commun.*, 298–299, 135–140.

Rao, M.S., Sanyal, B., Bhargavi, K., Vijay, R., Kityk, I.V., Veeraiah, N. (2014). Influence of induced structural changes on thermoluminescence characteristics of γ-ray irradiated PbO–Al_2O_3–SiO_2: Dy^{3+} glasses. *J. Mol. Struct.*, 1073, 174–180.

Roisine, G. (2018). Céramiques glaçurées de Bernard Palissy : à la recherche des secrets d'un maître de la Renaissance. PhD Thesis, Université Paris Sciences et Lettres, Paris.

Ronay, C. and Seff, K. (1985). Crystal structures of Pb6-A and $Pb_9(OH)_8(H_2O)_3$-A. Zeolite A ion exchanged with Pb^{2+} at pH 4.3 and 6.0 and evacuated. *J. Phys. Chem.*, 89, 1965–1970.

Rossano, S. and Mysen, B. (2012). Raman spectroscopy applied to earth sciences and cultural heritage. In *EMU Notes in Mineralogy, Volume 12*, Dubessy, J., Caumon, M.-C., Rull, F. (eds). Eotvos University Press, Budapest.

Roy, B.N. (1990). Infrared spectroscopy of lead and alkaline-earth aluminosilicate glasses. *J. Am. Ceram. Soc.*, 73, 846–855.

Rybicki, J., Rybicka, A., Witkowska, A., Bergmanski, G., Di Cicco, A., Minicucci, M., Mancini G. (2001). The structure of lead-silicate glasses: Molecular dynamics and EXAFS studies. *J. Phys. Condens. Matter*, 13, 9781–9797.

Saini, A., Khanna, A., Michaelis, V.K., Kroeker, S., González, F., Hernández, D. (2009). Structure–property correlations in lead borate and borosilicate glasses doped with aluminum oxide. *J. Non-Cryst. Solids*, 355, 2323–2332.

Sawvel, A.M., Chinn, S.C., Bourcier, W.L., Maxwell, R.S. (2005). Local structure of amorphous $(PbO)_x[(B_2O_3)_{1-z}(Al_2O_3)_z]_y(SiO_2)_y$ dielectric materials by multinuclear solid state NMR. *Chem. Mater.*, 17, 1493–1500.

Scholze, H. (1980). *Le verre. Nature, structure et propriétés*. Institut du verre, Paris.

Seward, T.P., Uhlmann, D.R., Turnbull, D. (1968). Phase separation in the system BaO-SiO_2. *J. Am. Ceram. Soc.*, 51, 278–285.

Shannon, R.D. (1976). Revised effective ionic radii and systematic studies of interatomic distances in halides and chacogenides. *Acta Crysta*, A32, 751–767.

Shannon, R.D. and Fischer, R.X. (2016). Empirical electronic polarizabilities of ions for the prediction and interpretation of refractive indices: Oxides and oxysalts. *Am. Mineral.*, 101, 2288–2016.

Shelby, J.E. (1983). Property/structure relationships in lead silicate glasses. *Glastech. Ber.*, 56, 1057–1062.

Shevchenko, M. and Jak, E. (2018). Experimental phase equilibria studies of the PbO–SiO_2 system. *J. Am. Ceram. Soc.*, 101, 458–471.

Shrikhande, V.K., Sudarsan, V., Kothiyal, G.P., Kulshreshtha, S.K. (2001). ^{29}SiMAS NMR and microhardness studies of some lead silicate glasses with and without modifiers. *J. Non-Cryst. Solids*, 283, 18–26.

Shrikhande, V.K., Sudarsan, V., Kothiyal, G.P., Kulshreshtha, S.K. (2007). Photoluminescence and structural studies on Na_2O-PbO-SiO_2 glasses. *J. Non-Cryst. Solids*, 353, 1341–1345.

Siidra, O.I., Zenko, D.S., Krivovichev, S.V. (2014). Structural complexity of lead silicates: Crystal structure of $Pb_{21}[Si_7O_{22}]_2[Si_4O_{13}]$ and its comparison to hyttsjöite. *Am. Mineral.*, 99, 817–823.

Singh, K.J., Kaur, S., Kaundal, R.S. (2014). Comparative study of gamma ray shielding and some properties of PbO-SiO_2-Al_2O_3 and Bi_2O_3-SiO_2-Al_2O_3 glass systems. *Radiat. Phys. Chem.*, 96, 153–157.

Sinn, E. (1979). Measurement of the shear viscosity in garnet high temperature solutions by a cylindrical rotation viscometer. *Krist. Tech.*, 14, 117–127.

Smart, R.M. and Glasser, F.P. (1974). Compound formation and phase equilibria in the system PbO-SiO_2. *J. Am. Ceram. Soc.*, 57, 378–382.

Smets, B.M.J. and Lommen, T.P.A. (1982). The structure of glasses and crystalline compounds in the system $PbO-SiO_2$, studied by X-ray photoelectron spectroscopy. *J. Non-Cryst. Solids*, 48, 423–430.

Sorokina, M.F., Kanunnikova, O.M., Gil'mutdinov, F.Z., Kozhevnikov, V.I. (1996). X-ray electron spectroscopy investigation of the structure of double lead silicate glasses. *Glass Ceram.*, 53, 11–13.

Stols-Witlox, M. (2011). *The Heaviest and the Whitest: Lead White Quality in Northwestern European Documentary Sources, 1400–1900*. Archetype Editions, London.

Suzuya, K., Kohara, S., Ohno, H. (1999). A reverse Monte Carlo study of lead metasilicate glass. *Jpn. J. Appl. Phys.*, 38–1, 144–147.

Takaishi, T., Takahashi, M., Jin, J., Uchino, T., Yoko, T. (2005). Structural study on $PbO-SiO_2$ glasses by X-ray and neutron diffraction and ^{29}Si MAS NMR measurements. *J. Am. Ceram. Soc.*, 88, 1591–1596.

Takamori, T. (1979). Solder glasses. In *Treatise on Materials Science and Technology, Volume 17, Glass II*, Tomozawa, M., Doremus, R.H. (eds). Academic Press, New York.

Tite, M.S., Freestone, I.C., Mason, R., Molera, J., Vendrell-Saz, M., Wood, N. (1998). Review article. Lead glazes in antiquity – Methods of production and reasons for use. *Archaeometry*, 40, 241–260.

Vogel, W. (1994). *Glass Chemistry*. Springer-Verlag, Berlin.

Volf, M.O. (1984). *Chemical Approach to Glass. Glass Science and Technology 7*. Elsevier, Amsterdam, New York.

Waghmare, U.V., Spaldin, N.A., Kandpal, H.C., Seshadri, R. (2003). First-principle indicators of metallicity and cation off-centricity in the IV-VI rocksalt chalcogenides of divalent Ge, Sn and Pb. *Phys. Rev. B*, 67, 125111.

Walsh, A. and Watson, G.W. (2005). The origin of the stereochemically active Pb(II) lone pair: DFT calculations on PbO and PbS. *J. Solid State Chem.*, 178, 1422–1428.

Wang, P.W. and Zhang, L. (1996). Structural role of lead in lead silicate glasses derived from XPS spectra. *J. Non-Cryst. Solids*, 194, 129–134.

Warren, B.E. and Loring, A.D. (1935). X-ray diffraction study of the structure of soda-silica glass. *J. Am. Ceram. Soc.*, 18, 269–276.

Watson, G.W., Parker, S.G., Kresse, G. (1999). Ab initio calculation of the origin of the distortion of alpha-PbO. *Phys. Rev. B*, 59, 8481–8486.

Welcomme, E., Walter, P., Bleuet, P., Hodeau, J.-L., Dooryhee, E., Martinetto, P., Menu, M. (2007). Classification of lead white pigments using synchrotron radiation micro X-ray diffraction. *Appl. Phys. A*, 89, 825–832.

Yeom, Y.H., Kim, Y., Seff, K. (1997). Crystal structure of zeolite X exchanged with Pb(II) at pH 6.0 and dehydrated: $(Pb^{4+})_{14}(Pb^{2+})_{18}(Pb_4O_4)_8Si_{100}Al_{92}O_{384}$. *J. Phys. Chem. B*, 101, 5314–5318.

Yoshihoto, M. and Soga, N. (1987). The effect of composition on crack propagation in lead silicate glasses. *J. Non-Cryst. Solids*, 95–96, 1039–1046.

Zachariasen, W.H. (1932). The atomic arrangement in glass. *J. Am. Chem. Soc.*, 54, 3841–3851.

4

Optical Properties and Coloration of Lead Silicate Glasses

Odile MAJÉRUS, Adèle MUNOZ-BONGRAND and Daniel CAURANT
IRCP, CNRS, Chimie ParisTech, France

Colored lead silicate glasses, whether bulky or thin (glazes), have generally been appreciated for the beauty of their color. In his commented translation of Neri, Dr. Merret wrote in 1665: "If this species of glass had the same solidity as the *cristallin* glass, it would be superior to all the others, due to the beauty of its color" (Bontemps 1868, p. 527). Moreover, the colorless lead silicate glasses of SiO_2-PbO-K_2O system have been known and used since the 18th century for their optical properties, being referred to as "flint glass" (this term, originally given in England to potassium and calcium white glass, was reserved for potassium and lead glass since the end of the 17th century. It now designates a dispersive optical glass of high refractive index).

This chapter reviews the physical bases of the main optical properties, with an emphasis on color. Then it presents an overview of the optical properties and applications of transparent SiO_2-PbO-M_2O (M, alkaline) glasses that are often said to take a yellow color. Furthermore, it discusses the color of lead glasses colored by transition ions, the case of "colored transparent glazes". The colors of these glasses differ from those of alkaline glasses colored by the same ions, and hypotheses are proposed to understand the origin. This chapter does not cover opacification and coloration by lead-based pigments (lead antimonates, arsenates or stannates), which can be added to alkaline glasses and also to lead glasses, considering that it involves

Lead in Glassy Materials in Cultural Heritage,
coordinated by Anne BOUQUILLON and Patrice LEHUÉDÉ.

an opacification and coloration whose origin is external to the lead glass. Moreover, the opacification of lead glasses and their possible coloration by semiconductor or metallic particles will not be covered either, as they involve the same agents and mechanisms as the silicate glasses in general, lead oxide not modifying the phenomena except for the increase in the refractive index of the glass and the role that it can play in redox equilibriums.

4.1. Physical bases of optical properties and the origins of glass color

4.1.1. *Complex refractive index and the main optical properties*

When traveling through a material medium such as glass, light is partly reflected, partly refracted and transmitted, and partly absorbed if absorbing species are present in the glass. These properties of the material medium in relation to light are summarized in the complex refractive index (Kittel 1983):

$$\sqrt{\varepsilon(\omega)} = n(\omega) + i\,\kappa(\omega)$$

where $\varepsilon(\omega)$ is the complex dielectric function, $n(\omega)$ is the refractive index and $\kappa(\omega)$ is the extinction coefficient.

NOTE.– A homogeneous and continuous material medium is considered here, whose electron density distribution can be considered uniform at the scale of wavelengths of UV–visible–IR domain. In this case, the complex refractive index does not depend on the wave vector *k* of light.

The wave vector k of light in vacuum changes into the wave vector $(n + i\kappa)k$ in the medium of index $n + i\kappa$. Considering the amplitude of the incident wave is $E_{y,inc} = E_{inc}e^{i(kx-\omega t)}$, the amplitude of the transmitted wave is $E_{y,trans} \propto e^{i((n+i\kappa)kx-\omega t)}$ which is:

$$E_{y,trans} \propto e^{-\kappa x}e^{i(nkx-\omega t)}$$

The refractive index $n(\omega)$ therefore reflects the decrease in the phase velocity of light due to the medium ($velocity = \frac{\omega}{k} = \frac{c}{n(\omega)}$), driving a change in the direction of light when crossing the interface (*refraction*). The dependence of n on the frequency ω is referred to as *dispersion*, a term used for light dispersion by a prism, but it causes also the chromatic aberration of lenses. The extinction coefficient $\kappa(\omega)$ corresponds to the amplitude *attenuation* of light traveling through a medium due to absorption phenomena. The two indices *n* and κ are also known as “optical constants” of the medium.

Under normal incidence, the amplitude of the reflected wave relative to the incident wave is:

$$\frac{E_{refl}}{E_{inc}} = \rho(\omega)e^{i\theta(\omega)} = \frac{n + i\kappa - 1}{n + i\kappa + 1}$$

Experimentally, the reflection coefficient is measured under normal incidence, being the square module of the amplitude ratio:

$$R(\omega) = \rho^2(\omega)$$

The phase information is therefore lost. However, Kramers and Kronig have shown that, due to the analytical form that can be given to $\varepsilon(\omega)$ (not detailed here), it is possible to mathematically transform $R(\omega)$ to calculate $\theta(\omega)$. These "Kramers-Kronig relations" can therefore be used to get $n(\omega)$ and $\kappa(\omega)$ from the quantitative measurement of the optical reflection spectrum $R(\omega)$.

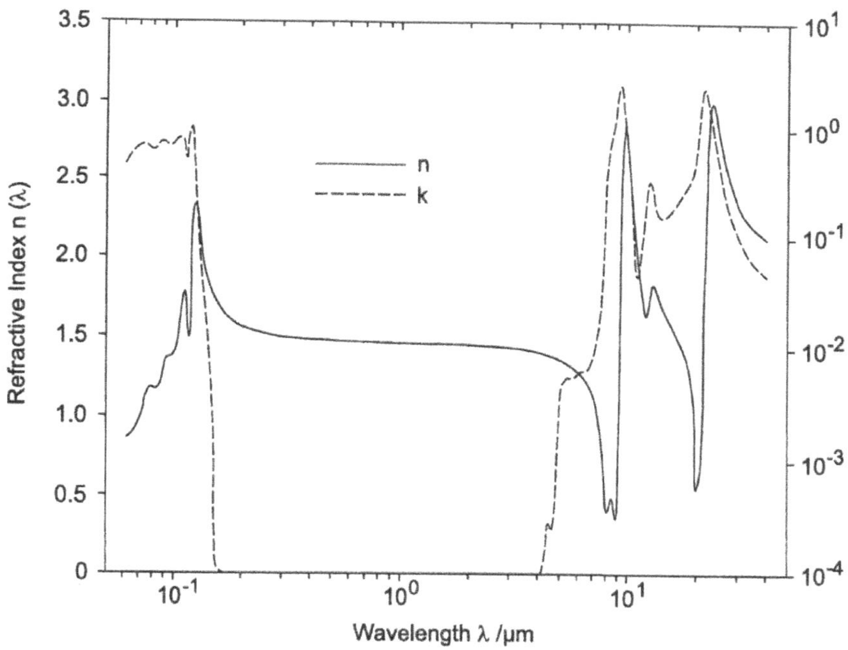

Figure 4.1. *Refractive index and extinction coefficient of the silica glass measured by optical reflection. Figure extracted from Lentes (1998, p. 23)*

Figure 4.1 provides the indices n and κ as a function of the wavelength λ of light from UV to IR for the silica glass. Absorptions take place in far UV (toward 150 nm) and IR (toward 9 μm). The nature of these absorptions varies: electronic for UV and vibrational for IR. They cause strong variations of the indices: as these absorptions occur, reflection and dispersion are both strong. However, in the extended visible domain (between 1 μm and 0.3 μm), the index κ is zero and the index n is quasi-constant around 1.5. The silica glass is therefore a perfectly transparent and little dispersive medium in the visible, near UV and near IR domains.

Colorless and transparent glasses are the glasses that contain no species absorbing in the visible and no diffusing species. Their use for optical applications requires more precise information on their refractive index and their dispersion. As already mentioned in Chapter 1, several characteristic optical properties can be defined, among which are the following:

– main refractive index n_d is the refractive index measured at 587.56 nm (He spectral line);

– reciprocal dispersive power or Abbe number ν_d is the ratio:

$$\nu_d = \frac{n_d - 1}{n_F - n_C}$$

where n_F and n_C are the refractive indices measured, respectively, at 486.13 nm (H spectral line) and at 656.27 nm (another spectral line of H). The lower the Abbe number, the higher the dispersive power.

Colored glasses contain species absorbing in one part of the visible spectrum. The absorption phenomenon takes out from the light a part of the visible spectrum; the transmitted light is no longer "white" but colored. The color that is finally perceived depends on the spectrum of the illuminating light (it is not always white light), on its absorption by the glass and on eye sensitivity. These data are gathered in the three colorimetric coordinates calculated from the transmission spectrum or from the reflection spectrum by postulating a standard illumination and a standard sensitivity for the human eye. There are various systems of colorimetric coordinates, each combining in three numbers the three perceptions, namely the *hue*, the *lightness* and the saturation (related to the existence of very dominant absorptions).

When the extinction coefficient κ remains small at least in one part of the visible domain, and in the absence of any diffusing species, the colored glass is referred to as "colored transparent". For a massive glass (object, stained glass, etc.), the color is perceived in transmission. For an enamel or a transparent glaze, light is first transmitted, then reflected in a more or less diffuse manner by the substrate (a metal or ceramic surface), then once again transmitted and gets out of the glass reaching our eye. The reflection by the substrate is therefore added to the perceived color. When trying to measure the colorimetric coordinates from a reflection spectrum, attention should be paid to choosing a perfectly reflective or perfectly diffusive substrate, meaning that it does not absorb in the visible.

Glass is no longer transparent when the extinction coefficient κ is very high in all the visible domain (case of chalcogenide glasses, for example) or when heterogeneities inside it cause the phenomenon of *diffusion*. Indeed, particles of index differing from that of glass and of size a that is very small or close to visible wavelengths will behave as diffusing objects, the complete behavior being given by the Rayleigh theory ($a \ll \lambda$) or by Mie theory ($a \sim \lambda$). Depending on the size of these particles and on the difference between their index and that of glass, the intensity and the spectrum of diffused light varies from pale bluish (very small sizes of the order of 1–10 nm, diffused intensity varying with $^1/_{\lambda^4}$) to "white" (larger sizes, 0.1–10 μm, the diffused intensity no longer varies with the wavelength). It can then be said that these glasses are opalescent or opacified, respectively.

4.1.2. *Coloring species in glasses*

The visible domain corresponds to the energies of the valence electrons. The absorptions in the visible are therefore always related to resonant electron excitations at the level of the most external electrons. These excitations can be of various natures and are related to various species.

4.1.2.1. *d-d or f-f transitions*

Transition ions, whose valence level is a partially filled d level (3d, 4d, etc.), are the most common species for glass coloration since ancient times (iron, copper, cobalt, manganese ions). Under the effect of the non-spherical symmetry of the electrical field generated by the oxygen neighbors (negatively charged), the level d is split into several levels. This splitting is due to the "crystalline field". The electron–electron interactions inside the d configuration add an additional splitting term. These splittings and the energies of various levels are described by the Tanabe Sugano diagrams for each configuration d^n (n = 1–9) placed in a crystalline field of octahedral or tetrahedral symmetry depending on the force of the crystalline field and on the electron-electron interactions. The energy differences between the higher d levels and the lowest d level are in the visible and near IR: visible light can therefore be absorbed by the transition of a d electron to a higher d level, which is referred to as "d-d transition". The absorption bands of d-d transitions are not intense, as they are forbidden by Laporte rule that imposes a change of parity between the departure orbital and the arrival orbital. On the other hand, they are very large, both because of the vibronic coupling related to the contribution of d orbitals to the metal-oxygen chemical bond and the distribution of environment and therefore crystalline field in the glasses.

Lanthanide ions, whose valence level is the partially filled f level, give rise to f-f transitions in the visible, not intense and very thin, as the f levels are core levels that participate to a little extent to the chemical bond and therefore to vibrations. The use of these ions for glass coloration appeared in the 20th century (see Chapter 7).

4.1.2.2. *Optical gap transitions and charge transfer transitions*

Light may excite the valence electrons of ligands (anions of oxides, sulfides, etc.) and place them in the lowest unfilled orbitals of metallic centers. In the solid, depending on the overlapping degree of valence orbitals, there may be a charge transfer transition (little or no overlapping) or an optical gap transition between the valence band and the conduction band (strong overlapping, electron delocalization). These transitions are fully authorized and the absorptions are therefore very intense.

In pure silica glass, the absorbance in far UV (toward 150 nm) corresponds to the optical gap transition between the top of the valence band, mainly formed of the 2p level of oxygen, and the bottom of the conduction band, mainly formed of the 3s-3p hybrid level of silicon.

In the alkaline silicate glasses, the introduction of non-bridging oxygens (denoted by NBO in what follows) isolated the ones from the others creates filled 2p levels above the valence band. The transition from these 2p levels leads to an absorbance in near UV (toward 350 nm).

In the alkaline silicates glasses that contain transition ions, these ions have their d levels below the lowest of the conduction band. The electron transition from the 2p levels of non-bridging oxygens toward the unfilled d levels of the metal ion to which these oxygens are linked is not an optical gap transition, as these levels are localized. It is properly speaking a ligand charge transfer transition toward the metallic center, as in a coordination complex.

Generally, the charge transfer transitions between the ligands and the d orbitals of the transition ions are in the near UV, but they can extend in the visible. A yellow coloration may then appear (case of glasses very rich in TiO_2, for example). The oxoanions are isolated molecular entities formed between very charged metal ions and their oxygen ligands: chromates, vanadates, etc. In these entities, the charge transfer transition is located in the visible and causes very intense colorations (yellow for chromate or vanadate oxoanions). Moreover, the sulfide anions can also be bound to metal ions substituting oxide anions. The charge transfer of S^{2-} toward the metal ion is itself shifted in the visible. It causes, for example, the amber coloration of glasses containing Fe^{3+} - S^{2-} chromophore formed by reduction of sulfate ions in glass melting (Chopinet 2002).

It is important to note here that the color of crystalline pigments, such as the lead-based ones (lead antimonates, arsenates, stannates) or metal chalcogenides (CdS, CdSe, etc.), results from the optical gap transition in these particles, to which the diffusion effect should be added.

4.1.2.3. *Plasmon resonances*

Plasmon resonances are collective resonant excitations of the free electrons at the surface of metal nanoparticles (gold, silver, colloidal copper). These resonances cause thin and strong absorptions in the visible, the domain of absorbed wavelengths depending on the size of the nanoparticles. This phenomenon is, for example, at the origin of copper red (Cu^0 nanoparticles) or silver yellow (Ag^0 nanoparticles) of stained glasses.

4.2. Optical properties and color of transparent SiO_2-PbO-M_2O glasses

SiO_2-PbO-M_2O glasses form a very ancient family of transparent glasses used in optics, referred to as "flint". Their general optical properties are a high refractive index and a strong dispersion in the visible domain (rather low Abbe number) (see Chapter 1). The high electron density of Pb^{2+} ions and the polarizability of their 6s orbital filled by two electrons explain the high refractive index. As for the strong dispersion, it is related to the existence of an absorbance in the near UV due to the O(2p) $\rightarrow$ Pb(6p) transition where the O(2p) level is modified by the overlapping with the 6s level of lead. It is the nature of this absorbance and the yellow color that it confers in the presence of metal impurities that will be briefly described in this section. A more detailed discussion of the literature is presented in Majérus et al. (forthcoming).

4.2.1. *Electronic structure and color of lead oxides PbO*

The colors of lead oxides result from the optical gap absorption between the top of the valence band and the bottom of the conduction band. Experimentally, the optical gap is indeed given at 1.9 eV for litharge α-PbO, red, and 2.5 eV for massicot β-PbO, yellow (Radhakrishnan et al. 1983). The electronic structure of PbO and the orbital composition of the valence and conduction bands were the subject of many studies and debate in the literature. It was less a matter of understanding the color, but rather the nature of the free pair whose stereochemical activity manages to distort the environment of Pb^{2+} ion in these oxides. A first interpretation was to assign the optical gap to a transition between the hybridized 6s-6p levels of lead, slightly overlapping with 2p orbitals of oxygens. These hybridized levels responsible for the optical gap were associated with the free pair. Their more or less significant overlapping in the direction z of the structure explained the different color in α-PbO and in β-PbO.

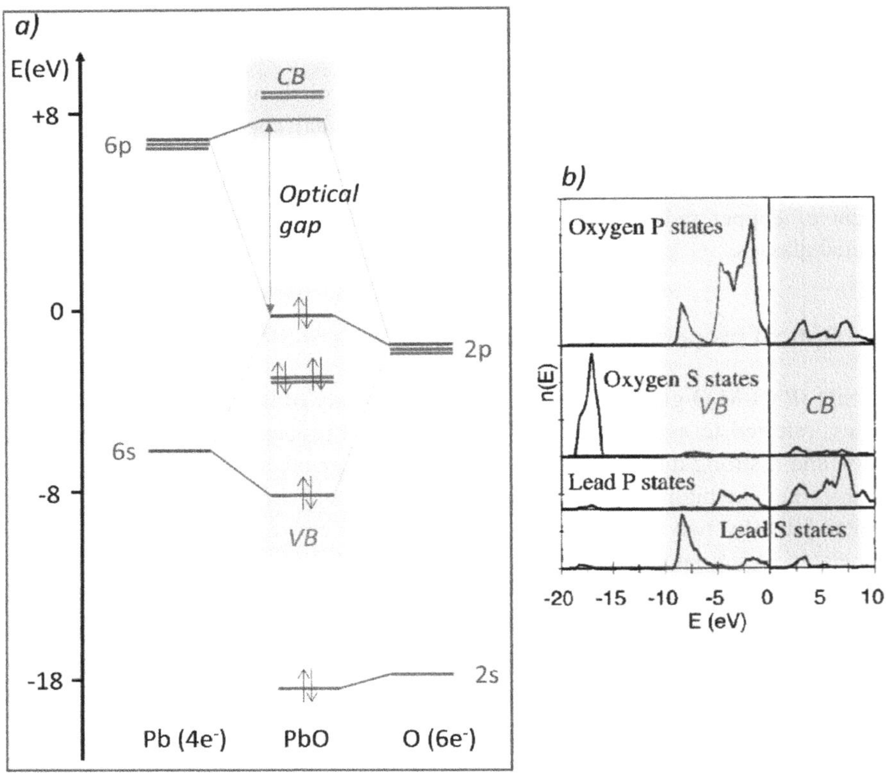

Figure 4.2. *(a) Very simplified diagram of the chemical bond in PbO compound according to the partial densities of states calculated by ab initio method (DFT) for α-PbO by Watson et al. represented in (b) (Watson and Parker 1999). For a color version of this figure, see www.iste.co.uk/bouquillon/lead.zip*

COMMENT ON FIGURE 4.2.– *The orbitals contributing to the top of the valence band, which can be assimilated to the "free pair" presenting a stereochemical activity, are represented in violet and the other orbitals in blue. In these calculations, the top of the valence band is essentially formed of* $2p_z$ *orbitals of oxygens, with a minor contribution of 6s and* $6p_z$ *of lead. VB = valence band. CB = conduction band.* e^- *= valence electrons.*

Ab initio calculations with DFT (Density Functional Theory) method (Terpstra et al. 1995; Watson and Parker 1999) have called into question the nature of the free pair and the contribution of various orbitals in the valence band. Indeed, internal 6s-6p hybridization did not allow understanding why the environment of Pb^{2+} is distorted in PbO, while it is perfectly spherical in PbS. The anion had to play a role (Du et al. 2014).

According to these calculations, which were experimentally validated in the 2000s (Payne et al. 2006), the top of the valence band in PbO oxide is mainly formed of 2p levels of oxygens, with a very small contribution of 6s and 6p orbitals of lead. The optical gap is therefore assimilated to a O(2p) → Pb(6p,6s) transition, where the O(2p) level is slightly overlapping with 6s and 6p orbitals of lead (Figure 4.2). This information is important for understanding the electronic transitions at the UV limit of the spectrum in glasses containing PbO.

4.2.2. *Electronic structure, optical properties and color of* SiO_2*-PbO (-*M_2O*) glasses*

The studies dedicated to the electronic structure of binary SiO_2-PbO glasses or to alkaline lead ternary glasses are relatively few. While the silica glass does not absorb below 9 eV, absorptions appear at about 6.7 eV, 5.3 eV and 4.2–4.6 eV when PbO is added to the silica (Ellis et al. 1979). The absorptions explain the significant increase of the refractive index and the dispersion of these glasses with respect to the silica glass in the near UV domain of the spectrum.

These low energy electronic transitions in the UV domain have been mainly explained by the internal 6s → 6p transition of Pb^{2+} in the studies published between 1970 and the beginning of 2000 (Bourdillon et al. 1978; Ellis et al. 1979; Robertson 1980; Gubanov et al. 1991; Weinstein et al. 2001). To our knowledge, these interpretations were not resumed and called into question until present. However, in the light of recent results on the electronic structure of PbO already mentioned, and of knowledge on the local order around Pb^{2+} in glasses, it is likely that these transitions have in reality a O(2p) → Pb(6p) character predominant over Pb(6s) → Pb(6p). Indeed, the local environment of Pb^{2+} ions in SiO_2-PbO glasses is close to the one in lead oxides (see Chapter 3, section 3.3.3.2). These environments of low coordination number and distorted reflect the strong interactions due to the overlapping of 6s and 6p orbitals of lead with the 2p orbitals of oxygens, similar to PbO oxide (Figure 4.2). A simplified electronic structure is therefore proposed for these glasses inspired from the PbO oxide in Figure 4.9(a).

Let us now consider the effect of these transitions on the color. It was often noted that beyond a certain lead content, SiO_2-PbO or SiO_2-PbO-M_2O glasses have in their thickness a yellow coloration (Hampton 1946; Stroud 1971; Cohen et al. 1973). The origin of this coloration was debated in the attempt to establish whether it was intrinsic and resulted from a shift and a thermal widening of near UV transitions, or it was related to impurities. By showing that a lead flint glass with a refractive index of 1.90 is colorless over 5 cm thickness if it is quite pure, Hampton settled the argument and established the essential role of impurities (Hampton 1946). These impurities can be d transition ions (typically Fe^{3+}) and also platinum if the glass is elaborated in platinum crucible.

The very significant sensitivity of lead glass coloration to the presence of impurities is however remarkable and not equivalent to that of alkaline glasses, as it was very clearly established by Stroud in a study related to 10 coloring species (V^{5+}, Cr^{6+}/Cr^{3+}, Mn^{3+}/Mn^{2+}, Ni^{2+}, Cu^{2+}/Cu^{+}, Ce^{4+}/Ce^{3+}, U^{6+} ions and Rh, Pd, Pt metals with their ionized species) (Stroud 1971). Moreover, Cohen et al. noted that the cutoff wavelength in near UV increases regularly with the PbO content: in other terms, lead glasses are all the more yellow (UV absorbance extending toward violet) as their PbO content increases. They therefore suggest that "the presence of variable concentrations of metal ion impurities is not the only source of yellow color in PbO-rich glasses" (Cohen et al. 1973).

The "mystery" therefore persists, as clearly formulated by Stroud and Cohen, on this yellow coloration that seems to originate *both* in the presence of impurities and in PbO content.

This point will be discussed in further detail in section 4.3 related to the color of SiO_2-PbO(-M_2O) glasses colored by transition ions.

4.3. SiO_2-PbO-M_2O glasses colored by transition ions

A remarkable property of lead glasses is their color in the presence of transition elements, and this is presented in most chapters of this book. Lead glazes have warm and intense colors, unlike their soda lime or alkaline borosilicate analogues colored with the same transition ions.

This fact is well known by the potters of varnished terracotta who desperately try to reproduce, with leadless commercial glazes, the ochre-yellow (with Fe^{3+}) and sap green (with Cu^{2+}) colors so characteristic of their ceramic art (Figure 4.3). In contrast, the same ions in the borosilicate glazes offer pale and cold substitutes to these colors, at best creamy-white and blue-green.

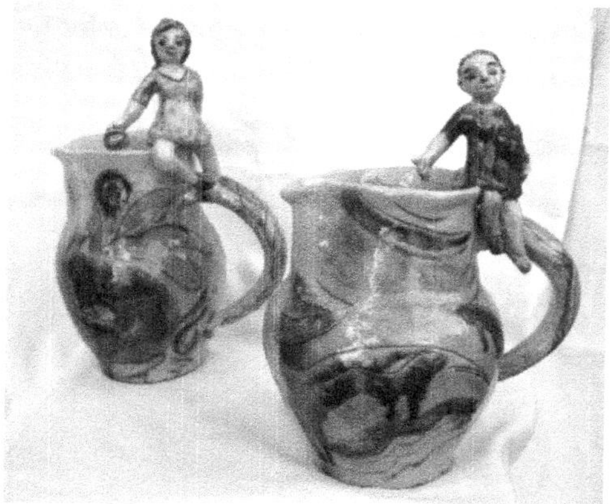

Figure 4.3. *Two varnished terracotta jugs by Agnès Cabaret, ceramist in Sarthe. The green and ochre colored glazes are based on a lead glazing mixture and copper oxide (green) and iron oxide (ochre). The transparent lead glaze covering the jugs is slightly colored in yellow by a little iron oxide. For a color version of this figure, see www.iste.co.uk/bouquillon/lead.zip*

a)

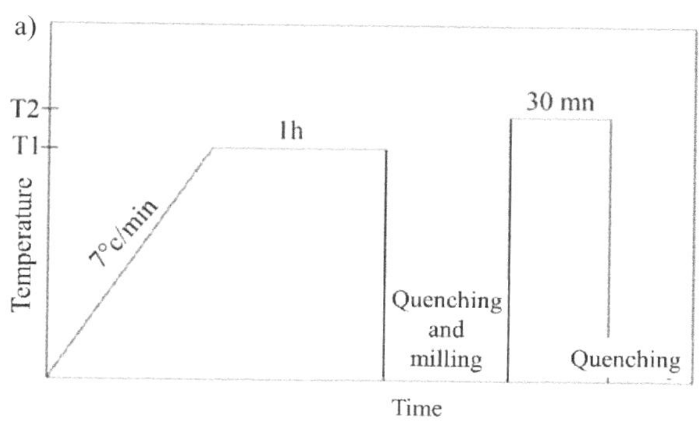

b)

Glass	T1	T2
PbI		
PbFe	1050	1050
PbCu		
NaPbI		
NaPbFe	1050	1150
NaPbCu		
NaI		
NaFe	1150	1300
NaCu		

Figure 4.4. *(a) Glass elaboration process with two meltings at temperatures T_1 and T_2, summarized in table (b). Soda glasses being more viscous than lead glasses, these temperatures had to be increased*

In order to study the origin of the specific color of lead glasses, nine glass compositions were elaborated in the laboratory by melting-quenching under air. These nine compositions correspond to three vitreous matrices: a lead-only matrix (% mol) (57 SiO_2–5 Al_2O_3–38 PbO), an identical matrix but where all PbO oxide was replaced by Na_2O (57 SiO_2–5 Al_2O_3–38 Na_2O) and a mixed lead and soda matrix (57 SiO_2–5 Al_2O_3–20 Na_2O–18 PbO), combined with two coloring oxides (Fe_2O_3, CuO or "colorless" glass due to the absence of coloring agent). The melting temperature had to be modified to take into account the viscosity variations, the melting temperatures being summarized in Figure 4.4. The colors and the names of these glasses are shown in the photographs in Figure 4.5.

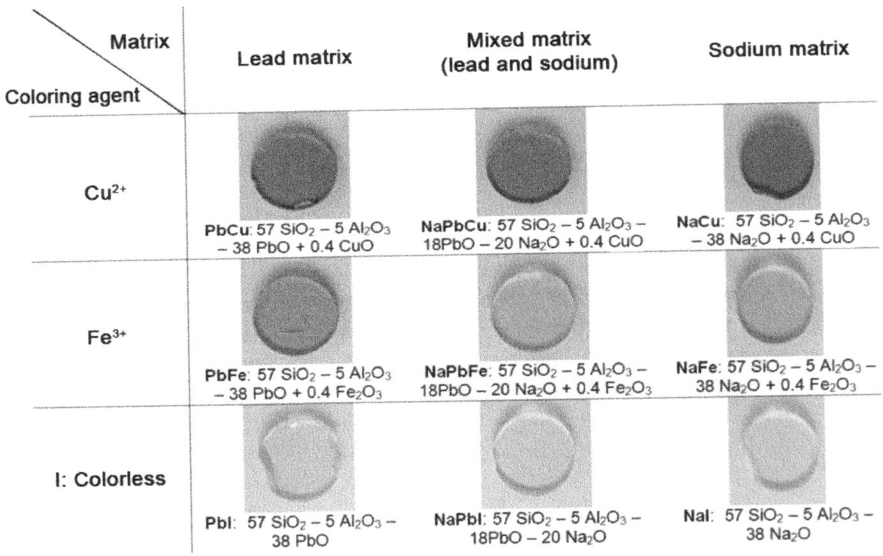

Figure 4.5. *Nominal molar compositions and photographs of nine glasses prepared in the laboratory; their names being specified in bold. The compositions were analyzed in the laboratory by EDX, and they are very close to these nominal compositions (deviation below 1% molar for each oxide). For a color version of this figure, see www.iste.co.uk/bouquillon/lead.zip*

The color of lead glass is stronger and greener (with Cu^{2+}) or more "honey-like" (with Fe^{3+}) than the soda glass. The study of these glasses by optical absorption spectroscopy and electron paramagnetic resonance (EPR) presented below can be used to propose an origin to this color change[1].

4.3.1. *Optical absorption spectroscopy and origin of color changes*

As a first step, the optical absorption spectra of colorless glasses are presented in Figure 4.6. As these glasses do not contain any coloring species (except for impurities visible by the small band at 27,000 cm^{-1} in NaI glass), their absorbance remains close to zero throughout the visible domain and it strongly increases in the near UV under the effect of the optical glass transition (top of the valence band, in fact constituted of isolated levels of NBO → bottom of the conduction band. It is important to note that rigorously speaking the term "optical gap" is not suitable to a glass, due to the disorder that prevents the orbital overlapping throughout the solid). When Na_2O is substituted by PbO, this optical gap transition strongly shifts toward the visible. The shift is of about 5,800 cm^{-1} (Table 4.1 and Figure 4.6). In the lead-only glass (PbI), absorption extends toward 450 nm, which explains the yellow coloration of this glass.

The optical absorption spectra of the colorless soda glass (NaI) and iron-colored glasses (NaFe, NaPbFe and PbFe) are represented in Figure 4.7. First of all, the comparison between the spectrum of NaI glass and the spectrum of NaFe glass makes it possible to identify the transitions related to the iron oxide introduced in the glass and attributable to the two degrees of oxidation of iron, namely Fe^{2+} (wide absorption band at 10 000 cm^{-1}) and Fe^{3+} (several absorption bands between 20,500 and 27,000 cm^{-1}). These absorption bands extending largely in the visible are due to d-d transitions whose attribution is indicated in the figure (for the details on these attributions, see Vercamer 2017; Munoz-Bongrand 2018). They are at the origin of the yellow-green color of the soda NaFe glass. Besides these d-d transitions, another effect of the addition of Fe_2O_3 is the shift of the optical gap absorption from UV to near UV and visible. This shift reflects the appearance of unfilled levels below the conduction band of the glass: unfilled d levels of Fe^{2+} and Fe^{3+} ions. The nature of the absorption in near UV therefore changes, passing from an optical gap transition

---

1. All the optical absorption spectra are presented here either in the form of absorbance normalized to the thickness of the glass slide (unit cm^{-1}), or in the form of absorbance normalized to the thickness and to the nominal molar concentration of coloring ion (unit L·mol^{-1}·cm^{-1}). A baseline was subtracted from all the spectra before normalization. For each glass composition, this baseline was established from the difference spectrum between the spectrum of a thick slide and the spectrum of a thinner slide of glass, following the procedure described in (Vercamer 2017; Munoz-Bongrand 2018).

in the colorless glass to a charge transfer transition in the complex formed by the transition ion and its oxygen ligands in NaFe glass (O(2p) → Fe(3d) transition).

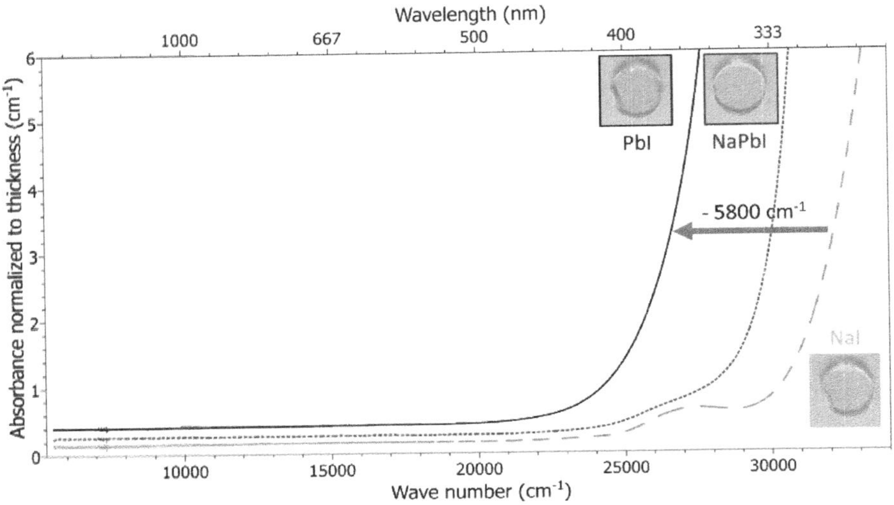

Figure 4.6. *Absorption spectra of three "colorless" glasses NaI, NaPbI and PbI, showing the shift of 5,800 cm^{-1} toward the visible of the charge transfer band when Na_2O is substituted by PbO. This shift is at the origin of the slightly yellow color of the PbI glass. The band at 27,000 cm^{-1} visible on the spectrum of NaI and NaPbI glasses is attributed to impurities. The dashed, dotted and continuous lines refer to the spectra of NaI, NaPbI and PbI glasses, respectively. For a color version of this figure, see www.iste.co.uk/bouquillon/lead.zip*

It is now interesting to compare the spectra of the three glasses doped with Fe_2O_3. These spectra are normalized to the thickness and to the concentration of Fe ions in the glass, so that their ordinate corresponds to the linear and molar extinction coefficient ε (linked to the above-mentioned linear extinction coefficient κ by the relation $\kappa = \varepsilon \times c$, where c is the concentration in mol·L^{-1} of ions in the glass). In the PbFe glass compared to the NaFe glass, the d-d band of Fe^{2+} ions at 10,000 cm^{-1} strongly decreases in absorbance and very slightly shifts toward high energies (shift of about + 300 cm^{-1}, not visible in Figure 4.7). But most of all, the charge transfer band shifts very strongly toward the visible by about 5,500 cm^{-1} (Table 4.1 and Figure 4.7). It extends this time up to over 600 nm, explaining the marked "honey-like" color of PbFe glass.

Finally, Figure 4.8 summarizes the spectra of NaI glass and of three glasses doped with CuO. These spectra show a d-d absorption band attributed to the $^2E \rightarrow {}^2T_2$ transition of Cu^{2+} ions in octahedral symmetry toward 12,500 cm^{-1}, and the charge transfer absorption in near UV. The d-d band is slightly asymmetric due to the Jahn–Teller effect, which causes an axial deformation of the octahedral complex CuO_6 and splits the transition into several sub-transitions. Here too, when Na_2O is replaced by PbO, this band shifts very lightly toward high energies (near + 500 cm^{-1}). But above all, the charge transfer band shifts strongly toward the visible by about 5,500 cm^{-1} as in the PbFe glass. These shifts toward the visible cause the modification of color between lead glasses and soda glasses, as the violet to blue parts of the spectrum become absorbed, which introduces a prominent yellow-orange component in the perceived color. The d-d bands vary very little and do not explain the change of color.

In any case, the spectrum of the mixed soda and lead glass is between the two other spectra and shows that these changes are related to the gradual substitution of Na_2O by PbO.

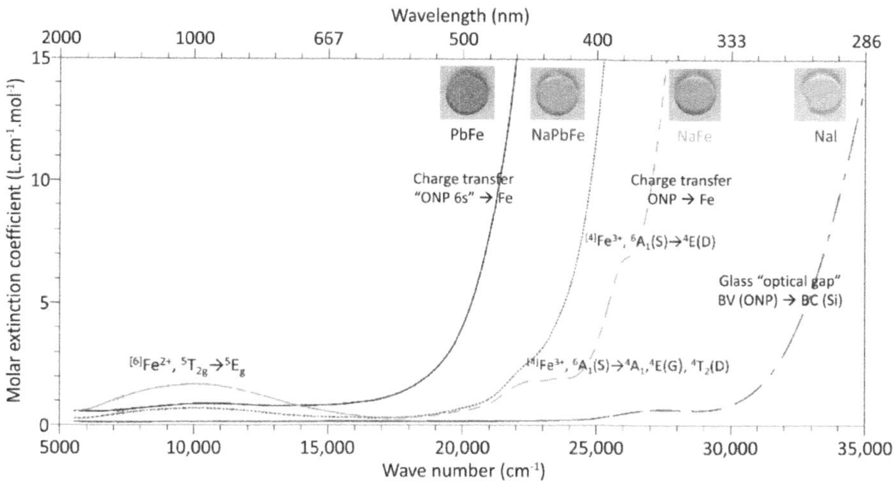

Figure 4.7. *Molar extinction coefficient ε of iron-doped glasses and comparison with the normalized spectrum of non-doped NaI glass. The dotted-dashed, dashed, dotted and continuous lines refer to the spectra of NaI, NaFe, NaPbFe and PbFe glasses respectively. For a color version of this figure, see www.iste.co.uk/bouquillon/lead.zip*

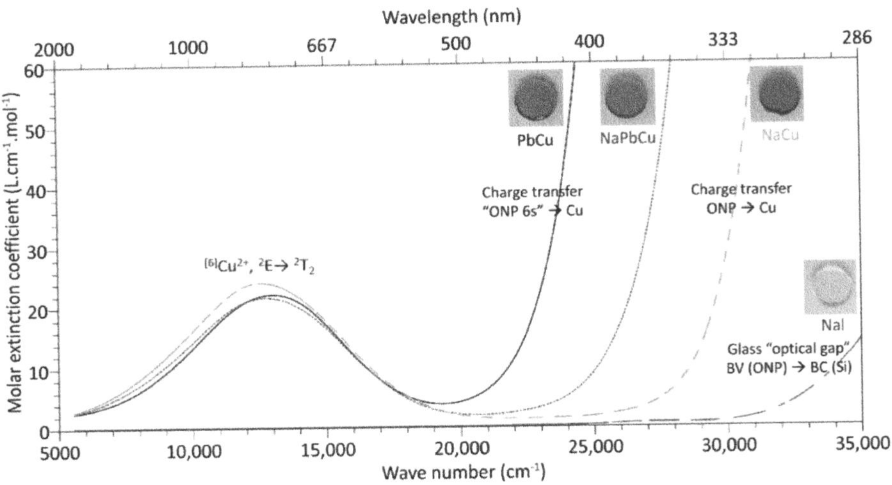

Figure 4.8. *Molar extinction coefficient ε of copper-doped glasses and comparison with the normalized spectrum of non-doped NaI glass. The dotted-dashed, dashed, dotted and continuous lines refer to the spectra of NaI, NaCu, NaPbCu and PbCu glasses, respectively. For a color version of this figure, see www.iste.co.uk/bouquillon/lead.zip*

	Pb	NaPb	Na	Na–Pb difference
Cu^{2+}	**24,500**	28,000	**31,000**	5,500
Fe^{3+}	23,000	26,300	28,500	5,500
/	**28,400**	31,200	**34,200**	5,800

Table 4.1. *Wavenumbers (in cm^{-1}) for which the absorbance normalized to thickness reaches A = 10 cm^{-1} for each of the glasses. The four values in bold are used for the energy level diagram in Figure 4.9*

The shifts toward the visible of near UV absorptions have been evaluated by the energy for which the absorbance normalized to thickness takes the value A = 10 cm^{-1} (transmission of 10^{-10}). These energies are summarized in Table 4.1 and the shifts by total substitution of Na_2O by PbO are calculated by the difference in the last column. In the PbFe and PbCu glasses, these shifts, of 5,500 cm^{-1}, can be due either to the reduction of energy of the first unfilled d levels of the metal ion, or to the increase in the energy of donor levels of ligands O(2p), or to the two combined effects. As there is such a shift also in the non-doped PbI glass (5,800 cm^{-1}), this proves that it is, in fact, due *only to the energy increase of donor levels*. The diagrams of the energy levels in Figure 4.9 summarize these conclusions.

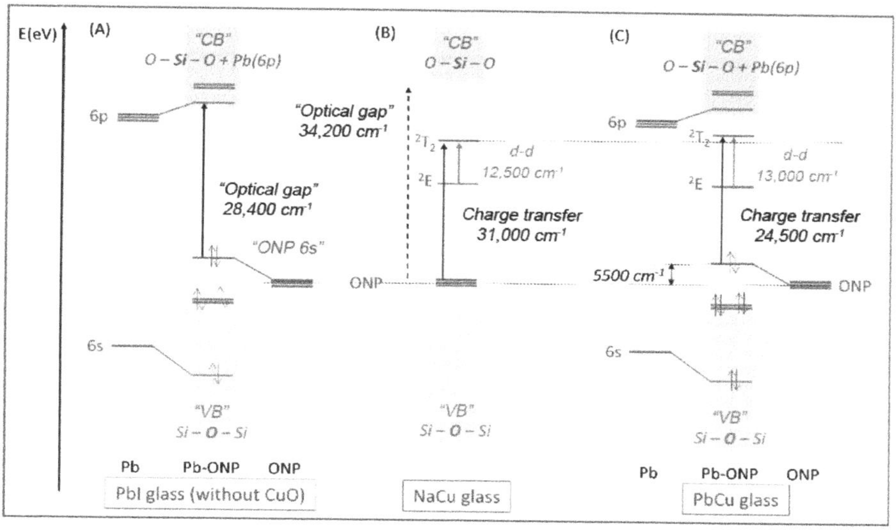

Figure 4.9. *Schematic energy level diagrams, using the transition energies experimentally determined on the optical absorption spectra of PbI (a), NaCu (b) and PbCu (c) glasses and summarized in Table 4.1. For a color version of this figure, see www.iste.co.uk/bouquillon/lead.zip*

In Figure 4.9, the terms "conduction band" (CB), "valence band" (VB) and "optical gap" are given in quotation marks to highlight the fact that these terms rigorously designate delocalized levels that do not actually exist in the glasses (because of the disorder, the electronic levels are rather localized). However, these terms are used by direct analogy with the crystals, as the nature of transitions is similar (same types of departure and arrival orbitals). In the diagram of the PbI glass, the donor level of the optical gap transition is formed of orbitals of NBO modified by the overlapping with the 6s level of Pb, denoted by "NBO 6s", and the acceptor level is formed of 6p orbitals of Pb modified by the overlapping with O(2p), as in the PbO crystal. We therefore consider that in PbI glass with 38% mol in PbO, the local order around Pb^{2+} ions and the Pb-O chemical bond are similar to those in PbO (see Figure 4.2). This is justified by the structural results presented in detail in section 3.3.3.2 and by the comparative examination of the optical absorption spectra conducted here. In the diagram of NaCu glass, the Cu^{2+} ions introduce their 3d levels under the "CB" so that the first absorption in UV is a charge transfer absorption between NBO(2p) and Cu(3d) in the quasi-molecular CuO_6 complex. In the diagram of PbCu glass, the charge transfer takes place this time between the NBO levels modified by the presence of lead (denoted by "NBO 6s" and represented in violet) and the Cu(3d) levels. The energy difference with respect to the same charge transfer in NaCu is of –5,500 cm^{-1}, pushing this transition in the violet-blue part of the visible spectrum.

These positions relative to the energy levels can be interpreted as follows at molecular scale. In the glass, Cu^{2+} ions are bound to 6 NBO, forming a quasi-molecular CuO_6 complex. This complex has excess negative charge of –4 (Cu^{2+} surrounded by 6 NBO carrying each a formal negative charge). This charge excess is compensated by alkaline ions or Pb^{2+} that are near the complex, so that the following formal equations can be written as:

$$\text{in NaCu: } CuO + 2Na_2O \xrightarrow{[SiO_2]} (CuO_6^{4-}, 4Na^+)_{SiO2\ glass}$$

$$\text{in PbCu: } CuO + 2PbO \xrightarrow{[SiO_2]} (CuO_6^{4-}, 2Pb^{2+})_{SiO2\ glass}$$

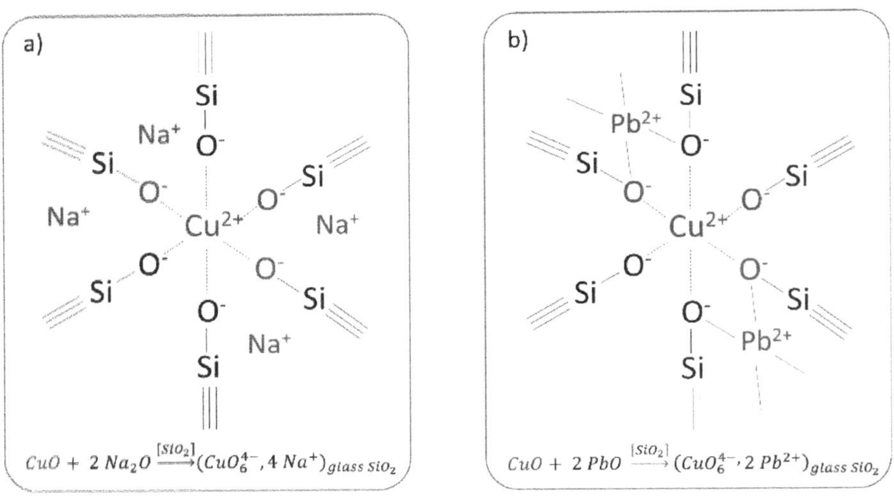

Figure 4.10. *Schematic representation of quasi-molecular CuO6 complexes of octahedral symmetry formed in the (a) NaCu and (b) PbCu glass. Each "O⁻" is a NBO playing the role of ligand and binding the Cu^{2+} ion to the silicate network. The bond between Pb^{2+} and O^- is represented by a line to indicate its covalent character related to the interaction between O(2p) and Pb(6s, 6p). For a color version of this figure, see www.iste.co.uk/bouquillon/lead.zip*

In these equations, CuO and Na_2O or PbO oxides are considered as modifiers of silica: each of their oxygens reacts with silica and forms two non-bridging oxygens. In the resulting quasi-molecular complex, these non-bridging oxygens are bound at the same time to the Cu^{2+} center and to the compensating Na^{+} or Pb^{2+} ions. These complexes are schematically represented in Figure 4.10, where the NBO brought by Na_2O, PbO or CuO are drawn each in a different color.

The charge transfer transition is the transition between the 2p level of these NBO having the highest energy (donor level) and the unfilled 3d level with the lowest energy of Cu^{2+}. In the $(CuO_6^{4-}, 2\,Pb^{2+})$ complex, the antibonding overlapping of O(2p) and Pb(6s) orbitals (as in PbO, Figure 4.2) forms a donor level denoted by "NBO 6s" that is significantly higher in energy than in the $\mathrm{CuO_6^{4-}, 4Na^{+}}$ complex and is represented in violet in the diagrams in Figures 4.2 and 4.9.

The same conclusions are obviously drawn for Fe^{3+} ions.

4.3.2. *Variation of the local environment of Cu^{2+} ions depending on the type of glass*

The local environment of Cu^{2+} and Fe^{2+} is slightly modified in the lead glass compared to the soda glass, as proven by the slight shift toward high energies of their d-d absorption band (for Fe^{3+}, the d-d transitions are in the charge transfer band of PbFe glass and therefore impossible to observe). This high energy shift reflects a stronger interaction between the 3d orbitals and the NBO ligands, that is, an increase of the crystalline field. In diagram (c) of Figure 4.9, it was represented by an increase in $^2T_2$ level, which is the one interacting the most with the ligands.

Cu^{2+} ion has a d^9 configuration with a total electronic spin S = ½, which makes it a paramagnetic species that can be studied by EPR.

The EPR-based study of Cu^{2+} ions in the two types of glass is presented in detail in Majérus et al. (forthcoming). The symmetry of the CuO_6 complex is slightly modified in the lead glass compared to the soda glass, but the crystalline field is slightly stronger and the Cu-O bond more covalent. These conclusions are also those of a recent study on the Cu^{2+} ions in various lead matrices (Andreev et al. 2020).

In summary, in this study related to two transition ions (Fe^{3+} and Cu^{2+}), it appears that the warmer and more intense colors of lead glasses are due to the prominent absorption of the violet-blue component of the visible spectrum in the presence of both lead and metal ion. In the quasi-molecular complex formed by the metal ion, its oxygen ligands and its charge compensators Pb^{2+}, the charge transfer transition is clearly shifted in the visible. This is due to the hybridization of 2p orbitals of oxygen ligands with the 6s orbital of Pb^{2+} ions that increases the energy of the donor level. There is a stronger interaction of this level with the d orbitals, which is reflected by a slightly stronger crystalline field and a more covalent metal-oxygen bond.

The conclusions of this study are confirmed by the recent results on the electronic structure in PbO presented in section 4.2.1, which show that the top of the valence band (comparable to the donor level here) is constituted by 2p orbitals of the oxygen in antibonding overlapping with Pb(6s) orbital. They are consistent with the studies of Hampton, Stroud and Cohen, highlighting both the role of impurities and the PbO content in the yellow coloration in transparent lead glasses.

4.4. References

Andreev, M.N., Ratnikov, D.S., Dolzhenko, V.D., Belousov, Y.A., Drozdov, A.A. (2020). Influence of macro- and microcomponent content on the color of M_2O-PbO-SiO_2 (M = Li, Na, K, Rb, Cs) glasses doped with copper. *Russ. Chem. Bull.*, 69, 697–703.

Bontemps, G. (1868). *Le guide du verrier.* Librairie des Arts et Métiers, Paris.

Bourdillon, A.J., Khumalo, F., Bordas, J. (1978). The reflection spectra of lead glass. *Philos. Mag. B*, 37, 731–738.

Chopinet, M.-H., Lizarazu, D., Rocanière, C. (2002). L'importance des phénomènes d'oxydoréduction dans le verre. *C.R. Chimie*, 5(12), 939–949.

Cohen, B.M., Uhlmann, D.R., Shaw, R.R. (1973). Optical and electrical properties of lead silicate glasses. *J. Non-Cryst. Solids*, 12(2), 177–188.

Du, Y., Ding, H.-C., Li, S., Savrasov, S.Y., Wan, X., Duan, C.G. (2014). Microscopic origin of stereochemically active lone pair formation from orbital selective external potential calculations. *J. Phys. Condens. Matter*, 26, 025503.

Ellis, E., Johnson, D.W., Breeze, A., Magee, P.M., Perkins, P.G. (1979). The electronic band structure and optical properties of oxide glasses II. Lead silicates. *Phil. Mag. B*, 40(2), 125–137.

Gubanov, V.A., Zatsepin, A.F., Kortov, V.S., Novikov, D.L., Friedman, S.P., Cherlov, G.B., Shchapova, U.V. (1991). Electronic states spectrum for lead silicate glasses with different short-range order structures. *J. Non-Cryst. Solids*, 127(3), 259–266.

Hampton, W.M. (1946). Colour of heavy lead silicate glass. *Nature*, 158, 582.

Kittel, C.K. (1983). *Physique de l'état solide.* Dunod/Bordas, Paris.

Lentes, F.T. (1998). Optical Properties. In *The Properties of Optical Glass*, Bach, H., Neuroth, N. (eds). Springer-Verlag, Berlin.

Majérus, O., Munoz-Bongrand, A., Caurant, D., Roisine, G., Binet, L., Touati, N. (forthcoming). Origin of the yellow color of lead silicate glasses. *J. Non-Cryst. Solids*.

Munoz-Bongrand, A. (2018). Influence de la composition de la matrice vitreuse sur la couleur des glaçures plombifères de Bernard Palissy. Internship report, Sorbonne Université, Paris.

Payne, D.J., Egdell, R.G., Walsh, A., Watson, G.W., Guo, J., Glans, P.A., Learmonth, T., Smith, K.E. (2006). Electronic origins of structural distortions in post-transition metal oxides: Experimental and theoretical evidence for a revision of the lone pair model. *Phys. Rev. Lett.*, 96(15), 1–4.

Radhakrishnan, S., Kamalasanan, N., Mehendru, P.C. (1983). Sensitization of photoconductivity in tetragonal lead monoxide. *J. Mater. Sci.*, 18, 1912–1916.

Robertson, J. (1980). Electronic structure of lead glasses and resonant covalence in glasses. *J. NonCryst. Solids*, 42(1–3), 381–392.

Stroud, J.S. (1971). Optical absorption and color caused by selected cations in high-density, lead silicate glass. *J. Am. Ceram. Soc.*, 54(8), 401–406.

Terpstra, H.J., De Groot, R.A., Haas, C. (1995). Electronic structure of the lead monoxides: Band-structure calculations and photoelectron spectra. *Phys. Rev. B*, 52(16), 11690–11697.

Vercamer, V. (2017). Spectroscopic and structural properties of iron in silicate glasses. Propriétés spectroscopiques et structurales du fer dans les verres silicatés. PhD Thesis, Université Pierre et Marie Curie, Paris.

Watson, G.W. and Parker, S.C. (1999). Origin of the lone pair of α-PbO from density functional theory calculations. *J. Phys. Chem. B*, 103(8), 1258–1262.

Weinstein, I.A., Zatsepin, A.F., Kortov, V.S. (2001). Effects of structural disorder and Urbach's rule in binary lead silicate glasses. *J. Non-Cryst. Solids*, 279(1), 77–87.

PART 3

History and Evolution of Lead Glasses

5

Lead in the Recipes of the Middle Ages and Renaissance

Marco VERITÀ[1] and Isabelle BIRON[2]

[1] *LAMA-Iuav, University of Venice, Italy*

[2] *C2RMF, IRCP, Paris, France*

5.1. The first written sources mentioning the use of lead

Excluding the Assyrian tablets of the seventh century BC that probably mention the indirect lead addition by glassmakers (Brill 1970), several medieval, Arab or European collections of recipes or technical treatises report the use of lead in glass making. The main such sources are, in a chronological order, the collection of recipes *Kitāb ad-Durra al-Maknūna* (compilation of texts from 8th to 10th centuries dating from 13th century) (Al-Hassan 2009), the *Syriac Alchemy* treatises (texts from eighthto ninth centuries) (Berthelot 1893[1]), Heraclius (*de coloribus et artibus Romanorum*, 10th century) (Merrifield 1849), Al-Bīrūnī (treatise on gemstones, 11th century) (Kahle-Bonn 1936) and of Theophilus (*De Diversis Artibus*, 11th–12th centuries) (Dodwell 1961, pp. 44–45) and *Jawâhar-nâma-i Nizâmi* (treatise on gemstones of the 12th century) (Porter and Pourjavady 2004). This chapter does not cover these first treatises, but presents several examples of lead glass recipes dating from the Middle Ages to Renaissance, coming from various places in Europe,

1. Work composed of an introduction and several Syriac and Arab alchemy treatises, according to the manuscripts of the British Museum and Cambridge. Syriac alchemy treatise contained in the ms. Mm. 6. 29 of the library of the University of Cambridge.

Lead in Glassy Materials in Cultural Heritage,
coordinated by Anne BOUQUILLON and Patrice LEHUÉDÉ.

and dedicated to various applications. Because of the important works of Luigi Zecchin and Cesare Moretti for the translation and publishing of written sources on the practices of Venetian glassmakers, the focus will be on Venice and Italy during the Renaissance.

5.2. Recipes of translucent and opaque lead glasses from the Middle Ages and Renaissance

Lead glass is considered in many treatises as ideal for colored glasses and false gemstones; it is "the foundation of any glass coloring art, as there are no proper colors unless lead glass is used"[2] (*Montpellier* H486 – 1536 (Zecchin 1987, p. 271)); it is "the raw material of any gemstone" (Sedacer, second half of the 16th century (Barthélémy 1985, p. 155)), and "when it comes to coloring, it is one of the most beautiful materials that is prepared in a furnace. It is employed to imitate all the gemstones of the East…"[3] (Neri 1612, p. 57); it is still recommended beyond the Middle Ages and the Renaissance for "the pastes intended for gemstones" (Haudicquer de Blancourt 1697)[4].

Most of these recipes are dedicated to the imitation of gemstones by manufacturing glass cabochons or false gemstones, to the fabrication of enamels and also of mosaic tesserae. This is an interesting specificity, as it represents a range of products requiring high quality in terms of colors, shades, brightness and shine. These are high added value ornamental objects. The recipes also concern rods for working on small objects using the oil lamp (beads, jewelry, figurines, etc.) and blown glass objects.

While the recipes mention various colors, the chemical analyses available for these lead-rich glasses show three colors: translucent yellow and green and opaque red. Glasses with high lead content generally have a yellow shade (see Chapter 4), which was used to produce yellow glass objects, and the addition of oxidized copper yields a green glass. An example are the yellow beads produced in Venice to make rosaries (*Paternostri*), whose production is also mentioned by Antonio da Pisa toward 1400 (Lautier and Sandron 2008). As for the use of lead in red glasses, it will be presented in further detail in Chapter 6.

2. "Il fondamento di tutta l'artenostra de vedri per far li colori, perchè... tu non farai mai boni colori se non li fai con il vedro di piombo".

3. L'arte vetraria (The art of glass), fourth book, chap. LXI in the original edition of 1612, p. 57.

4. Chapter LXXXII.

5.2.1. *Middle Ages*

Depending on the source, lead oxide is mentioned as calcined lead ashes, lead or white ceruse, litharge, lead calcine or calcined lead. Examples are the Syriac alchemy treatises (Berthelot 1893, p. 29) (eighth to ninth century – glass dyeing for the production of emerald with one part of ceruse [lead carbonate] and two parts of glass to be melted in a crucible), those of Theophilus[5] (Blanc 1980) (12th century – making rings by mixing ashes, salt, copper powder and lead) and Heraclius[6] (Merrifield 1849) (12th–13th – how to make a lead glass and color it. Burn good bright lead in a furnace, and once it cools down as powder, mix two parts of this powder with one part of sand, then put it in the furnace), and so on (Cannella 2006). In these sources, after the preparation of calcined lead, the latter is then mixed with silica in the form of crystal rock powder or sand, or with glass.

5.2.1.1. *Translucent glasses: imitation of gemstones, inlays and enamels*

Jean d'Outremeuse, in his fourth book of *Trésorier de Philosophie naturelle des pierres précieuses* (Treasurer of natural philosophy of gemstones) dedicated to "gems, colored glass, false gems in the Middle Ages", presents only two recipes using lead (Cannella 2006, pp. 143–146). The first one, difficult to interpret, concerns the fabrication of topazes and other lemon-colored stones. The second one is dedicated to emeralds and all the green stones. ("Moreover, take one hundred and a half ounces of ashes of burnt lead, nice and fine and one hundred drachme of crystal thus prepared, and transform everything into fine and clear powder in a clay pot. Burn this powder for one day and one night and then take it out of fire and the result will be a beautiful, clear and green block".) The amounts of lead estimated[7] in this last recipe may be up to about 87% lead oxide weight for about 13% silica; these contents are equivalent to those observed in yellow and green glasses in North-Eastern Europe (54–84% PbO, 34–9% SiO_2) (Wedepohl et al. 1995).

This small proportion of recipes of colored lead glass seems to reflect the small number of lead glass cabochons observed until present on the goldsmith's pieces of the Middle Ages. Those decorating the Limoges enamels of the Middle Ages from the 12th and 13th centuries are all made of the same soda–lime–silica glasses as those used for enamels (Biron et al. 1995, 2015, p. 209; Röhrs 2010). As for the medieval goldsmith's pieces, they are decorated with cabochons, or exclusively with

5. Book II, chap. XXXI.

6. De *coloribus et artibus Romanorum*, book III, chap. VIII – compilation of recipes extracted from ancient treatises. This third book dates from the 12th to 13th centuries, while the first two are more ancient.

7. Considering 1 ounce = approximately 25 g and 1 drachme = 1/8th ounce, around 3 g.

gems or at the same time with glass and precious or semiprecious stones. Stones often appear to be predominant in these mixtures (Biron and Cannella 2005; Joyner et al. 2006)[8], and are sometimes a minority, as on the Valasse cross (with 124 glass cabochons to 69 in stone)[9]. A larger number of analyses of these false gems would make it easier to appreciate the importance of glass and of the types of recipes employed during these periods.

A certain number of analyses prove however the relevance of written sources for the elaboration of translucent lead glass for green-blue inlays of Visigoth crowns and crosses (containing up to 70% lead oxide – see Figure 5.1) (Calligaro et al. 2001), glasses adorning two Celtic goldsmith's pieces of the 7th century (Bimson and Werner 1965) (Ardagh chalice and Tara brooch), a green Byzantine enamel from 11th to 12th century (43%)[10], yellow and green plique enamels of the first quarter of the 14th century (see Figure 5.2, 50% and 60–70%) (Biron et al. 2015, pp. 152–153), beads, rings, imitation of gemstones and mosaic tesserae found in North-Western Europe (from 54 to 84%) (Wedepohl et al. 1995).

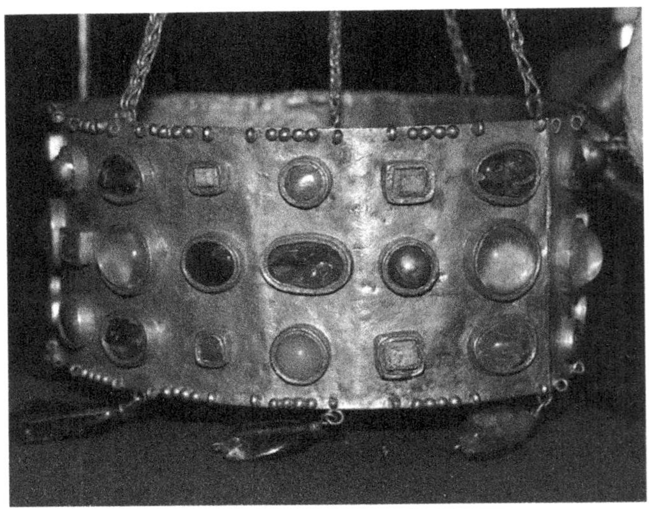

Figure 5.1. *Visigoth crown of the 7th century discovered at Guarrazar near Toledo in the 19th century. National museum of the Middle Ages and thermae in Cluny, inv. Cl. 2879 (H: 67.3 cm and D: 16.8 cm) (© C2RMF, T.Calligaro). For a color version of this figure, see www.iste.co.uk/bouquillon/lead.zip*

8. And C2RMF report no. 21283, pastoral staff known as of Saint Austreberthe, 11th century.

9. C2RMF report no. 29769, 2016, Valasse cross (11th and 12th century).

10. Inv. OA6458 of the Louvre museum, probably of non-Byzantine origin.

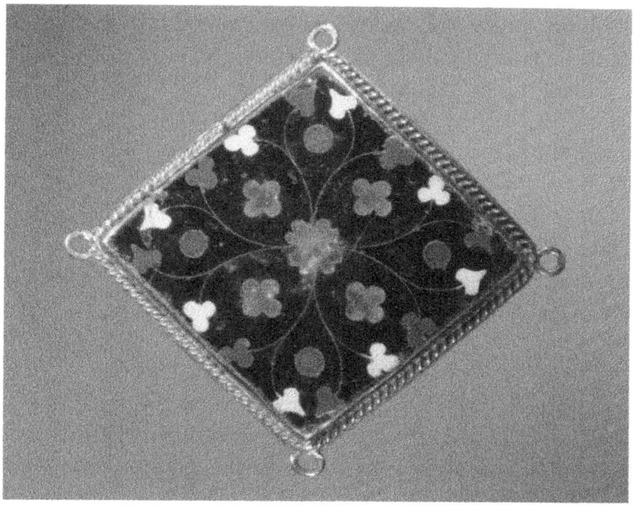

Figure 5.2. *Plique enamel plate preserved in Paris at the National museum of Middle Ages and thermae of Cluny (overall view and detail), inv. Cl. 23411 b, rhombus (H: 3.5 cm; L: 3.2 cm). Lead-rich transparent yellow and green glass (© C2RMF/D. Bagault). For a color version of this figure, see www.iste.co.uk/bouquillon/lead.zip*

5.2.1.2. *Opaque red glasses*

In the category of opaque glasses containing lead, red glasses have a remarkable place, with two large families of composition; one with low copper and lead contents (PbO < 9%), for which the coloring agent is in the form of metallic copper particles as such or mixed with copper oxide particles (Freestone et al. 2003, pp. 142–154). The other is colored by large cuprite dendritic crystals (Cu_2O), whose growth is facilitated by high copper and lead contents (PbO 10–57%) (Barber et al. 2009, pp. 115–127). Due to these particles and crystals, the glass is opaque. While the manuscripts of the Middle Ages and Renaissance mention the addition of calcined copper, and of calcined copper and iron to a colorless transparent glass in order to color it in opaque red (metallic copper), to our knowledge no recipe corresponds to cuprite lead glasses (Cannella 2006, pp. 168–176, pp. 306–316).

5.2.1.3. *Opaque glasses and white enamels*

The several medieval recipes concerning the melting of glass or white enamel recommend preparing a batch of (generally calcined) lead and tin with silica and/or glass powder. This mixture reduced into powder should either be directly used or it should be added to a glass melt or be fired (Cannella 2006, p. 296). Therefore, there are two opacifying processes, one involving the addition of independently synthesized crystals and the other the addition of a *corpo* (glass with a high concentration of opacifier crystals) to the melt.

According to the writings of Mappae Clavicula[11] (Phillipps 1847, p. 240; Smith and Hawthorne 1974) (9th–12th century), to obtain the white color, one should take "ten pounds of tin, one pound of lead reduced into powder, 10 pounds of Asian alum, 8.5 pounds of yellow sand; heat on fire; after burning, break and sieve the result of the burning; then add 8.5 pounds of sand sieved in the same manner and then 6.5 pounds of lead and 5 pounds of tin in the same manner".

Jean d'Outremeuse in the 14th century in the fourth book of his *Trésorier de Philosophie naturelle des pierres précieuses* (Treasurer of natural philosophy of gemstones) presents six recipes for making the white enamel and to make glasses in various colors, one of which is very close to the previous one (Cannella 2006, pp. 295–296): "take three pounds of calcined tin, one pound and a half of calcined lead, two ounces of mercury, one ounce and a half of glass, mill everything together with water in a vessel ... Put it to the furnace to coagulate and then let it cool. And mill in the thin powder that is found coagulated and work with water and paint the glass and then fire it once more".

Another method originating in the previous source: "should one want to dye the glass in white, take one ounce and one drachme and a half of burnt tin, one drachme of ceruse, and one half of drachme of glass and mix them with water and then paint the glass" (Cannella 2006, p. 296).

Finally, other rarer recipes do not use lead, but directly add tin to the glass melt (Cannella 2006, p. 298).

5.2.2. *Italian Renaissance*

The methods presented in this section originate in the articles on the history of Venetian glass of Luigi Zecchin (Zecchin 1987, pp. 103–122; Zecchin 1989, pp. 337–349; Zecchin 1990, pp. 213–226) and in various treatises of the Renaissance written by Venetian and Florentine glassmakers dating from the end of the 14th to the 17th century on raw materials, glass elaboration and coloration presented as follows:

– three booklets (end of the 14th to the 15th century) (*Trattatelli* – referred to by TR in what follows) preserved at the Archivio di Stato of Florence (Milanesi 1864);

11. N° CCLXXXI.

– *Montpellier*, 1536 (Bibliothèque de l'École de médecine, Montpellier, manuscript n. H.486, referred to by M), partially translated and commented upon by Luigi Zecchin (Zecchin 1987, pp. 247–276);

– *Anonimo*, 1560, book of recipes of an unknown Venetian glassmaker, probably a copy of two treatises of Murano glassmakers (referred to by A): the first part possibly dating from the second half of the 15th century to the beginning of the 16th century (Moretti and Toninato 2001, pp. 30–32);

– *Darduin*, collection of recipes belonging to the members of a family, whose workshop was established at the end of the 15th century (referred to by D). Some date from the end of the 17th century/beginning of the 18th century (Zecchin 1986);

– *Arte Vetraria*, the first printed book (1612) on glassmaking technology by the Florentine priest Antonio Neri (1576–1614), which contains a certain number of Venetian recipes (Neri 1612).

5.2.2.1. *Opaque glasses*

For glasses in general and enamels applied on glass (Venetian enameled glass; Verità and Biron 2015, 2017; Figure 5.3) and on metal[12] of the Renaissance, the presence of lead is strictly related to the production of opaque white, yellow and green colors, in which white crystals (tin oxide) or yellow crystals (lead antimonate and stannate) are used both as coloring agent and opacifier. Their fabrication process is presented below. Other colors may also contain lead, but it is then always introduced via opacifiers of white color to modify the hue.

– *Preparation of opaque white glass (lattimo)*: the recipes for preparing a *lattimo* glass describe first the elaboration of lead-tin calx[13] (lead and tin generally calcined together in a ratio close to 1:1). Comparable concentrations of glass frit and lead-tin calx and lower contents of silica were then mixed, calcined and molten. Manganese was added as a decolorizer agent (Verità et al. 2018). Chemical analyses confirm that *lattimo* glass of the 15th and 16th centuries was opacified by tin oxide crystals (cassiterite) dispersed in a soda–lime–silica glass containing lead. A new white glass known as *smalto* was employed in the last years of the 17th century (recipe D219) and rapidly replaced tin oxide for blown glasses. It was prepared by the addition of arsenic oxide, saltpeter and lead oxide to a traditional glass batch. The *smalto* was opacified by micrometric crystals of lead arsenate, producing an intensely opaque glass with better properties (viscosity, smooth surface of objects). Moreover, it was less expensive than the traditional *lattimo* glass. With the same components, but in

12. See Chapter 9 and section 5.2.

13. See section 5.2.

different quantities and ratios, the Venetian glassmakers started in 1693 the production of a new translucent dichroic glass (blue in reflected light, yellow in transmitted light) referred to as *girasole* (sunflower, recipe D214) (McCray and Kingery 1996).

– *Preparation of yellow pigments, lead antimonate or stannate*: the processes to prepare *anime* (souls, or yellow pigments) in a range of colors from lemon yellow to orange brown are reported in Venetian recipes, which refer to lead stannate, lead antimonate and lead and tin antimonate compounds. The *anime* were prepared in the glassworks. Lead stannate, for example, was obtained by melting a batch of lead-tin calx, silica, lead oxide and lead-silica glass (recipes from D52 to D60); the yellow crystals of lead stannate precipitate during cooling; the *anima* (pigment) thus obtained was then added to a colorless (or green) transparent soda–lime glass melt, then the whole was rapidly stirred and cooled (Moretti and Hreglich 1984).

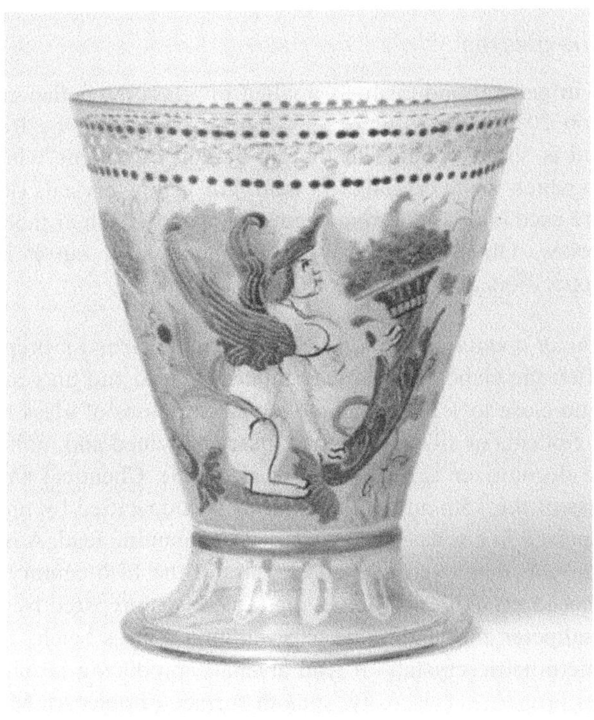

Figure 5.3. *Enameled goblet with two sphinxes facing each other, Venice, end of the 15th century to the beginning of the 16th century, Louvre museum R 51 (H: 11.8 cm, D: 9.5 cm) (picture C2RMF, A. Maigret). For a color version of this figure, see www.iste.co.uk/bouquillon/lead.zip*

5.2.2.2. *Transparent glasses*

Limoges painted enamels of the Renaissance show the use of lead not only for opacified white glass (similar to the white glass presented in section 5.2) but also in transparent glasses. Before 1530, all translucent colored glasses are systematically poor in lead (<1% involuntary addition), while after this date lead contents reach several percentages for certain colors and between 20 and 25% for others. These concentrations tend to increase toward the middle of the 17th century and become systematically high by the mid-19th century (17–60%) (Biron 2015, pp. 359 and 364–366). The addition of minium to the batch of silica sand or crushed stone and *salt alkali* (purified vegetal ash) is mentioned in the writings of B. de Vigenère (1895, p. 65).

Apart from the production of white and yellow pigments, the recipe books of Venetian glassmakers mention starting with the 14th century the use of lead for several applications concerning colored transparent glass. In general, a glass made with lead oxide and silica (lead-silica glass) was prepared in advance. It could be processed as such, or added to the traditional Venetian glass (soda-lime-silica glass [Na-Ca-Si] or sodium silicate [Na-Si]) to make a lead–soda–silica glass (Na-Pb-(Ca)-Si). The lead to silica ratio varies a lot (Pb/Si from 1/1 to 4/1), and lead oxide can be found in the final glass in the range: PbO 20–67%. The variable lead to silica ratios are justified by the necessity for the glassmaker to make softer glasses (with more lead) or harder glasses (with more silica) (Moretti and Toninato 1987).

In the *Montpellier* manuscript (Zecchin 1987, p. 271), recipe M134 describes a golden yellow lead-silica glass made by melting a batch of five parts of silica (the pebbles of the Ticino river give a glass that is less yellow than those of the Adige river which contain more iron) and 10 parts of lead oxide (but lead content may be increased to have a softer glass). Examples of lead glass recipes are presented in the following examples concerning various applications.

– *Imitation of gemstones*: there are recipes for imitating emeralds with a lead-silica glass (PbO/SiO_2 weight ratio: between 1.5/1 and 2/1) to which oxidized copper (CuO) is added (e.g. TRIII-24-31-32). Recipe TRIII-29 for yellow glass to imitate topaz is a lead-silica glass (PbO/SiO_2 ratio: 3/1) without addition of coloring agents. To imitate chalcedony[14] (A14-15; M43), lead silicate is added to the soda–lime glass, followed by addition of silver and other coloring agents. The recipes A71-79

14. Chalcedony is a mineral or a rock composed of quartz crystallites, whose size ranges between 50 and 100 nm, arranged in fibers. Its marbled appearance with veins in various colors (white, blue, brown) was imitated by the glassmakers.

for the imitation of all the gemstones use a basic lead silica glass (Pb/Si 2/1 or 2.5/1) or Pb-Si-Na glass (A81).

Recipe A71 presents the composition of the base glass for all the recipes of imitations of gemstones. "Take one pound of translucent chalcedony (silica) and one pound of minium, mill them finely and put them into a small crucible, heat them on a strong fire with bellows for 12 hours and that is all". The same is valid for recipes A73 (four ounces of rock crystal, eight ounces of alkaline salts, five ounces of calcined lead) and A77 (7 parts of ceruse, one part of quartz).

– *Rods for lamp working of beads and other small objects*: in *Darduin*, there are recipes for lead-silica glass colored in green using copper D84 (PbO/SiO_2 2/1) and D85 (PbO/SiO_2 3/1) that can be used, either for imitation of gemstones or for making low melting rods for lamp working (*tenera da giosar a lume*). Recipes D88-89 (green glass by adding copper), D92 (very detailed) and D93-95 (more or less yellow glass depending on the amount of soda glass added) can be used to make rods, beads and earrings, by adding lead-silica glass to *cristallo* glass cullet (very pure soda glass; common glass cullet in recipe D90). It is a soda–lead–silica glass with more or less calcium (Na-(Ca)-Pb-Si).

– *Mosaic tesserae*: in *Montpellier* (M36, *vetro tenero per musicho*), in order to make a glass easy to cut for mosaic tesserae, lead oxide is added to common soda–lime glass (PbO 50% in the final composition). For gold leaf tesserae (M108), two parts of soda–lime glass are mixed with one part of lead-silica glass (Pb/Si ratio is not mentioned). The addition of lead-silica glass to soda–lime glass in order to make mosaic tesserae is also found in the *Anonimo* manuscript dated 1560, Murano (A35-36, 40, 41, 42, 46, 50).

– *Opaque enamels on metal*: several recipes of opaque glass (A21-26) mention the addition of coloring agents to *lattimo* glass made with *lead-tin calx* (see the example of production presented in section 9.2).

– *Transparent blown glass* (more or less yellow): the recipe TRIII-52 of the third *Trattatello* gives information on how to make a lead glass that can be blown by adding 1/3 lead oxide to a batch of silica and purified soda plant ash. The recipe is nearly identical to the recipe A13 of the *Anonimo*, which gives more information. It explains that the glass color is yellow and suggests a further addition of lead glass to produce a more intense yellow ("*e se fusse tropo chiaro e che lo volessi più scuro agiongeli del vetro de Piombo sopraditto e lo farai venir de ce forze de color che vorai chiaro e scuro a tuo modo*"). This glass is also used for blowing and making drinking glasses ("*vasi diversi da bere o per altri vasi*"), or for making rods that are used for the manufacturing of beads. It must be carefully processed, as it is soft.

Further details can be found in *Neri*'s book, which features the same recipe and the same indications (N64): it is a glass that requires a lot of time to solidify and, therefore, it must be collected with the blowpipe several times, the metallic table on which gob is prepared must be wet (to cool the surface) and other suggestions.

The two recipes (TRIII-76 and 95) of the third *Trattatello* for making yellow-green glass correspond to this category of blown glass. For the coloration, to 100 parts of frit for making transparent colorless soda-lime glass, 20 parts of lead oxide (final concentration of lead in final glass: PbO 17%) and oxidized copper (CuO) are added. Without lead, the same recipe gives (TRIII-77) a dark green glass. Yellow-green potash-lime-silica glass pieces containing lead (PbO 25–28%) and colored by copper (CuO 2.5%) were found by analyzing samples of a 14th-century stained glass window of the Basilica of Santa Croce in Florence (Verità et al. 2019). This glass probably originates in the glassworks of Northern Europe, but the presence of recipes for making this yellow-green glass in the *Trattatelli* of Florence suggests a probable local production.

– *Colorless blown glass*: it is not before the last years of the 17th century that the Venetian glassmakers started to use lead for colorless transparent glass, with entirely new compositions for Murano. The glass was prepared from a batch of minium, potassium nitrate, silica and arsenic. This lead glass, with several modifications, will be used in Murano until the end of the 19th century, without ever replacing the traditional soda–lime glass. Apparently, lead content was of the order of 10% PbO for the blown glass and 40% PbO for the rods and gemstone imitations (Moretti and Toninato 2001). At the end of the 19th century, the blown glass is once again soda–lime glass, while the enamels for metals, rods, beads, etc., are made with a glass containing high lead contents.

5.3. Conclusion

In general, the chemical analyses of lead glasses from the Middle Ages and Renaissance are well in agreement with the recipes for enamels, glass rods and mosaic tesserae. On the other hand, given the current state of our knowledge, the analyses of glass cabochons or glass imitation of gemstones are an exception. Caution is however recommended related to this hypothesis, as it is susceptible of being biased given the still low number of analyses that are therefore non-representative for the whole production of the Middle Ages, and particularly for the Early Middle Ages.

These recipes show that the glassmakers of the Middle Ages and Renaissance knew how to exploit at least some of the properties of lead glasses, either transparent or opaque white. In these two cases, lead is dissolved in the vitreous matrix and confers on it all the properties described in detail in this book (Chapters 1 and 3). Depending on the type of object, it is easy to deduce what properties of the lead glass were sought by the glassmakers of the Middle Ages and Renaissance. Indeed, their color brightness and shine are required for the imitations of gemstones, jewelry and enamels. Moreover, their density renders the glass cutting and polishing easier. On the other hand, the increase in the thermal expansion coefficient of the glass induced by the addition of lead, bringing it close to that of the metallic substrate[15] and thus limiting the mechanical stresses between these materials, was not exploited by the enamellers. The translucent Limoges painted enamels are the proof – the same copper plate can be enameled by lead glass or by glasses with and without lead without any fracture due to stresses. As for the rare examples of other lead-rich translucent enamels on metal, these are always isolated cases that are quite peripheral in their production.

Compared to soda–lime–silica glasses of Antiquity and to the potassium or sodium silica glasses of the Middle Ages, the lower melting temperatures of the lead glasses[16] made it possible to use smaller furnaces that were particularly well adapted to the activity of small workshops dedicated to the fabrication of gemstones (Wedepohl and Baumann 1997), jewelry and enamels. The lead-potash glasses made during the Middle Ages in North-Western Europe are often small objects such as rings, beads and mosaic tesserae (Bezborodov 1975; Wedepohl et al. 1995; Wedepohl and Baumann 1997).

5.4. References

Al-Hassan, A.Y. (2009). An eighth century Arabic treatise on the colouring of glass: Kitāb al-Durra al-Maknūna (The Book of the Hidden Pearl) of Jābir ibn Ḥayyān (c. 721–c. 815). *Arabic Sciences and Philosophy*, 19, 121–156.

Barber, D., Freestone, I., Moulding, K. (2009). Ancient copper red glasses: Investigation and analysis by microbeam techniques. In *From Mine to Microscope – Advances in the Study of Ancient Technology*, Shortland, A., Freestone, I., Rehren, T. (eds). Oxbow Books, Oxford.

Barthélémy, P. (1985). L'Œuvre de Guillaume Sedacer. PhD Thesis, Université de Liège, Liège.

---

15. For example, the thermal expansion coefficient of the copper substrate of enamels is 16.6. 10^{-6} K^{-1}, that of the painted enamels of Limoges for a soda–lime silicate without lead is 7.95×10^{-6} K^{-1}, and that of silicate containing 25% lead weight is 11.31×10^{-6} K^{-1}.

16. Estimated at 740–760°C for a glass containing 65–84% lead oxide and 35–16% silicon oxide according to Wedepohl and Baumann (1997). Please see the binary PbO-SiO_2 diagram in Chapter 3, Figure 3.6.

Berthelot, M. (1893). *La chimie au moyen âge 2 : L'alchimie syriaque*. Imprimerie Nationale, Paris.

Bezborodov, M.A. (1975). *Chemie und Technologie der antiken und mittelalterlichen Gläser*. Philipp von Zarben, Mayence.

Bimson, M. and Werner, A.E. (1965). Scientific examination of ancient glass. In *Annales du 3e congrès de l'Association internationale pour l'histoire du verre, Damas, 1964*. AIHV, Liège.

Biron, I. (2015). Les émaux cloisonnés sur or byzantins du IXe au XIIe siècle. In *Les émaux sur métal d'Occident, du IXe au XIXe siècle, histoire, technique et matériaux*, Biron, I. (ed.). Éditions Faton, Dijon.

Biron, I. and Cannella, A.F. (2005). Identification of false gems on objects from the Middle Ages. In *Annales du 16e congrès de l'Association internationale pour l'histoire du verre. Londres, 2003*. AIHV, Nottingham.

Biron, I., Dandridge, P., Wypyski. M.T. (1995). Le cuivre et l'émail : technique et matériaux. In *L'œuvre de Limoges, émaux limousins du Moyen Âge*, Taburet-Delahaye, E., Boehm, B. (eds). RMN, Paris.

Blanc, A. (ed.) (1980). *Théophile prêtre et moine : essai sur divers arts*. Picart, Paris.

Brill, R.H. (1970). The chemical interpretation of the texts. In *Glass and Glassmaking in Ancient Mesopotamia. An Edition of the Cuneiform Texts which Contain Instructions for Glass-makers with a Catalogue of Surviving Objects*, Oppenheim, A.L., Brill, R.H., Barag, D., von Saldern, A. (eds). Corning Museum of Glass, Corning.

Brun, N. (1991). Étude de verres opaques celtiques et gallo-romains. PhD Thesis, Université de Paris-sud, Paris.

Brun, N. and Pernot, M. (1992). The opaque red glass of Celtic enamels from continental Europe. *Archaeometry*, 34(2), 235–252.

Calligaro, T., Dran, J.C., Poirot, J.P. (2001). Estudio por accelerator de las gemas del Museo Nacional de la Edad Media, Cluny. In *El Tesoro visogodo de Guarrazar*, Perea, A. (ed.). CSIG, Madrid.

Cannella, A.F. (2006). *Gemmes, verre coloré et fausses pierres précieuses au Moyen Âge*. Le quatrième livre du "Trésorier de philosophie naturelle des pierres précieuses de Jean d'Outremeuse". Bibliothèque de la faculté de philosophie et lettres de l'université de Liège, fascicule CCLXXXVIII. Librairie Droz, Geneva.

Challet, V. (1992). *Les celtes et l'émail*. CTHS, Paris.

Dodwell, C.R. (1961). *Theophilus. De Diversis Artibus. The Various Colours of Opaque Glass*. Thomas Nelson and Sons Ltd, London.

Freestone, I.C., Stapleton, C.P., Rigby, V. (2003). The production of red glass and enamel in the Late Iron Age, Roman and Byzantin periods. In *Through a Glass Brightly – Studies in Byzantin and Medieval Art and Archaeology Presented to David Buckton*, Entwistle, C. (ed.). Oxbow Books, Oxford.

Haudicquer de Blancourt, J. (1697). *De l'art de la verrerie.* Jean Jombert, Paris.

Héraclius (1849). De coloribus et artibus romanorum. In *Original Treatises on the Arts of Painting*, Merrifield, M.P. (ed.). John Murray, London.

Joyner, L., Freestone, I., Robinson, J. (2006). Crowning glory: The identification of gems on the head reliquary of St Eustace from the Basle Cathedral Treasury. *J. Gemmol.*, 30(3/4), 169–182.

Kahle-Bonn, P. (1936). Bergkristall, Glas und Glasflüsse nach dem Steinbuch von el-Bērūnī. *Zeitschrift der deutschen morgenländischen Gesellschaft*, 90, 322–356.

Kunckel, J. (1752). Ars vitraria experimentalis. In *L'art de la verrerie de Neri, Merret et Kunckel.* Durand, Pissot.

Lautier, C. and Sandron, D. (2008). *Antoine de Pise, l'art du vitrail vers 1400.* CTHS, Paris.

McCray, W.P. and Kingery, W.D. (1996). The unusual optical properties of two Venetian glasses. *Glass Technol.*, 37(2), 57–68.

Merrifield, M.P. (1849). *Original Treatises on the Arts of Painting.* John Murray, London.

Milanesi, G. (1864). *Dell'arte del vetro per mosaico. Tre trattatelli dei secoli XIV e XV, ora per la prima volta pubblicati.* Romagnoli, Bologna.

Moretti, C. and Hreglich, S. (1984). Opacification and coloring of glass by the use of anime. *Glass Technol.*, 25, 277–282.

Moretti, C. and Toninato, T. (1987). Cristallo e vetro di piombo ricettari dei '500, '600, e '700. *Rivista della Stazione Sperimentale del Vetro*, 17, 31–40.

Moretti, C. and Toninato, T. (2001). *Ricette vetrarie del Rinascimento. Trascrizione da un manoscritto anonimo veneziano.* Marsilio, Venice.

Neri, A. (1612). *L'Arte Vetraria, libro IV.* Giunti, Florence.

Phillipps, T. (1847). Mappae Clavicula – A treatise on the preparation of pigments during the Middle Ages. *Archaeologia*, 32, 183–244.

Porter, Y. and Pourjavady, N. (2004). Le quatrième chapitre du Javāher-nāme-yeNeẓāmī. In *Sciences, techniques et instruments dans le monde iranien (X^e^–XIXe siècle)*, Pourjavady, N., Vesel, Ž. (eds). Peeters Publishers, Louvain.

Röhrs, S. (2010). The vitreous decoration of the British Museum Ciborium, the scientific analysis. In *The Heritage of "Maître Alpais"*, LaNiece, S., Röhrs, S., McLeod, B. (eds). The British Museum, London.

Smith, C.S. and Hawthorne, J.G. (1974). Mappae Clavicula: A little key to the world of medieval techniques. *Trans. Am. Philos. Soc.*, 64(4), 1–128.

Verità, M. and Biron, I. (2015). Enamels on Venetian Renaissance glass. An analytical and technical approach. *Glass Technol. – Part A*, 56(6), 177–190.

Verità, M. and Biron, I. (2017). Renaissance Venetian enamelled glass. An analytical approach to distinguish between genuine, façon de Venise and fakes. In *Annales du 20e Congrès AIHV (Fribourg 2015)*. AIHV, Romont.

Verità, M., Biron, I., Cagnini, A., Porcinai, M., Barbe, F. (2018). The so-called Venetian enamelled copper artworks of the Italian Renaissance: The technology and provenance of the enamels – An analytical approach. *Archaeometry*, 60(2), 233–254.

Verità, M., Bracci, S., Porcinai, S. (2019). Analytical investigation of 14th century stained glass windows from Santa Croce Basilica in Florence. *Int. J. Appl. Glass Sci.*, 10, 1–12.

de Vigenère, B. (1895). Emaillerie au XVIe siècle. In *L'art de l'émail de Limoges ancien et moderne : traité pratique et scientifique*, Meyer, A. (ed.). H. Laurens, Paris.

Wedepohl, K.H. and Baumann, A. (1997). Isotope composition of medieval lead glasses reflecting early silver production in central Europe. *Mineralium Deposita*, 32, 292–295.

Wedepohl, K.H., Krueger, I., Hartman, G. (1995). Medieval lead glass from northwestern Europe. *J. Glass Stud.*, 37, 65–82.

Zecchin, L. (1986). *Il Ricettario Darduin, un codice vetrario del Seicento, trascritto e commentato*. Arsenale, Venise.

Zecchin, L. (1987). *Vetro e vetrai di Murano*, volume 1. Arsenale, Venice.

Zecchin, L. (1989). *Vetro e vetrai di Murano*, volume 2. Arsenale, Venice.

Zecchin, L. (1990). *Vetro e vetrai di Murano*, volume 3. Arsenale, Venice.

6

The First Lead Glasses

Bernard GRATUZE
IRAMAT-CEB, CNRS, University of Orléans, France

6.1. Introduction

This chapter dedicated to the history of lead used as raw material in the glassmaking industry focuses on the period from the beginning of glassmaking, by the middle of the second millennium BC until the 17th and 18th centuries (see Figure 6.1), when lead crystal glass appears (Dungworth and Brain 2013).

In Western Europe, the "official" beginning of lead glass is often associated with the invention in England of the colorless glass known as *flint glass*. Indeed, in 1674 George Ravenscroft filed a patent that would give him exclusivity for its fabrication in England. "Flint glass", a colorless glass composed of silica, lead oxide and potash, is traditionally, and rightfully so, associated with the names of Ravenscroft, and/or Da Costa (Moretti 2005; Brain 2008), who had invented this type of glass, as it will be seen in Chapter 7. However, since the origins of the glass history, lead was part of the raw materials used by the glassmakers, either on purpose, as an ingredient of the recipe, or incidentally, by means of other raw materials. In 1877, Eugène Péligot (1877) highlights the controversies related, in his opinion, to the confusion between flint glass as such and the use of lead, more generally, as raw material by the glassmakers. Flint glass, as potash and lead silicate, is according to Péligot the most beautiful vitreous material that can be produced.

Lead in Glassy Materials in Cultural Heritage,
coordinated by Anne BOUQUILLON and Patrice LEHUÉDÉ.

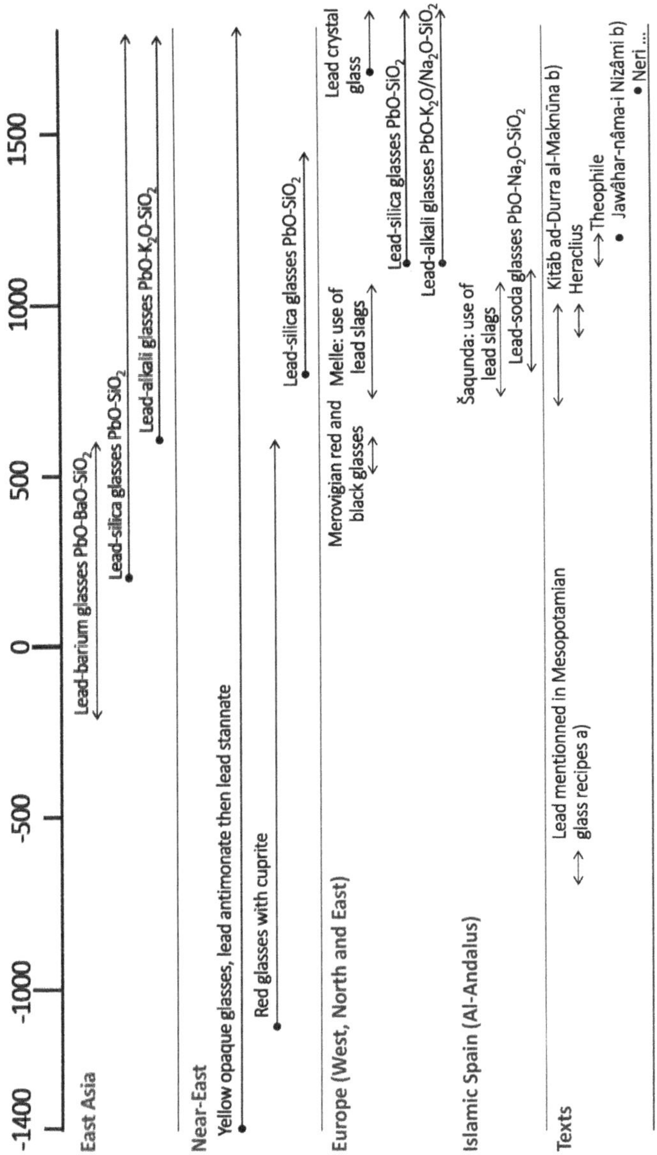

Figure 6.1. *Simplified chronology of the fabrication of lead glasses and the use of lead in glassmaking. (a) (Brill 1970); (b) Schibille, Chapter 8*

In his opinion, this very specific glass was not known by the ancient people, even if they used lead in various glass compositions. Péligot mentions the "hematin" glass used by the Gauls for enameling their weapons, as well as the enamels and false gemstones whose recipes, described in the treatises of Heraclius (10th century) and Theophilus (11th–12th century), explicitly refer to the use of lead for their fabrication. He also mentions one of the first archaeomeric studies conducted in 1787 by Fougeroux de Bondaroy (1787) on the so-called Virgil's mirror that was part of the treasury of Saint-Denis. This yellowish-green mass of glass of about 15 kg was polished on the two faces and had an oval shape. Fougeroux de Bondaroy estimated that half of the weight of this glass consisted of "lead lime" (litharge or lead oxide), while the other half consisted of vitrifiable earth (silica). He compared this glass with the English *flint glass* and the strass, invented about 40 years previously (see Chapter 7) and concluded that lead glasses were fabricated long before the invention of crystal glass. This conclusion is probably at the origin of the controversy that Péligot mentioned in 1877. This piece, no longer available, that can be dated between the last quarter of the 9th century and the first half of the 12th century, could be compared with the Islamic lead glasses (see Chapter 8).

Since the pioneering works of Auguste-Denis Fougeroux de Bondaroy and Eugène Péligot, many studies contributed to a refinement of our knowledge on the fabrication of the first lead glasses, and more generally on the use of lead by the glassmaking industry. The history of this development covering slightly over three millennia is the focus of this chapter. The exploration starts in Egypt and Mesopotamia, cradles of the glassmaking industry at the middle of the second millennium BC. The next points of interest are the whole Middle East and the Eastern Mediterranean for the period of the first millennium BC. Recipes developed by glassmakers from East Asia and South Asia between the middle of the first millennium BC and the middle of the first millennium AD will then be examined. The exploration will then bring us back to the West, with a focus on the recipes elaborated by European glassmakers between the middle of the first millennium AD and the 17th century. Islamic lead glasses are beyond the scope of this chapter, being explored in Chapter 8.

6.2. Glasses of the Eastern Mediterranean from the second and first millennia BC

There is very little evidence of the first development phases of the glassmaking industry between the second half of the third millennium and the second half of the second millennium BC. This craftsmanship, rooted somewhere between Syria and Mesopotamia (Henderson 2013), is probably related to that of faïence (Angelini et al. 2019). A quasi-industrial development of this activity starts in the second millennium BC in Mesopotamia and Egypt. A complex operational chain is rapidly established: it involves primary workshops that elaborate the raw glass, and secondary workshops that process it. Since the 14th century BC, similar to the copper ingots, the raw glass is commercialized over long distances, as proven by the discovery of Egyptian glass ingots in the Uluburun shipwreck (Pulak 2008). Chemical analyses conducted on soda-lime glasses of this period regularly evidence the presence of lead at contents varying from several parts per million to several percents, even several tens of percents for yellow glasses (Brill 1999; Shortland and Tite 2000; Shortland 2002). At this period of time, lead is therefore present in glasses, either as impurity brought by the raw materials or tools, or as an ingredient associated with the preparation of coloring materials.

6.2.1. *Yellow and green glasses*

The studies published by Brill and Shortland on the first Egyptian glasses (Brill 1999; Shortland and Tite 2000; Shortland 2002) show in fact that lead presence is closely related to the presence in the vitreous matrix of particles of a yellow pigment: lead antimonate. For green glasses, which contain this colorant, lead oxide contents vary from 0.3 to 2% (in weight) and for the yellow glasses from 1 to 7%. It can be noted that in yellow glasses lead oxide is in excess with respect to the stoichiometry of lead antimonate. It is however not added on purpose to the glass melt, being a result of the colorant fabrication process. For other color glasses fabricated during this period, lead content rarely exceeds a thousand parts per million (0.1%) and it is most often below 100 parts per million.

These lead contents of yellow glasses are predominant until the beginning of the 1st millennium BC. From then on, the pigment contents, and consequently the lead oxide contents of yellow glasses, increase progressively (Brill 1999) until reaching 20–30%. Lead content may sometimes reach 40%, and even more, but these high values seem most of the time related to a strong degradation of the vitreous matrix. Lead then becomes the second component of this type of glass, right after silica (45–60%).

6.2.2. *Red glasses*

Starting with the first millennium, increasingly higher lead contents are systematically used as components of red glasses with cuprite (Cu_2O), and less systematically of red glasses with metallic copper (Cu). At Hasanlu (Iran), red glasses with cuprite, dated between 1100 and 800, contain 7% of PbO, while those found at Nimrud (Irak), dating from the seventh century BC, have 10–23% content (Freestone 1987; Brill 1999; Barber et al. 2009). It can go up to 30% and even more at the beginning of AD (Hughes 1972; Brun and Pernot 1992). Lead contents of the red glasses with metallic copper are generally lower and far more variable.

Unlike yellow glasses, in which lead is one of the main components of the pigment, lead is not an essential component for the fabrication of red glasses, as proven by the Egyptian red glasses of the 15th and 14th centuries BC, which do not contain lead (Freestone 1987; Brill 1999; Barber et al. 2009). Lead oxide is therefore added on purpose to this type of glass in order to facilitate its elaboration, on the one hand, and improve its appearance and make it easier to process, on the other hand (Freestone 1987). The studies conducted show indeed that starting with 10%, lead facilitates the formation of Cu^{+} rather than of Cu^{2+} thus allowing the nucleation of a larger number of cuprite crystals. Beyond 15% lead also enhances the growth and size of these crystals. The increase in the number and size of crystals thus contributes to the glass opacity and coloration. Moreover, the presence of lead limits devitrification, facilitates its engraving and renders the glass brighter (see Chapter 1), making it look more similar to gemstones (Freestone 1987). These studies also show that, in contrast with cuprite glasses, less benefits are obtained by the addition of lead oxide to the melt for the fabrication of red glasses with metallic copper (Freestone 1987). For these red glasses, the presence of lead is indeed highly variable, and it is not correlated with the dating of glasses. It seems to depend more on traditions and recipes specific to various workshops. In contrast, for red glasses with cuprite, the addition of lead at contents above 15% becomes systematic during the first millennium BC, to the detriment of recipes without lead, which are hardly ever used.

Red glasses with cuprite and, to a lesser extent, those with metallic copper, seem to be the first glasses in which the glassmakers purposefully incorporate lead oxide, in order to improve its fabrication and its appearance. With contents ranging between 20 and 35 to 40% lead oxide, this element, similar to the yellow glasses, is the second component of glass, immediately after silica (40–50%). For this reason, red glasses with cuprite could be considered the first lead glasses produced in the Mediterranean region.

Throughout the first millennium BC, there is a more or less constant increase in lead contents in glasses, other than yellow or red. However, this increase is not a reflection of the voluntary addition of this element, whose contents are most often below percent, seeming rather the consequence of the addition of metallic colorants from metallurgical wastes (lead bronze, for example) and from the incidental recycling of yellow or red colored glasses. Two cases of voluntary addition of lead oxide to the glass are however worth mentioning: antimony-decolorized glasses from the end of the Hellenistic period to the first century BC (Paynter and Jackson 2019), as well as white tin oxide-opacified glasses dating also from the first century BC (Rolland 2017).

6.2.3. *Hellenistic glasses containing antimony*

A quasi-systematic presence ranging between 2 and 3% lead oxide is noted in the antimony-decolorized glasses produced at the end of the Hellenistic period. These glasses differ from the ancient antimony-decolorized glasses (produced since the end of the first century AD), whose lead content is rarely above 300 ppm. The works of Paynter and Jackson (2019) highlight the role of the base glass composition, and more specifically that of calcium content, on the effects of a 2% lead oxide addition for calcium antimonate opacified glasses and antimony decolorized glasses. While this addition does not essentially modify the dispersion of calcium antimonate crystals in the glasses with low lime content (Egyptian compositions), it favors their dispersion in the glasses with high lime content (Syrian-Palestinian compositions). The presence of lead would have therefore been advantageous for Syrian-Palestinian glassmakers that fabricate white opaque glass, as it improved the glass quality, and perhaps the opacity, by producing finer and better dispersed crystal glasses (see Chapter 1).

For the Hellenistic colorless glasses, which contained lower contents of calcium and aluminum oxides, two hypotheses may explain the addition of lead oxide to the antimony oxide-decolorized glasses. The purpose of this addition may have been to render this material easier to refine and process (engraving, polishing, see Chapter 1). Likewise, while the use of lead is not necessary for the fabrication of white pigments (calcium antimonate), it may have been incidentally or purposefully combined with antimony at the end of the Hellenistic period. Thanks to its chemical properties, facilitated by oxidation and combination with oxide phases of other elements present as impurities, lead is widely used during the metal refining operations (silver cupellation, for example). It may have been part of the operational chain implemented for the preparation of compounds used for glass opacification and discoloration. The association of lead and antimony in certain glasses at the end of the Hellenistic period may consequently be simply due to the method of preparation of antimony compounds.

6.2.4. *Tin oxide–opacified glasses*

In Western Europe, during a short period around the beginning of the first century BC (Tène D), it can be noted that antimony compounds are replaced by tin compounds for the production of yellow opaque glasses (lead stannate replaces lead antimonate) and white opaque glasses (tin oxide replaces calcium antimonate) (Tite et al. 2008; Rolland 2017). This mutation, concerning essentially the Celtic glass adornment, does not influence the lead concentrations of yellow glasses, which remain in the range between 15 and 40%. On the other hand, white opaque glasses, colored/opacified by tin oxide, feature a systematic presence of 2–5% lead oxide, while the presence of lead remains exceptional in glasses colored/opacified with calcium antimonate. It is difficult to decide between these two hypotheses:

– either it is a deliberate addition of lead oxide to facilitate the dispersion of tin oxide in the glass, as evidenced for antimony (Paynter and Jackson 2019);

– or the lead addition is related to the fabrication of white pigment of tin oxide (see Chapter 5), with the purpose, once more, to facilitate its incorporation into the glass.

6.3. Lead glasses in Asia starting with the second half of the first millennium BC

While the use of lead by the glassmakers of the Mediterranean Basin seems limited to very specific cases, the situation is quite different in East Asia (China, Korea and Japan). Toward the end of the third century BC, lead appeared to be one of the main components of the glasses manufactured in this wide geographical area (Gan 2009a, 2009b). Three large families of silica glass are of particular interest in this chapter: glasses with lead and barium (Pb-Ba) oxides, glasses with lead oxide properly speaking (lead silicate, Pb-Si) and lead–potash glasses (Pb-K).

6.3.1. *Lead–barium (Pb-Ba) glasses*

Silica glasses with lead and barium oxides appeared in Central China toward the end of the third century BC. There is a relatively large variability of compositions of these glasses and several subgroups were identified in this family (Gan 2009a, 2009b; Cui et al. 2011; Wang and Jackson 2014; Dong et al. 2015; Henderson et al. 2018). According to Henderson et al. (2018), their composition is characterized by the following minimal and maximal contents: 20.5–67.1% SiO_2, 12.3–54.5% PbO and 0.16–20.3% BaO, while other classifications (Wang and Jackson 2014) propose minimal contents for the barium oxides (BaO > 9%) and lead oxides (PbO > 20%).

Their contents in alkaline oxides, K_2O and Na_2O, range between 2 and 5%. The main other oxides, CaO, Al_2O_3 and MgO, have contents of percent order and the other elements present in these glasses are essentially colorants such as iron, copper or cobalt. This type of glass mainly used for making adornments and objects other than vessels (beads, disks, funerary objects, etc.) is also found in Korea and Japan (Gan 2009b; Tamura and Oga 2014). Its production gradually decreases between the second and sixth centuries AD.

6.3.2. *Lead glasses in the strict sense (Pb-Si)*

At the beginning of AD, a second group of lead glass appears (Gan 2009b). It is essentially composed of lead oxide (35–75%), silica (35–75%) and less than 5% alkaline oxides ($Na_2O + K_2O < 5\%$). It is found throughout the first millennium and at the beginning of the second millennium AD in China, Japan and Korea (Gan 2009b; Wang and Jackson 2014; Henderson et al. 2018). While this glass probably appeared in Southern China, it was very rapidly found in Korea (first century BC to first century AD) and in Japan (sixth century). Its later fabrication in Japan and Korea is not excluded. The first blown glass table glassware with this composition appears between the third and sixth centuries. This glass is a precursor to Merovingian glasses of the sixth century, Islamic lead glasses that appear in the Eastern Mediterranean in the ninth century (see Chapter 8), lead glass adornments in Eastern Europe of the 10th and 13th centuries (Siemianowska et al. 2019), and even of the ninth century (Bezbodorov 1975) and lead glasses of the modern era in Western Europe (Moretti 2005).

6.3.3. *Lead–potash glasses (Pb-K)*

Pb-Si glasses seem to gradually evolve toward lead–potash glasses, in which lead oxide is partly replaced by potash. It is not clearly established when these glasses appeared. Their lead oxide contents range between 35 and 50% and their potash contents between 7 and 15%. Being probably initially fabricated also in China, these types of glass can be found during the first millennium in Korea, in Japan starting with the 12th century and in Vietnam toward the 16th and 17th centuries (Gan 2009b). Once more, certain compositions of this group of glass (Kang and Yun 2012) are precursors to those of European lead-potash glasses of the 10th and 13th centuries (Bezbodorov 1975; Siemianowska et al. 2019) and of the lead–alkali glasses of the modern age, and except their high rate of impurities, those of *flint glass* (Moretti 2005; Brain 2008; Dungworth and Brain 2013). With its lead–silica glasses and its lead–potash glasses, East Asia seems to have had a pioneering role in the elaboration of various families of lead glass that appeared in the Mediterranean Basin and in Eastern Europe in the ninth century.

With respect to lead glasses, the situation in South Asia and South-East Asia resembles that observed in the Mediterranean Basin. For the various glass families identified in this wide geographical area, the references considered are the following: (Lankton and Dussubieux 2006, 2013; Dussubieux and Gratuze 2013) and the special volume of the *Archaeological Research in Asia* (Carter et al. 2016). The use of lead is in this region, too, correlated with coloration recipes and pigment preparation modes. The quasi-totality of glasses that contain lead oxide at contents above 2 or 3% are yellow or green glasses characterized by the presence of lead stannate: this ingredient is the yellow pigment employed by the glassmakers of this geographical area, where antimony is not used. In contrast with the Mediterranean Basin, lead is not used for the fabrication of red glasses, with either cuprite or metallic copper. The low lead concentrations measured in red glasses (PbO > 1%) rather indicate that this element was incidentally introduced by means of metallurgical wastes employed as a source of copper.

6.4. Medieval lead glasses in Western Europe

6.4.1. *The Merovingian period (end of the fifth century to the first third of the eighth century)*

From the beginning of AD until the end of Antiquity, the use of lead by Mediterranean and European glassmakers did not evolve compared to the previous period (see section 6.2). For the glasses of this period, the presence of lead, and of other metallic impurities, is rather considered as an indicator of the glass recycling rate (Freestone 2015). This situation appears to change at the end of Antiquity. A clear increase in lead contents in the productions of western glassmaking workshops can be noted starting with the Merovingian period. One of the characteristics of glassmaking in this period is the massive fabrication of beads, whose complexity and variety of shapes and patterns increase, reaching its peak between the last third of the sixth century and the first third of the seventh century AD (Pion 2014). While lead, with contents ranging between 20 and 60%, is still one of the main oxides of yellow glasses, it also becomes one of the integral components of certain categories of red glasses colored by metallic copper, of black glasses, colored by iron, and to a lesser extent of white glasses.

6.4.1.1. *Yellow and white opaque glasses*

Yellow opaque glasses of this period are essentially obtained by the addition of lead stannate, yellow opacifying pigment that gradually replaces lead antimonate at the end of the fifth century AD (Tite et al. 2008). While there is still a residual presence of yellow glasses with lead antimonate, due to the recycling of ancient mosaic tesserae, the latter are rare. Yellow glasses with lead stannate however differ from yellow glasses with lead antimonate by higher lead contents. These on average range between 30 and 50% (reticella beads and glasses) while they rarely exceed 20–30% for glasses with lead antimonate. Lead and tin contents of yellow opaque glasses appear to constantly increase between the fifth and seventh century. While yellow glasses with lead stannate from the beginning of the Merovingian age are identical to those of the end of Antiquity, the addition of increasingly higher amounts of colorant can be noted during the sixth century. The coloring batch sometimes represents more than half of the mass of glass of these objects (mostly monochrome or polychrome beads; Mathis et al. 2022).

As already noted for white opaque glasses colored and opacified with tin oxide at the end of the Celtic period (Tène D), in the Merovingian period there was a quasi-systematic addition of lead oxide in white opaque glasses with tin oxide (up to 5% and even 10% PbO for tin oxide contents ranging between 3 and 20%), while it remains nearly absent from white glasses opacified with calcium antimonate. It may be that, similar to calcium antimonate (Paynter and Jackson 2019), this addition of lead oxide facilitates the dispersion of tin oxide crystals in the glass.

6.4.1.2. *Red and black glasses*

The two major innovations of the glassmakers in this period related to the use of lead concern the fabrication of red and black glasses between the end of the fifth century and the last third of the sixth century, or the MA1 and MA2 (Mathis et al. 2022). Until the end of Antiquity, unlike the red glasses with cuprite and to a lesser extent with metallic copper (see section 6.2), lead was not part of the ingredients used for the fabrication of black glasses, colored by iron oxide, where it is most often only present in the state of impurity (Cosyns 2011; Cagno et al. 2014).

Starting with the end of the fifth century, a glass containing very significant amounts of lead oxide (25–60%) and iron oxide (2–19%) appeared in North-Western Europe. This composition, which is also characterized by the absence of alkaline or alkaline-earth elements ($Na_2O + MgO + K_2O + CaO < 4\%$) and by variable alumina contents ($2.5 < Al_2O_3 < 9\%$), was only used for the fabrication of black and red glass beads. Slight composition differences are observed between red and black glasses (Mathis et al. 2022), the former containing less iron and alumina than the latter. These differences remain however minor and are probably related to the colorant materials. With an average silica and lead oxide content of about 80% (SiO_2 + PbO = 78% for black glasses and nearly 86% for red glasses), this glass could be obtained by mixing litharge (lead oxide) with silica. Minor elements and traces may be caused by migrations from the crucibles. Due to the low contents in lime, antimony and barium, the hypothesis of the addition of slags resulting from lead metallurgy becomes less probable. The use and therefore the fabrication of this glass seem to suddenly stop from the last quarter of the sixth century. Other red and black glasses without lead oxide, as well as red and black glasses with intermediate compositions (Mathis et al. forthcoming), circulate also in North-Western Europe in the Merovingian period. The latter probably reflect the batch of leadless glasses and lead glasses.

6.4.2. *From the Carolingian period (middle of the eighth century to the 10th century) to the beginning of the modern period*

6.4.2.1. *Glasses resulting from the recycling of metallurgical slags*

During the eighth century, another group of lead glass emerges in Western Europe. First identified at Novgorod (Russia), then at Haithabu (Germany) upon the study of glass smoothers (Scapova 1992; Wedepohl 2001), this specific composition was then linked to the productions of a workshop probably located in the region of Melle (France, Deux-Sèvres) (Gratuze et al. 2017b). Until the discovery of a set of blown glass goblets of this composition (Gratuze et al. 2014), this lead glass, issued from the recycling of slags, had longtime seemed specific to Carolingian glass smoothers.

This use of lead slags, produced during the processing of galena for obtaining silver, whose only equivalent known at that time was recently identified at Šaqunda in Spain (Schibille et al. 2020), had two advantages for the glassmakers in Melle. On the one hand, it offered them a substitute product to cope with the shortage of soda–lime glass encountered in that period throughout Western Europe (Angelini et al. 2019), and, on the other hand, it allowed them to save fuel. The systematic presence of excess soda in this type of lead glass, with respect to slags, highlights the fact that these glassmakers were probably recycling ancient soda–lime glasses. The lead contents of Melle glasses remain moderate and highly variable: 1.5–16% of PbO for the goblets and 8–26% of PbO for the smoothers (Gratuze et al. 2014, 2017b). Strictly speaking, these glasses can be hardly named lead glasses, as shown by the very strong variability of their lead contents, and therefore rather related to slag recycling. Indeed, lead is not introduced as such in the glass batch, being indirectly added by the use of metallurgical waste resulting from silver production.

6.4.2.2. *Lead and lead–alkali glasses*

Several families of lead and lead–alkali glasses, strictly speaking, were identified in Europe between the 9th and 14th centuries (Wedepohl et al. 1995; Stephan et al. 1997; Mecking 2013; Siemianowska et al. 2019). Their fabrication is mentioned by two medieval authors, Heraclius (Merrifield 1849) and Theophilus (Dodwell 1961). These glasses, mainly present in Central and Eastern Europe, seem to have been used mostly for the production of adornment elements (rings, beads). Unlike the Spanish lead glasses (Schibille et al. 2020), they are more rarely used for table glassware or for glazing that until the 12th century were most often produced from alkali-lime glasses (Bezbodorov 1975; Wedpohl et al. 1995; Brill 1999). It is important to note however the fabrication of vessels from strongly colored glass with high lead content in certain countries in North-Western Europe (Germany, England, Ireland, Sweden and Finland) between the 12th and the 14th centuries (Wedepohl 2000; Krueger 2006). While most of these glasses have an emerald-green shade, there are also bright yellow and sometimes brown-red vessels. The fabrication of these vessels from lead-colored glass seems to come to an end in Europe in the 14th century. Moreover, in the 13th and 14th centuries, smaller objects, such as knife-handles, were fabricated from enamel (lead-soda glass opacified with tin and colored or not with cobalt), such as the one of Saint-Denis (La Gabrielle 2017, Inv. 16.1680.5., p. 54).

The studies conducted on adornment elements (Mecking 2013; Siemianowska 2019) show the presence of various regional groups of lead–alkali glasses. The most widespread are lead glasses having lime-potash as a fluxing agent, while lead glasses with soda or soda-potash as fluxing agent are less common. Regional specificities are highlighted:

– in Western Europe (France, England) and in Northern Europe (Scandinavia), there are almost exclusively lead glasses *str.s.*;

– in Central Europe (Germany, Austria, Czech Republic, Slovakia), the ratio of lead glasses *str.s.* and lead-alkali glasses (essentially lime-potash glasses) is 2/3–1/3 and is reversed in Eastern Europe (Poland, Russia), where the latter become predominant and represent more than 2/3 of the objects.

While lead and lead-alkali glasses in Central and Eastern Europe appear to be the result of local production (Siemianowska 2019), the origin of some of the oldest lead glass *str.s.* elements of adornment (9th–10th centuries) found in Scandinavia (Sode et al. forthcoming), England (Bayley 2009) and France (Velde and Gratuze 2009) is still being questioned. Unlike other types of glass, the chemical compositions of lead glasses *str.s* vary little, and isotopic analyses are the only means to identify their origin and differentiate between local productions and/or import from Central Europe or the Islamic world. According to Mecking (2013), a possible explanation for the significant production of these lead glass elements of adornment in Central and Eastern Europe, between the 10th and 14th centuries, is the wide availability of lead, related to increasing silver extraction during this period (Wedepohl and Baumann 1997).

6.5. European lead glasses, from the beginning of the modern period until the invention of crystal glass

The quasi-exclusive use of lead glasses for the fabrication of adornment elements or for enameling (see Chapter 9) appears to continue between the 15th and 17th centuries. Various recipes of lead glasses, some of which clearly mention the fabrication of beads or the imitation of gemstones, are described in several manuscripts of this period (Moretti 2005). According to these recipes, the proportions of lead in the glasses range from 22 to 28% (anonymous collection from the 15th/14th centuries), to about 56% ((Neri et al. 2001), but original from 1612), and reach 66% (recipe from 1635 of Brunoro's collection). It is important to note, both for Kunckel and Merret (Neri et al. 2001), that this type of glass is not always suitable for the fabrication of vessels, due to its "excessive brittleness". In the first half of the 17th century, these glasses appeared to have a too wide working range and a too low hardness for the fabrication of table glassware or glazing.

According to the compositions calculated by Cesare Moretti from these ancient recipes, these glasses also contain between 2 and 10% soda and slightly less than 2% potash. These two alkali are probably brought by cullet[1], as the recipes do not explicitly mention the addition of alkaline fluxing agent. They require only the addition of cullet or of frit of common glass or "cristallo" type in the Venetian sense of the term. In most of these recipes, lead is introduced in the form of litharge. The glass obtained is described as being transparent and having a golden-yellow shade. It appears to be especially appreciated for its refractive index higher than the one of the soda-lime glass, which makes it brighter (see Chapter 1). It is interesting to note that most of these recipes draw attention on the possibility of metallic lead precipitation in the vitreous mass. This precipitation is related to the absence of oxidizer, such as saltpeter, in the melt. According to these recipes, this compound does not appear to have been systematically used in Venice before the end of the 17th century.

---

1. Cullet designates all the glass debris added to the composition to facilitate melting. There are two types of cullet. The first one, also named groisil, results from recycled vessels, and the second one, called factory cullet or glass frit, results directly from other phases of fabrication of the raw glass, either from the workshop in question or from another workshop.

For the 15th and 16th centuries, the data on European lead glasses *str.s.* come mainly from the studies resulted from the excavations on San Salvador (Bahamas) and Nueva Cadiz (Venezuela) islands where small rolled-up beads of green and amber shades were discovered (Brill et al. 1987). These beads, attributed to the first stages of contacts with the Europeans, are dated from the beginning of the 16th century. Identical beads were also signaled in Columbia and Peru. The analyses conducted by Brill show that they were fabricated from a lead glass *str.s.* ($19.5 < SiO_2 < 27.5$ and $67.4 < PbO < 74.9$) and that they were colored by copper for the red glasses and by iron for the amber-shaded glass. The basic recipe for these glasses is equivalent to one part of sand to three parts of litharge in weight, or one part of litharge and one part of sand in volume (Brill et al. 1987): it closely resembles the recipes published one century later by Brunoro (Moretti 2005) for the fabrication of beads. Isotopic analyses show that lead may come from Spanish mines, which supports the hypothesis of their fabrication in Andalusia (Brill et al. 1987). Beads of lead glass *str.s.* or lead-alkali glass have also been identified on various African sites (Truffa Giachet 2019): at Old Buipe, (Ghana, 15th–18th centuries), at Alinguel and Toum-bounto (Senegal, 17th–19th centuries). On the Alinguel site, beads of lead glass *str.s.* and lead-alkali glass may come from levels dated as 11th–13th centuries. For the European 15th and 16th centuries, the available analyses also show the use of lead-alkali glasses for adornment elements of table glassware, from white opaque glass or red glass, which may resemble enameling (see Chapter 9). Lead contents of white patterns vary from 10 to 20% for tin contents from 7 to 15% (Guerit et al. 2020).

From the middle of the 17th century, lead glass recipes increase in frequency and number (Moretti 2005; Kerssenbrock-Krosigk 2008). They even clearly mention the addition of a significant amount of saltpeter, calcined scale (potassium tartrate) and sometimes borax, during the first decade of the 17th century. These formulations, which correspond to those of a lead-potash glass (PbO 17 to 23% and K_2O 18 to 22%), are then close to those developed by Ravenscroft in England around 20 years before (Brain and Dungworth 2005, 2009; Moretti and Hreglich 2009) (Chapter 7). The archaeological and archaeometric data thus highlight a clear intensification of lead glass production and the fabrication of the first pieces of table glassware and glazing with this type of glass from the second half of the 17th century (Dungworth 2005; Müller and Stege 2009; Biron et al. 2010, 2011; Gratuze and Arles 2012; Valente et al. 2016).

According to the book entitled *Glass of the Alchemist, Lead Crystal-Gold Ruby, 1650-1750* (Kerssenbrock-Krosigk 2008), the invention of Ravenscroft is part of an overall development that revolutionized European glassworks from the second half of the 17th century. The recent works of Lefrancq (2020) on the glassworks of

Verdun (Meuse) mention a correspondence, dating from 31 March 1675, between Jean-Tilman d'Heur, director of this glassworks, and his uncle Henri Bonhomme, who was the owner. In this letter, he reveals the secret of the lead glass of Bernard Perrot: "The secret of Perrotte in Orleans is linked to the use of ceruse: 100 pounds of saltpeter, 100 pounds of ceruse, 200 pounds of sand, a little bit of manganese. It is important to note that this secret makes a very nice and white glass, which sounds like silver; but it contains cord and can only be used for making molded vases". Considering the formula of saltpeter is KNO_3, that of ceruse $(PbCO_3)_2{\cdot}Pb(OH)_2$ and that of sand SiO_2, this yields a glass of average composition K_2O 14%, SiO_2 60% and PbO 26%. This composition is close to those of glasses fabricated by Bernard Perrot between 1685 and 1695: stained glass roses of the transept of Orleans cathedral: K_2O 12 and 13%, SiO_2 59 and 55%, PbO 25 and 29% (Gratuze and Arles 2012; Gratuze et al. 2017a). It is also close to that of molded portraits of Louis XIV: K_2O 12%, SiO_2 47%, PbO 35% (Biron et al. 2010, 2011). It is also strangely similar to those of glasses attributed to Ravenscroft (K_2O 13.3%, SiO_2 58.1% and PbO 28.6%) corresponding to group 5 of Brain and Dungworth (2005) and to recipe 5 of Moretti and Hreglich (2009).

While the use of this type of glass for the fabrication of glassware seems to be widespread from the end of the 17th century, its use in other domains, such as stained glass, may pose several problems to vitraillists, accustomed to HLLA glasses (*high-lime low-alkali*, category of glasses developed in Europe at the end of the medieval period for the fabrication of glazing and bottles) for the application of black stains. The remarks of Pierre le Vieil concerning the glasses of Bernard Perrot that his grandfather, Guillaume, used for the roses of the transept of Orleans cathedral (Le Vieil 1774) echo here the reserves of Kunckel and Merret for certain applications of this type of glass.

At the same time with lead, other elements such as arsenic and gold entered the range of ingredients used by the glassmaker at the end of the 17th century. They allowed the glassmakers to develop new coloration recipes from a lead glass base. This explains the presence, alongside monochrome beads of lead glass from the previous centuries, of polychrome beads such as the cornalines of Aleppo fabricated from red lead glasses (gold-colored) and white lead glasses (opacified with arsenic). It is only by the middle of the following century (1746) that Georg Frederich Strass, a jeweler from Strasbourg, gave his name to a transparent and colorless lead glass. This glass, whose composition is close to that of Ravenscroft's crystal glass, but slightly richer in lead, will be used for making jewels imitating the diamond at a lesser cost. The middle of the 18th century is the period when the first large crystal glass-making workshops in Europe emerge and large-scale development of lead crystal production starts (see Chapter 7).

6.6. Conclusion

Since the beginning of glassmaking industry lead as oxide was part of the range of ingredients used by the glassmakers. Its use is very different depending on period or region.

In the Mediterranean basin and Europe, for over two millennia, it was mainly used as an ingredient for the fabrication of colored and/or opacified glasses. It is either an integral component of the colorant as in the case of yellow glasses opacified with lead antimonate or lead stannate, or an additive: it makes it easier to obtain a color by facilitating the nucleation and crystallization of the coloring agent in the case of cuprite, or its dispersion in the vitreous mass in the case of white opacifiers, as tin oxide. Lead is only occasionally used as the main component of the glass for the fabrication of black and red elements of adornment in the Merovingian period. It is only at the end of the first millennium AD that the first lead glasses *str.s.* and lead-alkali glasses appeared in the Islamic world and in Eastern Europe. Moreover, it was the time when the first recipes for the fabrication of lead glasses were published. In Europe, until the 17th century, they are mainly used for the fabrication of adornment elements and the imitation of gemstones. They were only exceptionally employed for the fabrication of table glassware and glazing elements, in contrast with the Islamic world, where they were widely used for this type of productions.

In East Asia, lead glasses *str.s.* and lead-alkali glasses emerge at the end of the first millennium BC. Probably developed in China, their recipes evolve throughout the first millennium AD and are gradually adopted throughout East Asia, where they persist for a large part of the second millennium.

Several hypotheses can be considered for the late appearance of this type of glass in Europe. These recipes may have traveled along the silk route (Henderson 2013, p. 109), being brought from Asia to the West by the intermediary of the Islamic world. In this case, there was a gradual passage from the East to the West of this type of recipes, whose use in Western Europe seems to remain anecdotal and limited to the fabrication of adornment elements, until their formalization by Ravenscroft in the last quarter of the 17th century. A further hypothesis is that of a separate development of this technology in Asia, Europe and the Near East. The similarity of compositions between the glasses produced in these various parts of the world may be due only to the physicochemical properties of the SiO_2-PbO/SiO_2-PbO-K_2O systems (Shevchenko and Jak 2018) and to the resourcefulness of certain glassmakers. The processing of lead ores for silver production may in fact lead during galena reduction phases to the production of a vitreous slag of lead composition *str.s.* if the galena gangue is siliceous. The observation of these vitreous slags by the glassmakers, at a time when raw glass production started in Europe, may have also been the element triggering the production of this type of glass.

6.7. References

Angelini, I., Gratuze, B., Artioli, G. (2019). Glass and other vitreous materials through history. In *The Contribution of Mineralogy to Cultural Heritage*, Artioli, G., Oberti, R. (eds). Mineralogical Society, Twickenham.

Barber, D., Freestone, I., Moulding, K. (2009). Ancient copper red glasses: Investigation and analysis by microbeam techniques. In *From Mine to Microscope – Advances in the Study of Ancient Technology*, Shortland, A., Freestone, I., Rehren, T. (eds). Oxbow Books, Oxford.

Bayley, J. (2009). Early medieval lead-rich glass in the British Isles – A survey of the evidence. In *Annales du 17e Congrès de l'Association internationale pour l'histoire du verre (Antwerp)*, Janssens, K., Degryse, P., Cosyns, P., Caen, J., Van't Dack, L. (eds). University Press, Brussels.

Bezborodov, M.A. (1975). *Chemie und Technologie der antiken und mittelalterlichen Gläser*. P. von Zarben, Mayence.

Biron, I., Gratuze, B., Pistre, S. (2010). Étude en laboratoire d'objets en verre attribués à Bernard Perrot. In *Bernard Perrot (1640–1709). Secrets et Chefs-d'œuvre des Verreries royales d'Orléans*, Klinka-Ballesteros, I., de Valence, C. (eds). Aimery Somogy, Paris.

Biron, I., Gratuze, B., Lehuédé, P., Pistre, S. (2011). Étude en laboratoire d'objets attribués à Bernard Perrot. *Bulletin de l'AFAV 2011, 25e rencontres de l'AFAV*.

Brain, C. (2008). Vitrum Saturni: Lead glass in Britain. In *Glass of the Alchemists: Lead Crystal-Gold Ruby, 1650–1750*, von Kerssenbrock-Krosigk, D. (ed.). Corning Museum of Glass, Corning.

Brain, C. and Dungworth, D. (2005). English seventeenth-century crystal glass study: Phase 1. In *16th congrès de l'Association internationale pour l'histoire du verre. Londres, 2003*. AIHV, Nottingham.

Brain, C. and Dungworth, D. (2009). Late 17th century English crystal glass. In *Annales du 17e Congrès de l'Association internationale pour l'histoire du verre (Antwerp)*, Janssens, K., Degryse, P., Cosyns, P., Caen, J., Van't Dack, L. (eds). University Press, Brussels.

Brill, R.H. (1970). The chemical interpretation of the texts. In *Glass and Glassmaking in Ancient Mesopotamia. An Edition of the Cuneiform Texts which Contain Instructions for Glassmakers with a Catalogue of Surviving Objects*, Oppenheim, A.L., Brill, R.H., Barag, D., von Saldern, A. (eds). Corning Museum of Glass, Corning.

Brill, R.H. (1999). *Chemical Analyses of Early Glasses*. Corning Museum of Glass, Corning.

Brill, R.H., Barnes, I.L., Tong, S.S.C., Joel, E.C., Murtaugh, M.J. (1987). Laboratory studies of some European artifacts excavated on San Salvador Island. In *Columbus and his World: Proceedings of the First San Salvador Conference*, Gerace, D.T. (ed.). Bahaman Field Station, Fort Lauderdale.

Brun, N. and Pernot, M. (1992). The opaque red glass of Celtic enamels from continental Europe. *Archaeometry*, 34(2), 235–252.

Cagno, S., Cosyns, P., Izmer, A., Vanhaecke, F., Nys, K., Janssens, K. (2014). Deeply colored and black-appearing Roman glass: A continued research. *J. Archaeol. Sci.*, 42, 128–139.

Carter, A.K., Abraham, S.A., Kelly, G.O. (2016). Asia's maritime bead trade. *Archaeological Research in Asia*, 6.

Cosyns, P. (2011). The production, distribution and consumption of black glass in the Roman Empire during the 1st–5th century AD. An archaeological, archaeometric and historical approach. PhD Thesis, Vrije Universiteit, Brussels.

Cui, J.F., Wu, X.H., Huang, B.L. (2011). Chemical and lead isotope analysis of some lead-barium glass wares from the Warring States period, unearthed from Chu tombs in Changde City, Hunan Province, China. *J. Archaeol. Sci.*, 38, 1671–1679.

Dodwell, C.R. (1961). *Theophilus. De Diversis Artibus, Book II, XII, "The Various Colours of Opaque Glass"*. Thomas Nelson and Sons Ltd, London.

Dong, J.Q., Li, Q.H., Liu, S. (2015). The native development of ancient Chinese glassmaking: A case study on some early lead–barium silicate glasses using portable XRF spectrometer. *X-Ray Spectrometry*, 44, 458–467.

Dungworth, D. (2005). The scientific study of late seventeenth-century glassworking at Silkstone, England. In *Annales du 16e congrès de l'Association internationale pour l'histoire du verre (Londres, 2003)*. AIHV, Nottingham.

Dungworth, D. and Brain, C. (2013). Seventeenth- and eighteenth-century English lead glass. In *Modern Methods for Analysing Archaeological and Historical Glass*, Janssens, K. (ed.). John Wiley and Sons Ltd, Chichester.

Dussubieux, L. and Gratuze, B. (2013). Glass in South Asia. In *Modern Methods for Analysing Archaeological and Historical Glass*, Janssens, K. (ed.). John Wiley and Sons Ltd, Chichester.

Fougeroux de Bondaroy, A.-D. (1787). Sur la fusion de différentes substances vitrifiables, et particulièrement sur un verre connu sous la dénomination de Miroir de Virgile. In *Histoire de l'académie royale des sciences, année 1877. Avec les Mémoires de Mathématique et de Physique, pour la même année. Tirés des registres de cette académie.* Imprimerie Royale, Paris.

Freestone, I.C. (1987). Composition and microstructure of early opaque red glass. In *Early Vitreous Materials*, Bimson, M., Freestone, I.C. (eds). British Museum, London.

Freestone, I.C. (2015). The recycling and reuse of Roman glass: Analytical approaches. *J. Glass Stud.*, 57, 29–40.

Gan, F. (2009a). Origin and evolution of ancient Chinese glass. In *Ancient Glass Research Along the Silk Road*, Gan, F., Brill, R.H., Tian, S. (eds). World Scientific Publishing, Singapore.

Gan, F. (2009b). The Silk Road and ancient Chinese glass. In *Ancient Glass Research Along the Silk Road*, Gan, F., Brill, R.H., Tian, S. (eds). World Scientific Publishing, Singapore.

Gratuze, B. and Arles, A. (2012). Composition des verres et des plombs de la rose du transept sud de la cathédrale Sainte-Croix d'Orléans. *Bulletin de l'AFAV 2012*, 61–72.

Gratuze, B., Guérit, M., Simon, L., Villaverde, L., Barbier, E. (2014). Des verres carolingiens de compositions atypiques à Bressuire et Faye-sur-Ardin (nord de la région melloise, Deux-Sèvres). *Bulletin de l'AFAV 2014*, 114–120.

Gratuze, B., Arles, A., de Valence, C. (2017a). L'origine des verres de la rose du transept sud de la cathédrale Sainte-Croix d'Orléans. Complémentarité entre textes et analyses. In *Cathédrale d'Orléans, Patrimoine en région Centre-Val de Loire*, Ministère de la Culture et de la Communication (ed.). Presses Universitaires François Rabelais, Tours.

Gratuze, B., Guerrot, C., Foy, D., Arles, A., Téreygeol, F., Baron, S. (2017b). Les galets de verre au plomb carolingiens issus des scories de Melle : élaboration et distribution. In *Mine, métal, monnaie, Melle. Les voies de la quantification de l'histoire monétaire du haut Moyen Âge*, Bompaire, M., Sarah, G. (eds). Droz, Geneva.

Guérit, M., Gratuze, B., Joyeux, P. (2020). De nouvelles formes en verre à Orléans, de la seconde moitié du XVe à la première moitié du XVIe siècle : regards croisés entre typologie et analyses. In *Le verre du VIIIe au XVIe siècle en Europe occidentale – Actes du colloque international de l'AFAV tenu à Besançon (décembre 2016)*, Pactat, I., Munier, C. (eds). Presses universitaires de Franche-Comté, Besançon.

Henderson, J. (2013). *Ancient Glass – An Interdisciplinary Exploration*. Cambridge University Press, New York.

Henderson, J., An, J., Ma, H. (2018). The archaeometry and archaeology of ancient Chinese glass: A review. *Archaeometry*, 60(1), 88–104.

Hughes, M.J. (1972). A technical study of opaque red glass of the Iron Age in Britain. *Proc. Prehist. Soc.*, 38, 98–107.

Kang, H.-T. and Yun, E.-Y. (2012). Chemical compositions and lead isotope ratios of some glass beads from Seokga-tap, Gyeongju. *Conservation and Restoration of Cultural Heritage*, 1(1), 3–8.

Kerssenbrock-Krosigk, D.V. (2008). *Glass of the Alchemists. Lead Crystal-gold Ruby, 1650–1750*. Corning Museum of Glass, Corning.

Krueger, I. (2006). Post-medieval colored lead glass vessels. *J. Glass Stud.*, 48, 225–241.

Lagabrielle, S. (2017). *Le verre, un Moyen Age inventif, Catalogue de l'exposition Verre en Moyen Âge*. Musée national du Moyen Âge – RMNGP, Paris.

Lankton, J.W. and Dussubieux, L. (2006). Early glass in Asian maritime trade: A review and an interpretation of compositional analyses. *J. Glass Stud.*, 48, 121–144.

Lankton, J.W. and Dussubieux, L. (2013). Early glass in southeast Asia. In *Modern Methods for Analysing Archaeological and Historical Glass*, Janssens, K. (ed.). John Wiley and Sons Ltd, Chichester.

Le Vieil, P. (1774). *L'Art de la peinture sur verre et de la vitrerie, par feu M. Le Vieil.* L.F. Delatour, Paris.

Lefrancq, J. (2020). La verrerie de Verdun (Meuse) : ± 1664-1678. La direction des Bonhomme/Bounam et d'Heur de Liège. *Bulletin de l'AFAV 2020, 34e rencontres de l'AFAV*, 87–94.

Mathis, F., Gratuze, B., Vrielynck, O. (forthcoming). Étude de composition et de provenance des perles en verre mérovingiennes du cimetière de Bossut-Gottechain (Belgique, Wallonie). In *Un Monde en mouvement : la circulation des hommes, des biens et des idées à l'époque mérovingienne (V^e^–VIIIe siècle) Actes du 40e colloque de l'AFAM, Nantes 2019*, Henigfeld, Y., Peytremann, E. (eds).

Mecking, O. (2013). Medieval lead glass in Central Europe. *Archaeometry*, 55(4), 640–662.

Merriefield, H.E. (1849). *Original Treatises on the Arts of Painting.* John Murray, London.

Moretti, C. (2005). English lead crystal. A critical analysis of the formulation attributed to George Ravenscroft and points not yet clear on the process for the manufacture of "flint" glass. In *Annales du 16e congrès de l'Association internationale pour l'histoire du verre (Londres, 2003).* AIHV, Nottingham.

Moretti, C. and Hreglich, S. (2009). English lead crystal and Ravenscroft's formulation: Additional considerations on raw materials utilized and on batch melting temperatures. In *Annales du 17e Congrès de l'Association internationale pour l'histoire du verre (Antwerp)*, Janssens, K., Degryse, P., Cosyns, P., Caen, J., Van't Dack, L. (eds). AIHV, Loosdrecht.

Müller, K. and Stege, H. (2009). Material analysis of colourless lead glasses from a late 17th century glasshouse site in Groningen (The Netherlands). In *Annales du 17e Congrès de l'Association internationale pour l'histoire du verre (Antwerp)*, Janssens, K., Degryse, P., Cosyns, P., Caen, J., Van't Dack, L. (eds). AIHV, Loosdrecht.

Neri, A., Merrett, C., Cable, M. (2001). *The Art of Glass. The World's Most Famous Book on Glass-making.* Sheffield University Press, Sheffield.

Paynter, S. and Jackson, C. (2019). Clarity and brilliance: Antimony in colourless natron glass explored using Roman glass found in Britain. *Archaeol. Anthropol. Sci.*, 11, 1533–1551.

Péligot, E.M. (1877). Sur la composition du verre et du cristal chez les anciens. In *Bulletin de la Société d'Encouragement pour l'Industrie Nationale, 3e série, volume 4, Février 1877*, Péligot, E.M., Laboulaye, C. (eds). Bouchard-Buzard, Paris.

Pion, C. (2014). Les perles mérovingiennes : typo-chronologie, fabrication et fonctions. PhD Thesis, Université libre de Bruxelles, Brussels.

Pulak, C. (2008). The Uluburun shipwreck and late Bronze Age trade. In *Beyond Babylon: Art, Trade, and Diplomacy in the Second Millennium B.C.*, Aruz, J., Benzel, K., Evans, J.M. (eds). The Metropolitan Museum of Art, New York.

Rolland, J. (2017). L'artisanat du verre dans le monde celtique au second âge du fer : approches archéométriques, technologiques et sociales. PhD Thesis, Université Paris 1, Paris.

Scapova, J.L. (1992). Un lissoir de Novgorod, réflexions sur la verrerie médiévale. *Acta Archéologica*, 62, 231–243.

Schibille, N., De Juan Ares, J., Casal García, M.T., Guerrot, C. (2020). Ex novo development of lead glassmaking in early Umayyad Spain. *Proc. Natl. Acad. Sci.*, 117(28), 16243–16249.

Shevchenko, M. and Jak, E. (2018). Experimental phase equilibria studies of the PbO-SiO_2 system. *J. Am. Ceram. Soc.*, 101, 458–471.

Shortland, A.J. (2002). The use and origin of antimonate colorants in early Egyptian glass. *Archaeometry*, 44, 517–530.

Shortland, A.J. and Tite, M.S. (2000). Raw materials of glass from Amarna and implications for the origins of Egyptian glass. *Archaeometry*, 42(1), 141–151.

Siemianowska, S., Pankiewicz, A., Sadowski, K. (2019). On technology and production techniques of early medieval glass rings from Silesia. *Archaeometry*, 61, 614–646.

Sode, T., Gratuze, B., Lankton, J. (forthcoming). The glass beads of Ribe, evidence of local glass bead making and long-distance trade. In *Glass as a Trade Commodity in the Early Middle Ages – Cologne and the European North, Workshop, 20–22 March, 2018 in Ribe*, Messal, S. (ed.).

Stephan, H.-G., Wedepohl, K.H., Hartmann, G. (1997). Mittelalterliches Glas aus dem Reichskloster und der Stadtwüstung Corvey. *Germania*, 75(2), 673–715.

Tamura, T. and Oga, K. (2014). Distribution of lead-barium glasses in ancient Japan. *Crossroads*, 9(Special issue), 63–82.

Tite, M.S., Pradell, T., Shortland A.J. (2008). Discovery, production and use of tin-based opacifiers in glasses, enamels and glazes from the late Iron Age onwards: A reassessment. *Archaeometry*, 50, 67–84.

Truffa-Giachet, M. (2019). Étude archéométrique des perles en verre d'Afrique de l'Ouest : vers une meilleure compréhension des dynamiques techniques et commerciales à l'époque des empires pré-coloniaux. PhD Thesis, Université de Genève, Geneva.

Valente, F., Coutinho, I., Medici, T., Brain, C., Vilarigues, M. (2016). A group of early English lead crystal glass goblets found in Lisbon. *J. Glass Stud.*, 58, 211–225.

Velde, B. and Gratuze, B. (2009). Vaisselle, verre plat et petits objets en verre : aspects analytiques. In *Une résidence des comtes d'Angoulème autour de l'an Mil, Le castrum d'Andone (Ville-joubert, Charente). Publication des fouilles d'André Debord (1971–1995)*, Bourgeois, L. (ed.). CRAMH, Caen.

Wang, K.-W. and Jackson, C. (2014). A review of glass compositions around the South China Sea region (the late 1st millennium BC to the 1st millennium AD): Placing Iron Age glass beads from Taiwan in context. *Journal of Indo-Pacific Archaeology*, 34, 51–60.

Wedepohl, K.H. (2000). The change in composition of medieval glass types occurring in excavated fragments from Germany. In *Annales du 14e Congrès de l'Association internationale pour l'histoire du verre, Venezia-Milano 1998*. AIHV, Loosdrecht.

Wedepohl, K.H. (2001). The composition of glass from the Carolingian and post-Carolingian period in Central Europe. In *Il Colore nel Medioevo*, Dell'Acqua, F., Silva, R. (eds). Istituto storico lucchese, Lucca.

Wedepohl, K.H. and Baumann, A. (1997). Isotope composition of Medieval lead glasses reflecting early silver production in Central Europe. *Mineralium Deposita*, 32, 292–295.

Wedepohl, K.H., Krueger, I., Hartmann, G. (1995). Medieval lead glass from northwestern Europe. *J. Glass Stud.*, 37, 65–82.

7

Lead in Glasses: Recent Times

Patrice LEHUÉDÉ
C2RMF, IRCP, Paris, France

7.1. The adventure of lead crystal glass

7.1.1. *The creation of crystal glass*

As previously seen (Chapter 6), lead glass had been present long before the end of the 17th century; however, this was presumably the time when lead crystal glass was "invented". The starting point was a patent obtained by the Englishman George Ravenscroft in 1674 for a "specific glass resembling rock crystal" (Brain 2003). Rock crystal (quartz) had for a long time been a reference for the glassmakers, who admired its clarity and colorlessness. This is how Venice, over one century earlier, had introduced the "cristallo", a more colorless glass than the norm of the time, by using purified raw materials (Verità 2008). The problem with Ravenscroft's patent is that it does not mention the use of lead (Dungworth and Brain 2005), which may have been absent from the first glasses fabricated by Ravenscroft based on his patent. Moreover, these glasses proved to have extremely low durability, as the crizzling[1] phenomenon appeared within several months. This problem appears to have been solved by the introduction of lead into the composition or in any case by increasing its content: the analyses that could be conducted show effectively a lead content that gradually increased up to

1. It is a classical manifestation of glass alteration, which is reflected by the formation of a network of cracks in the altered layer.

Lead in Glassy Materials in Cultural Heritage,
coordinated by Anne BOUQUILLON and Patrice LEHUÉDÉ. © ISTE Ltd 2023.

34% (oxide weight) in the years following the patent filing (Moretti 2003; Dungworth and Brain 2004).

7.1.2. *The development of lead crystal glass*

Lead glasses production, which was peripheral until 1674, grew rapidly, first in Great Britain, where nearly 100 workshops fabricated this type of glass before the end of the century (Klein and Lloyd 1984, p. 107), then more widely throughout Europe (Charleston 1960). It would be interesting to know the origin of this rapid growth in the last quarter of the 17th century. This development may very well be explained by the mastery of the elaboration of this type of glass. Indeed, the incidental introduction of a reducing agent in the composition, carbon for example, triggers the reduction of the lead oxide in the glass into metallic lead, which, given its high density, falls to the bottom of the crucible (Watts 1990; Moretti 2003). But metallic lead rapidly attacks the refractory material of the crucible, generating a hole through which melting glass may flow (see Chapter 1). This is probably the reason why the previous lead glass productions were generally at small scale, which did not justify the use of large crucibles. The use of closed crucibles, associated with coal as fuel, was a solution to avoid the reducing action of fumes; adding oxidizers such as potassium nitrate (Moretti 2003) to the composition also contributed to preventing this reduction during glass elaboration.

The raw materials used to obtain the crystal glass were milled flint (the latter being the term that gave the name to this type of glass) or white sand, minium (Pb_3O_4) or litharge (PbO), tartar (potassium tartrate $KHC_4H_4O_6$) or potassium carbonate and nitre (potassium nitrate), also known as saltpeter (Moretti 2003; Dungworth and Brain 2005). The choice of raw materials containing few impurities, among which iron, rendered this glass significantly less colored than other glasses of the same period (Brain and Dungworth 2005; Dungworth and Brain 2005). The fuel employed was coal, wood being reserved for the British navy in that period.

In its development in Europe, lead crystal glass was most of all appreciated for its lack of color and its shine, in comparison with leadless glasses of those times; it was rather dedicated to luxury articles, which is still true nowadays. Many crystal glassworks developed in Europe as follows: Saint-Louis, Baccarat, Sèvres, Montcenis, Val Saint-Lambert, Portieux, Clichy, Choisy le Roi, Pantin, Legras, Vonêche, Zoude, Waterford, etc.

There was significant progress in the 19th century, particularly due to the advancement of chemistry and also of other domains (the technical evolution of furnaces allowed a higher melting effectiveness, and therefore a higher quality of the crystal glass, and higher efficiencies). For example, synthetic raw materials become widespread, such as sodium or potassium carbonate (Chopinet 2008). The composition of the glass and of the raw materials employed can be determined by chemical analysis, which clearly enhances production regularity and quality. Finally, in the 19th century or at the beginning of the 20th century, new chemical elements were discovered or isolated and used, such as chromium, uranium, cadmium, rare earths and selenium.

7.2. New colorants of lead glass

These new elements allow the glassmakers and the ceramists to widen the range of available colors. The progress achieved concerns all types of glasses, and particularly lead glasses (including crystal glass) and lead glazes, and will accompany the colored crystal glass fashion starting with the mid-19th century. Generally speaking, the colorations obtained with lead glasses are not only more intense than those obtained with soda-lime-silica matrices (Stroud 1971), but they can also be different (see the case of copper, Chapter 4).

7.2.1. *Chromium*

Chromium started to be widely used in solution in glasses to obtain a green color only at the beginning of the 19th century (Phillips 1981, p. 257; Verger 2015). It is in the form of Cr^{3+} that it gives a green coloration (Bamford 1977, p. 46). This coloration is stable, and the only difficulty of use is related to the fact that Cr_2O_3 crystals are very poorly soluble in glass. The pigments containing chromium associated with other elements (Al, Zn, etc.) can be used to obtain many other hues: green, pink, blue-green, brown and black (Verger 2015). Chromium in the form of Cr^{6+} is no longer used due to its high toxicity.

7.2.2. *Nickel*

Nickel brings a brown coloration (Bamford 1977, p. 42). Moreover, it has been used for the discoloration of lead crystal glass, often in association with cobalt traces: an even less colored glass is thus obtained.

7.2.3. *Uranium*

Uranium in the form of U^{6+} gives to the lead glass a beautiful yellow shade that may turn to green under reducing conditions that favor the formation of U^{4+} (Bamford

1977, p. 116). Other colors can be obtained in lead glasses and especially in glazes: ivory, orange and red, according to the glaze composition (Kantz 1934; Phillips 1981, p. 258). Uranium-based coloration is interesting in the history of art (Skelcher 2020): it was very widespread (Klein and Lloyd 1984, p. 175), but for a very limited period of time, since the 1830s until the 1940s (Strahan 2001), which allows dating the objects that use this element. Moreover, it can be easily identified (it often gives a green fluorescence under UV lighting and ionizing radiation generated by uranium can be readily detected by a Geiger counter). It can be found in Europe, but also in Japan, United States, etc. It is often found in common or ornamental objects (drinking glasses, bottles, vases, candlesticks, trays, beads, jewels, etc.), under the name of *vaseline*, but it can also be observed in less common objects, such as in a restoration from the 19th century of Limoges enamels from the tabernacle of Saint-Aignan (P. Lehuédé, unpublished data). Uranium has also been used in glazes, often at significant content (15% or more), where it produces variable bright colors including red and black (Weyl 1951, p. 207; Strahan 2001), depending on the composition of the matrix, temperature and burning atmosphere. The risks posed to health have obviously led to banning this element starting with the 1940s.

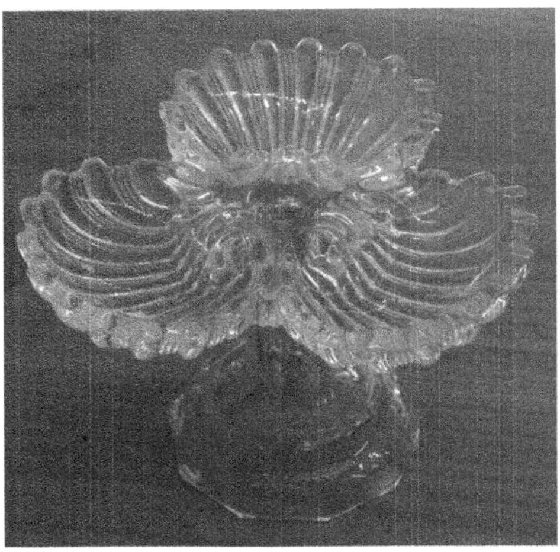

Figure 7.1. *Vaseline saltcellar, Saint-Louis museum, toward 1860, photo P. Lehuédé*

7.2.4. Selenium

Selenium was used as colorant in glasses since the end of the 19th century, but due to its high volatility it was difficult to control its use (LaCourse et al. 1980). It generates a pink coloration in soda–lime glasses. The color shifts toward gold-topaz in the case of lead glasses (Weyl 1951, p. 286). Given the various oxidation states possible for selenium, lead crystal glass may take a pink shade in the absence of oxidizer, and an amber shade in the presence of oxidizer (Humbert and Colin-Seigner 2013).

7.2.5. *Rare earths*

Coloration by rare earths is recent: the separation of various rare earths practically started at the beginning of the 20th century, but it was applied at industrial scale only after the Second World War, their cost remaining quite significant. Only the most abundant of these rare earths and the most interesting due to their properties are used: cerium, neodymium, praseodymium (Weyl 1951, p. 218). They can be used to obtain a large range of colors, including violet (by neodymium) or lemon yellow (by cerium, often associated with titanium) (Alakin et al. 1961). Moreover, cerium in the form of Ce^{3+} or Ce^{4+} has an absorption band in UV, thus giving UV-protection properties that are used in some modern glasses (Bamford 1977, p. 69). This band is wide and spills a bit in the visible domain, bringing a yellow coloration.

This range of colors is often further broadened by using a mixture of rare earths or by associating a rare earth to another colorant. Rare earths lead to optical absorption spectra with specific narrow bands (Banford 1977, p. 69; Shelby 2015) of relatively low intensity, which implies that a significant content of rare earth is required for a clearly visible coloration (generally several percents). The thinness of bands often results in a variation of the color of these glasses depending on the nature of lighting, as in the case of neodymium (Weyl 1951, p. 223; Mesnil 2012; Bellows 2016). Finally, rare earths often generate fluorescence under UV (Mesnil 2012; Shelby 2015; Bellows 2016).

7.2.6. *Coloration by colloidal particles*

This type of coloration has been well known for a very long time, particularly for copper, gold and silver. As already noted, the presence of lead was not necessary to facilitate the precipitation of copper Cu° nanoparticles; lead glasses colored in red by these nanoparticles could however be found (see Chapter 6). A very nice illustration in this sense is the rose of the Cathedral of the Holy Cross in Orleans. At least one part of the glass was provided by Bernard Perrot; the order dates from 1687, very shortly after Ravenscroft's patent. The red parts are lead glasses containing copper very probably in the form of Cu° nanoparticles (Gratuze 2012).

Coloration by gold nanoparticles has also been practiced for glasses and lead glasses for a very long time (Dararutana et al. 2008), at least 600 years, but was used to a little extent in Europe, as it was difficult to control its fabrication. Nevertheless, the color thus obtained being greatly appreciated and impossible to obtain by other processes in that period, the works of Cassius, inventor of "purple of Cassius", and of Kunckel allowed the production of ruby glass at a larger scale (Franck 1984) at the end of the 17th century. The process is adapted to lead glasses and particularly to lead crystal glass especially in the 19th century, certain crystal glassworks becoming specialized in this field, as Baccarat with its red ruby, but also for the enamels on glass (Beltrán et al. 2021) used for the stained glasses at the beginning of the 20th century.

Silver, in the form of nanoparticles, gives a yellow color to the glass, as in silver yellow, commonly used for stained glasses. Coloration by silver nanoparticles can also be used with the lead glass, color homogeneity being difficult to master, as proven by the praise brought by Bontemps to the crystal glassworks of Pantin at the Paris Universal Exposition in 1867 (Bontemps 1868, p. 545).

Cadmium, generally in the form of cadmium sulfoselenide, has been used since the second half of the 19th century in soda–lime glasses: beautiful colors are thus obtained, from *kaiser yellow* to red, depending on the proportion of sulfur and selenium (Weyl 1951, p. 271). These colorations can also be found in lead glasses or in lead glazes (Weyl 1951, p. 274; Eppler and Eppler 2000, p. 169), but the low thermal stability of CdS_xSe_{1-x} makes it difficult to use this chromophore. It can however be found in enamels for glass painting (Beltrán 2020), where it is employed at low temperature. Moreover, due to its toxicity, this compound is no longer used, except for glazes in the form of pigment inserted in zirconium silicate microcrystals (Matthes 2010, p. 92).

7.3. The new opacifiers

New opacifiers have been developed during the 19th century, at least in Europe; it is the case of calcium fluoride that precipitates under cooling when fluorspar is introduced in the composition (Péligot 1877, p. 424). This opacifier is especially used for lamp globes. It continues to be commonly used nowadays, not only for lead glasses. It appears to have already been used in China in the 12–13th centuries (Zhou 2019).

As for the lead arsenate, while it has been known since the 17th century (Moretti and Hreglich 2009; Biron and Chopinet 2013), it was widely used in the 19th and 20th centuries for lead glass (Lehuédé et al. 2015), but the nature of the opacifier proves to be more complex than what is generally admitted: in at least some number of cases, white $KNaPb_8(AsO_4)_6$ particles are at the origin of opacification. These particles can be obtained at very small dimensions (of about 0.1 μm); the specific optical properties they then confer are highly appreciated: opalescence, which is perceived as a yellowish shade in transmitted light and bluish shade in reflected light (Biron and Chopinet 2013).

7.4. The new processes of crystal glass decoration

7.4.1. *Engraving*

It has been practiced since Antiquity, especially on rock crystal, then it had its moment of glory on the Bohemian crystal glass (which, it is important to note, contains no lead), and finally made a comeback in the 19th century and during the Art Nouveau period; it was employed on glass, but also on lead crystal glass (de Langle 2002). Brilliant cutting on glasses involving several glass layers of different colors, possibly with one or several layers of opal glass, can be used to obtain new decorative effects that are highly appreciated when this technique is practiced by skillful artisans (see Figures 7.2 and 7.3). Crystal glass, softer than a soda–lime–silica glass, is quite suitable for engraving and polishing.

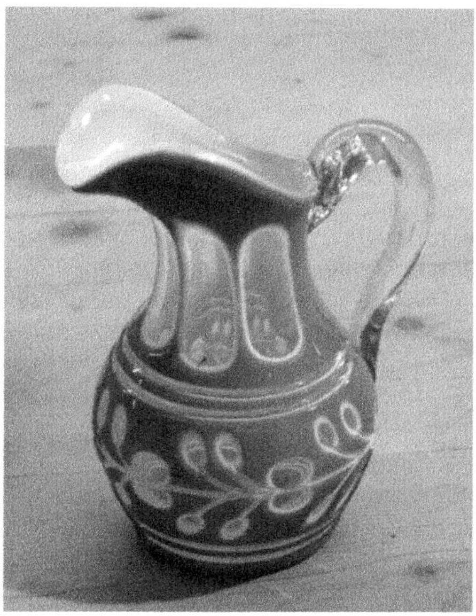

Figure 7.2. *Example of "overlay" technique (glass in several layers): colorless crystal glass doubled by white opaque and blue opaque crystal glass. Saint-Louis museum, toward 1850, photo P. Lehuédé. For a color version of this figure, see www.iste.co.uk/bouquillon/lead.zip*

On the other hand, acid etching (a hydrofluoric acid-based mixture), discovered at mid-19th century, is rapidly developing; it allows the engraving of part of surfaces unprotected by a varnish (generally bitumen of Judea), by immersing them in an acid mixture, which uncovers the underlying layers, and creates relief effects (see Figure 7.3). The state of the resulting surface may appear to be polished or satiny, depending on the composition of the bath employed (Spierings 1993).

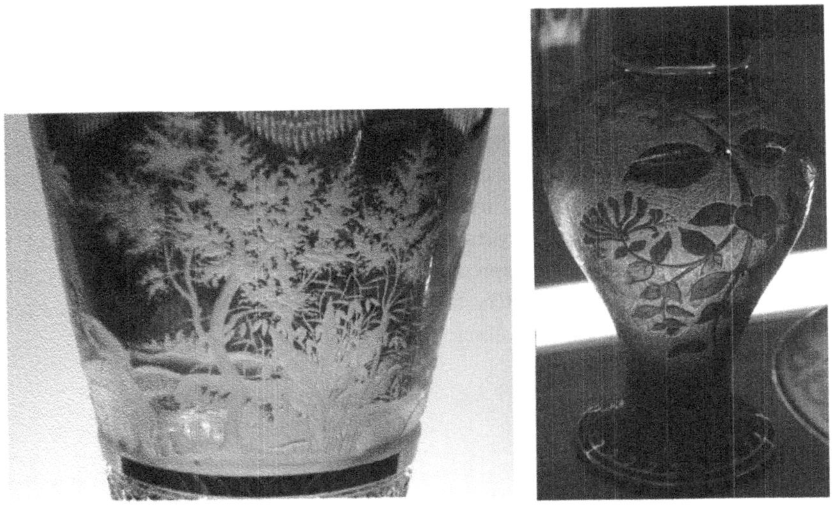

a) b)

Figure 7.3. *Examples of engraving. (a) Brilliant cutting (colorless crystal glass doubled by red crystal glass), by H. Winkler toward 1867. Saint-Louis museum, toward 1867, photo P. Lehuédé. (b) Acid engraving (turquoise crystal glass, doubled by amethyst crystal glass). Saint-Louis museum, toward 1900, photo P. Lehuédé. For a color version of this figure, see www.iste.co.uk/bouquillon/lead.zip*

Finally, sandblasting (propelling a stream of sand or corundum) is another method for engraving glass in a localized manner: it was rather employed for flat glass; the use of a mask makes it possible to recurrently engrave patterns or images at the glass or crystal glass surface.

7.4.2. *Iridescence*

Another method for surface decoration is iridescence, which was widely used at the end of the 19th century and at the beginning of the 20th century by great glassmakers such as Tiffany and Loetz. This complex technique has scarcely been scientifically studied (Jembrih et al. 2000; Mäder et al. 2005). It involves high temperature depositing of a 50–100 nm thick layer of tin oxide on a lead-rich glass. The presence of silver (0.1–0.4% oxide weight) in the glass under the tin oxide layer seems essential, as it was detected in all the cases, though its role remains mysterious. Colorants such as cobalt are sometimes also detected. The iridescent appearance is due to interferences between a light beam reflected at the air/SnO_2 layer interface and other light beams reflected at the SnO_2 layer/underlying glass interface. The iridescent areas generally have a satiny appearance and the observed colors vary quite little with the angle of incidence, in contrast to what is observed in smooth thin layers (see Figure 7.4).

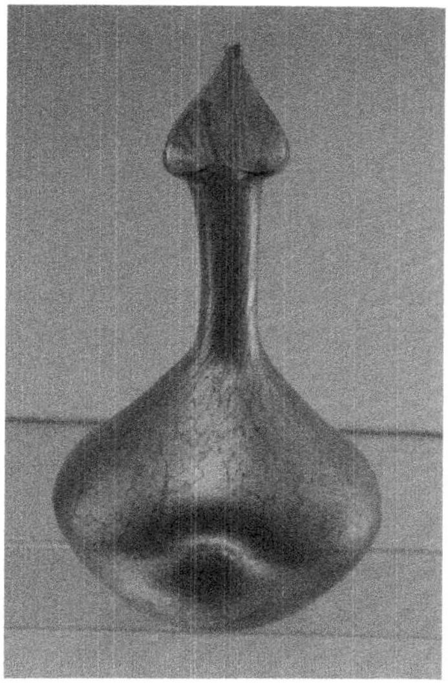

Figure 7.4. *Example of iridescent glass: Loetz, Bröhan-Museum, Glaskunst 1889–1939, toward 1900, photo P. Lehuédé. For a color version of this figure, see www.iste.co.uk/bouquillon/lead.zip*

Fine observations of iridescent surfaces show that the tin oxide layer presents some sort of wrinkling (Neuwirth 1986), which is most certainly responsible for the satiny appearance of the surface and the low sensitivity of the interference color to the angle of incidence. This wrinkling is similar to the one sometimes observed on the bath-side (which is tin enriched) of modern float glass, after thermal quenching in the air. Bath-side (or tin-side or tin bath side) is the term used by glass makers involved in the float glass process, and known by the glass community. It is attributed in this case to a difference of expansion coefficient between the surface of SnO_2-rich glass and the underlying glass: the surface contracts less than the underlying glass during cooling and the resulting stress leads to surface wrinkling. In the case of float glass, this phenomenon causes a defect known as "bloom" (Pantano et al. 1993; Frischat 2002), which manifests as a whitish diffusing appearance of the surface.

7.5. New glassmaking techniques

While artists were able to create magnificent objects using new colors, new opacifiers and new decoration techniques, new products and new implementation techniques played an important role in the accomplished expression of their creativity.

7.5.1. *Strass*

This is a very high lead content glass (often over 50%), which gives a spectacular shine to objects (Péligot 1877, p. 460). It was widely used in jewelry (it was invented in 1746 by a jeweler from Strasbourg, whose name was Strass). Metallic oxides can be added to the glass to give it the desired color. Imitations of gemstones such as emerald, ruby or sapphire, can thus be obtained.

7.5.2. *Flashed glass*

This technique as such is not new: it was used, for example, for the red glass of stained glasses where coloration by copper in the form of Cu° nanoparticles is very intense and therefore requires the use of this glass under a very small thickness (a fraction of a millimeter) to preserve satisfactory light transmission. In order to maintain the mechanical properties of glass, this red glass is hot sticked on a colorless glass of significantly greater thickness. As already mentioned, in the case of crystal glass, this technique was widely used starting with the 19th century, often associated with engraving in order to bring out various colors or various opacities.

7.5.3. *Millefiori glass*

G. Bontemps rediscovered the technique of filigree glass at the crystal glassworks of Choisy-le-Roi in 1838, then that of millefiori soon afterwards (Cappa 2001, p. 194). This forming technique has undoubtedly been brought to its peak in paperweights, as in the example shown in Figure 7.5. It has its origin in Greek and Roman mosaic glasses (de Langle 2003) and filigree glasses developed at Murano in the 16th century (Verità et al. 2018). These effects are obtained by hot immersion of fine opaque white or other color glass rods in a colorless glass batch. During forming, these rods can be twisted, thus giving to the object its specific appearance. The artists of the 19th century adapted these techniques to crystal, and all the great crystal glassworks competed to skillfully achieve paperweights presented during universal expositions (Dufrenne et al. 2005, p. 43) that were highly appreciated by the collectors.

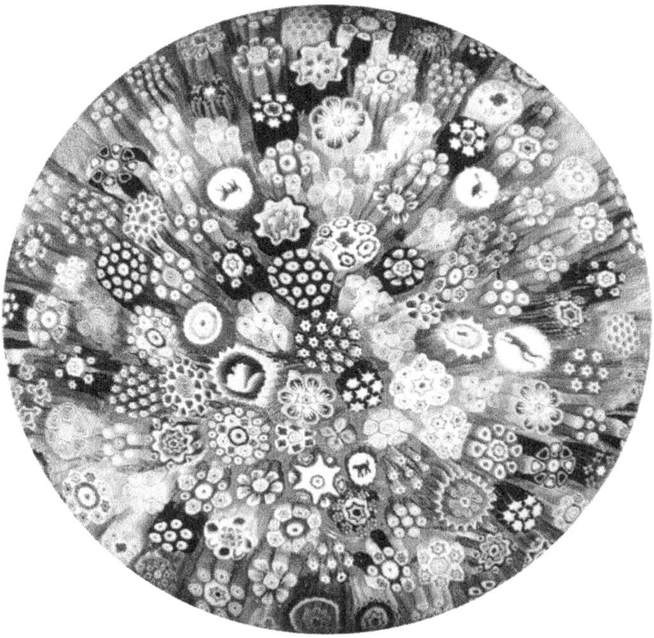

Figure 7.5. *Paperweight fabricated at Baccarat by the mid-19th century*[2]. *For a color version of this figure, see www.iste.co.uk/bouquillon/lead.zip*

7.5.4. *Pâte de verre*

Strictly speaking, this is not a new technique either, but it was used a lot by certain crystal glassworks such as Daum in the 20th century and artists such as Décorchemont or Argy-Rousseau (Klein and Lloyd 1984, pp. 221 and 230), as it allows the fabrication of objects of complex shape, such as small statues. It involves filling a mold of refractory plaster, negative of the object to be obtained, with crystal glass fragments, then melting the crystal glass at 900°C for several days. After

2. Wikimedia Commons: https://commons.wikimedia.org/w/index.php?curid=82022669 [Accessed on 20 October 2020].

cooling, the plaster is broken and small finishes can be conducted in order to perfect the state of the surface according to the artist's expectations. The crystal glass used can be colorless or colored, or a mixture in order to obtain a color gradient. This type of object highlights the intrinsic quality of the crystal, but withstands no defect. Lead crystal glass, because of its low viscosity, is particularly suitable for this technique.

7.5.5. *Lampworking glass, glass fibers*

The greatest forming facility brought by the presence of lead oxide in the composition was effectively used in the technique of "Nevers lampworking glasses". These glasses are lamp processed to form small objects, often figurines of opaque glass with several colors. This technique requires a quite low forming temperature (600°C), which is facilitated by the presence of PbO frequently evidenced in these objects (Lahlil 2008). These glasses were mainly produced during the 17th–19th centuries, at Nevers and also in other cities.

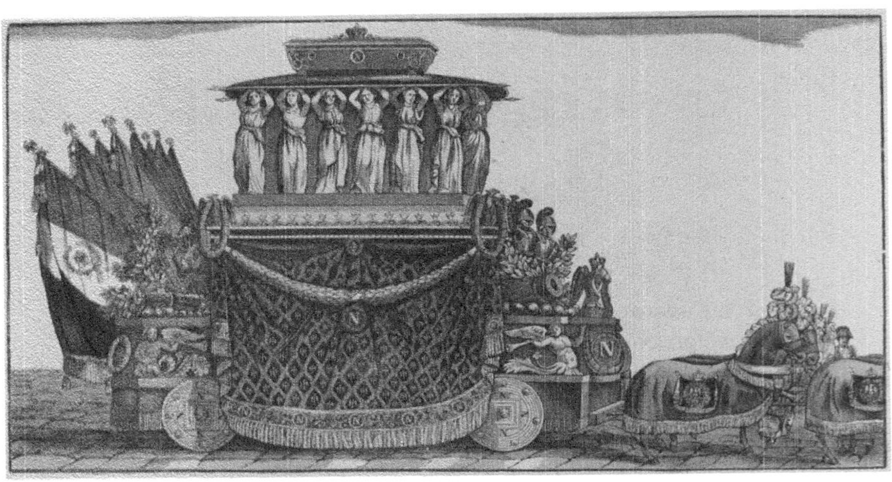

Figure 7.6. *Lithograph representing the funeral carriage of Napoleon during the transfer of ashes to Les Invalides in 1840*[3]

Glass fibers have been widely used for several decades, for thermal isolation and also for strengthening polymers or optical fibers, but already in the 19th century

3. Available at: http://www.collections.musee-bretagne.fr/ark:/83011/FLMjo139565 [Accessed 15 November 2020].

there are several examples of applications in the field of heritage materials. The oldest known presumably corresponds to the cloth used for decorating the carriage for the "retour des cendres" (return of ashes) of Napoleon to Les Invalides in 1840 (see Figure 7.6). This cloth consisted of a colored silk warp and a weft of gold-colored glass fibers (see Figure 7.7). According to the analyses, the glass used for these fibers had an 8% PbO content (oxide weight); as for its color, it resulted from a mixture of iron, manganese and copper oxides. The diameter of these fibers varies between 11 and 26 µm (Challes 2018; Lehuédé 2018).

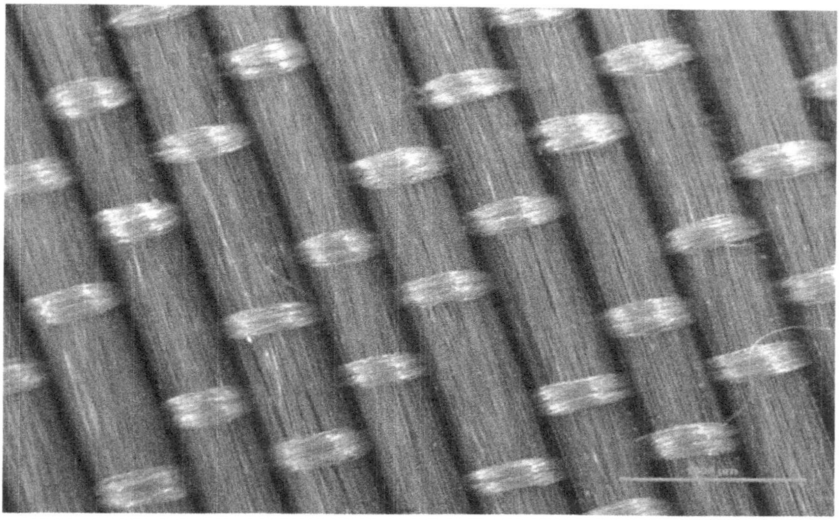

Figure 7.7. *Optical microscope photograph (reflected light) of a fragment of "retour des cendres" cloth. The weft is of glass fibers while the warp is of silk. Photo by P. Lehuédé. For a color version of this figure, see www.iste.co.uk/bouquillon/lead.zip*

A further example of glass fiber use in the artistic domain is the lion attacking a snake by René Lambourg (*Lion et serpent*) dated 1850–1855, and preserved at the Arts et Métiers museum in Paris. The lion's fur and mane, and grass on which it stands, are made from glass fibers. The analysis conducted on the fibers constituting the grass shows that it is a lead glass (with about 35% PbO) colored by some copper.

7.6. Conclusion

Glassmaking technique has for a long time evolved in an empirical manner: the development of lead glass by Ravenscroft is a very good illustration. But starting with the

18th century, the remarkable progress of sciences and techniques had a significant influence on the field of glass and glazes. The artists knew how to integrate very rapidly these advances in order to create new artworks, playing with the innovations in the field of color, opalization, chemical attacks, states of surface, forming processes, etc. This is particularly obvious in the Art nouveau period; it concerns glass in general, including lead glass, and also glazes. Though less exuberant, the following period took advantage of scientific and technical innovations, such as coloration using rare earths. The responsiveness of artists and glassmakers is stunning: the period of time between a scientific discovery and its application was generally very short.

The great Expositions played an important role in the development of arts and techniques in the 19th century: they offered craftsmen, artists and manufacturers the opportunity to present their know-how and innovations, and were a powerful motivation for the research and development of techniques.

7.7. References

Alakin, A.I., Nikitin, B.N., Tsarevskaya, N.P. (1961). The use of rare earths for coloring glass *Glass Ceram.*, 18, 144–146.

Bamford, C.R. (1977). *Colour Generation and Control in Glass. Glass Science and Technology*, volume 2. Elsevier, Amsterdam.

Bellows, C. (2016). Photochromism in rare-earth oxide glasses. PhD Thesis, Université d'Alfred, Ontario.

Beltrán, M., Schibille, N., Brock, F., Gratuze, B., Vallcorba, O., Pradell, T. (2020). Modernistenamels: Composition, microstructure and stability. *J. Eur. Ceram.*, 40, 1753–1766.

Beltrán, M., Schibille, N., Gratuze, B., Vallcorba, O., Bonet, J., Pradell, T. (2021). Composition, microstructure and corrosion mechanisms of Catalan Modernist enameled glass. *J. Eur. Ceram.*, 41, 1707–1719.

Biron, I. and Chopinet, M.-H. (2013). Colouring, decolouring and opacifying of glass. In *Modern Methods for Analysing Archaeological and Historical Glass*, volume 1, Janssens, K. (ed.). Wiley, Chichester.

Bontemps, G. (1868). *Le guide du verrier.* Librairie des Arts et Métiers, Paris.

Brain, C. (2003). The technology of 17th century flint glass. *Glass Technol.*, 43C, 357–361.

Brain, C. and Dungworth, D. (2005). English seventeenth-century crystal glass study: Phase 1. In *Annales du 16e congrès de l'Association internationale pour l'histoire du verre. Londres, 2003*. AIHV, Nottingham.

Cappa, G. (2001). *Le génie verrier de l'Europe*. Éditions Martaga, Brussels.

Challes, C. (2018). *Étoffes de verre et de soie*. Mémoire d'études, École du Louvre, Paris.

Charleston, R.J. (1960). Lead in glass. *Archaeometry*, 3, 1–4.

Chopinet, M.-H. (2008). Chimie industrielle et innovations dans les compositions verrières, fin XVIIIe–XIXe siècle. In *Actes du colloque "Les Innovations Verrières et leur Avenir" de Nancy*. Verre et Histoire, Paris.

Dararutana, P., Dutchaneephet, J., Chetanachan, P., Wathanakul, P., Sirikubrat, N. (2008). Effects of gold nanoparticles on the fabrication of red colored high refractive index lead glass. *Advanced Material Research*, 55–57, 601–604.

Dufrenne, R., Maës, J., Maës, B. (2005). *La Cristallerie de Clichy : une prestigieuse manufacture du XIXe siècle*. Association "La rose de Clichy", Clichy-la-Garenne.

Dungworth, D. and Brain, C. (2004). The "invention" of lead crystal glass. 34th International Symposium, Zaragoza [Online]. Available at: https://www.researchgate.net/publication/28309145_The_%27Invention%27_of_Lead_Crystal_Glass [Accessed January 30, 2021].

Dungworth, D. and Brain, C. (2005). Investigation of late 17th century crystal glass. *Centre for Archaeology*, 21.

Eppler, R.A. and Epple, D.R. (2000). *Glazes and Glass Coatings*. The American Ceramic Society, Westerville.

Frank, S. (1984). Gold ruby glass. *Glass Technol.*, 25(1), 47–50.

Frischat, G.H. (2002). Tin ions in float glass cause anomalies. *C. R. Chimie*, 5, 759–763.

Gratuze, B. (2012). Les premiers verres "cristal au plomb" produits en France par Bernard Perrot : contemporains de ceux de Ravenscroft ? In *Annales du 19e congrès de l'Association internationale pour l'histoire du verre – Piran, 2012*, Lazar, I. (ed.). AIHV, Koper.

Humbert, F. and Colin-Seigner, B. (2013). Rédox et couleurs. USTV – Atelier "Rédox" – Nancy, 21–22 March [Online]. Available at: https://www.ustverre.fr/site/ustv/Redox/Humbert.pdf [Accessed 17 December 2020].

Jembrih, D., Schreiner, M., Peev, M., Krejsa, P., Clausen, C. (2000). Identification and classification of iridescent glass artifacts with XRF and SEM/EDX. *Mikrochim. Acta*, 133, 151–157.

Kantz, K. (1934). The effect of glaze composition upon the colors produced by sodium uranate. *J. Am. Ceram. Soc.*, 17, 8–10.

Klein, D. and Lloyd, W. (1984). *The History of Glass*. Orbis Publishing Limited, London.

LaCourse, W.C., Ott-Teneyck, M., Ukwu, B. (1980). Selenium in oxide glasses – Methods for increase retention. *American Glass Review*, 8, 6–9.

Lahlil, S. (2008). Redécouverte des procédés d'opacification des verres à l'antimoine à travers l'Histoire. Étude des antimoniates de calcium. PhD Thesis, Université Pierre et Marie Curie, Paris.

de Langle, C. (2002). Verre et gravure : une histoire et des musées. *Verre*, 8(6), 50–57.

de Langle, C. (2003). Des parures pour le verre : du verre mosaïqué au règne de l'or et de l'émail. *Verre*, 9(2), 50–56.

Lehuédé, P. (2018). Fragment de soierie du Retour des cendres de Napoléon (1840, Paris). Report, C2RMF, Paris.

Lehuédé, P., Collin, M., Dubus, M. (2015). Techniques d'opacification de l'émail au XIXe siècle : nouvel éclairage apporté par le microscope à effet de champ. *Technè*, 42, 92–100.

Mäder, M., Jemdrih-Simbürger, D., Neelmeijer, C., Schreiner, M. (2005). IBA of iridescent art nouveau glass – Comparative studies. *Nucl. Instrum. Methods Phys. Res. B*, 239, 107–113.

Matthes, W.E. (2010). *Émaux et Glaçures céramiques : plus de 1100 formules*. Eyrolles, Paris.

Mesnil, J.J. (2012). Terres rares et coloration du cristal [Online]. Available at: https://www.ustverre.fr/site/ustv/Nice2012/Mesnil.pdf [Accessed 20 December 2020].

Moretti, C. (2003). The development of lead crystal in England. A critical analysis of the formulation attributed to George Ravenscroft and points not yet clear on the process for manufacture of "flint" glass. *Rivista della Stazione Sperimentale del Vetro*, 6, 19–30.

Moretti, C. and Hreglich, S. (2009). Les verres opaques : la technologie des verriers vénitiens (XV–XXe siècle). In *Actes du colloque "Les Innovations Verrières et leur Avenir" de Nancy*. Verre et Histoire, Paris.

Neuwirth, W. (1986). *LoetzAustria 1900*. Selbstverlag der Verfasserin, Vienna.

Pantano, C.G., Bojan, V., Verità, M., Geotti-Bianchini, F., Hreglich, S. (1993). Tin profiles in the bottom surface of float glass: Manufacturing and heat treatment effects. Fundamentals of glass science and technology ESG. *Rivista della Stazione Sperimentale del Vetro*, 23, 285–290.

Péligot, E. (1877). *Le verre, son histoire, sa fabrication*. G. Masson, Paris.

Phillips, P. (1981). *The Encyclopedia of Glass*. Crown Publishers, New York.

Shelby, J.E. (2015). Color in commercial glasses. Glass processing course [Online]. Available at: https://www.lehigh.edu/imi/teched/GlassProcess/Lectures/Lecture04_Shelby_ColoredGlass.pdf [Accessed 19 June 2020].

Skelcher, B. (2020). Uranium glass [Online]. Available at: www.glassassociation.org.uk [Accessed 6 June 2020].

Spierings, G.A.C.M. (1993). Review: Wet chemical etching of silicate glasses in hydrofluoric acid based solutions. *J. Mat. Sci.*, 28, 6261–6273.

Strahan, D. (2001). Uranium in glass, glazes and enamels: History, identification and handling. *Studies in Conservation*, 46(3), 181–195.

Stroud, J.S. (1971). Optical absorption and color caused by selected cations in high-density, lead silicate glass. *J. Am. Ceram. Soc.*, 54, 401–406.

Verger, L. (2015). Les oxydes de chrome dans les pigments et les décors de porcelaine. PhD Thesis, Université Pierre et Marie Curie, Paris.

Verità, M. (2008). Venetian innovations in glassmaking and their influence on the European glass history. In *Actes du colloque "Les Innovations Verrières et leur Avenir" de Nancy*. Verre et Histoire, Paris.

Verità, M., Zecchin, S., Tesser, E. (2018). Venetian filigree glass along the centuries: Some technological considerations. In *Study Days on Venitian Glass Venitianfiligrana Glass through the Centuries*, Barovier Mentasti, R., Tonini, C. (eds.). Instituto Veneto di Scienze, Lettere ed Arti, Venice.

Watts, D.C. (1990). Why George Ravenscroft introduced lead oxide into crystal glass. *Glass Technology*, 31(5), 208–212.

Weyl, W.A. (1951). *Coloured Glasses*. Society of Glass Technology, Sheffield.

Zhou, Y., Jin, Y., Wang, K., Sun, J., Cui, Y., Hu, D. (2019). Opaque ancient K_2O–PbO–SiO_2 glass of the Southern Song Dynasty with fluorite dendrites and its fabrication. *Herit. Sci.*, 7, 56.

8

Early Islamic Lead Glass

Nadine Schibille
IRAMAT-CEB, CNRS, University of Orléans, France

8.1. Introduction[1]

Islamic lead glass as a distinct compositional type was first recognized by Sayre and Smith in 1961. Because of the increasing volume of analytical and isotopic data of Islamic glass finds from a wide range of sites and geographical areas, several sub-types of Islamic lead glasses have since been identified and characterized (Brill 1989; Krueger 2014; Pollak 2017; Sayre and Smith 1961; Schibille et al. 2020; Shindo 2006).

Lead silica glass in the strict sense, that is, a mixture of lead oxide and silica, has lead oxide levels of at least 57% and typically contains very low concentrations of alkali metals (Table 8.1). All other elements, with the exception of deliberately added colorants and/or opacifiers, can be considered as contaminants contributed by silica or the lead component. Soda ash lead glass, in contrast, additionally contains significant levels of alkali and alkaline earth metals, and the lead concentrations range from about 10%–50% (Table 8.11). Accordingly, these soda ash lead glasses are made from three constituents, namely a soda-rich plant ash, silica and lead. These two main compositional categories – lead silica glass and soda ash lead glass – can be further subdivided based on accessory elements that reflect the use of distinct raw materials and production processes. It is important to stress that lead in

1. This article is part of a project entitled "Glass Routes: Mapping the First Millennium Glass Economy", funded by the ERC under the European Union's Horizon 2020 research and innovation programme (ERC-CoG-2014, Project ID: 647315).

Lead in Glassy Materials in Cultural Heritage,
coordinated by Anne Bouquillon and Patrice Lehuédé. © ISTE Ltd 2023.

these glasses serves as a fluxing agent and forms part of the primary glass matrix, but in combination with other metals (copper, tin, antimony), lead has also been used as coloring and opacifying agents since the late Bronze Age (Shortland 2002; Turner and Rooksby 1959). In this case, lead is decidedly not part of the base glass, even though some lead may have gone into solution.

		Na_2O	MgO	Al_2O_3	SiO_2	Cl	K_2O	CaO	Fe_2O_3	CuO	PbO
Islamic lead silica (n = 25)	Mean	0.31	0.2	0.81	28.4	0.20	0.26	0.59	0.27	**1.12**	**67.7**
	SD	0.31	0.1	0.61	4.2	0.11	0.20	0.48	0.19	0.80	4.8
Šaqunda slag glass (n = 15)	Mean	0.63	0.38	**4.73**	33.6	0.09	1.45	**4.16**	**1.72**	0.07	**52.2**
	SD	0.24	0.05	0.44	1.1	0.01	0.14	0.63	0.26	0.02	2.7
Andalusian soda ash lead glass (n = 55)	Mean	**6.61**	1.13	1.06	40.9	**1.82**	0.75	2.04	0.56	0.26	**42.8**
	SD	0.91	0.24	0.26	2.3	0.27	0.27	0.49	0.13	0.43	3.6

Table 8.1. *Mean composition and standard deviations (SD) of the three main Islamic high lead glass groups discussed in this article, given as wt%. Data sources: for Islamic lead silica glass (Brill 1999; Brill and Stapleton 2012; unpublished data); for Šaqunda (Schibille et al. 2020); for Andalusian soda ash lead glass (Duckworth et al. 2015; De Juan Ares and Schibille 2017a; De Juan Ares et al. 2021)*

Currently, it is not clear where and how Islamic lead glass originated and to what extent the finds from different regions are related in terms of their technologies and/or raw materials. High lead glass is not very common in the ancient Mediterranean (see Chapter 6). It first appeared in the Islamic world in the 8th and 9th century. In contrast, various types of high lead glasses are known from China, Korea and Japan, dating to the Han (206–220 CE), Sui (581–618 CE) and Tang Dynasties (618–907 CE) (Brill 1999; Gan 2009; Liu et al. 2012; Henderson et al. 2018). Since the composition of Chinese lead silica glass (as opposed to lead barium glass) is essentially the same as that of early Islamic lead silica glass from the Near East and Egypt, it has been proposed that the technique originated in East Asia and spread along the Silk Roads to Western Asia and further to the Mediterranean coast (Brill 1989; Brill and Stapleton 2012; Wypyski 2015; Henderson et al. 2018). An *ex nihilo* development of high lead silica glass has recently been documented in al-Andalus in the late 8th or early 9th century CE (Schibille et al. 2020). This was followed by the appearance of soda ash lead glass in al-Andalus shortly afterwards (Duckworth et al. 2015; De Juan Ares and Schibille 2017a).

Complementing Chapter 6, this article focuses on the advances that have been made in the analytical study of high lead glasses attributed to Islamic glassmaking based on archaeological context and/or style, from its earliest occurrence in the 8th and 9th centuries to the 12th century. I will begin by discussing glass containing mainly lead and silica found in the eastern Islamic regions of Egypt, Greater Syria, Mesopotamia and Central Asia, followed by glass from the Iberian Peninsula, where in recent years a multitude of different kinds of lead glasses have turned up in the archaeological record. The objective of this review is twofold. First, the chemical composition of sub-groups of Islamic high lead glass thus far identified will be characterized and differentiated according to their geographical and chronological contexts. By investigating glass finds from secured excavations and stratigraphic contexts, I hope to unravel the chronology of Islamic lead glasses and narrow down questions of technological transfer versus local innovations. The second objective concerns the origin of raw materials, especially lead. This involves the evaluation of the available lead isotope data of some of the glass samples in relation to the isotopic composition of lead ore deposits in the Near East, northern Africa and the Iberian Peninsula. These data may clarify the possible origins of the various types of lead glass and how they evolved.

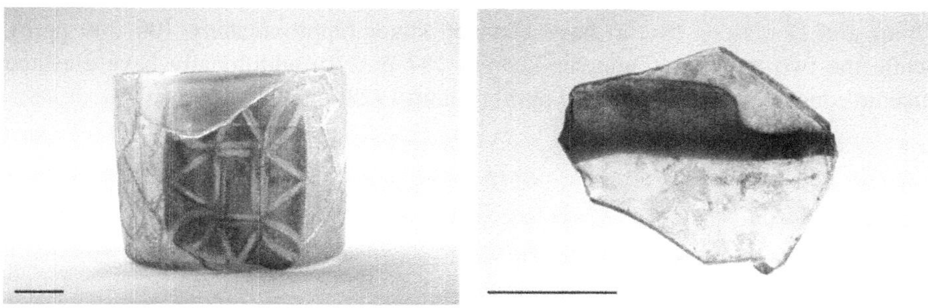

Figure 8.1. *Emerald green lead silica glass from Samarra: (a) Sam 800.2/Lamm 18; (b) Sam.I.44/Lamm 270. Museum für Islamische Kunst der Staatlichen Museen zu Berlin – Preußischer Kulturbesitz/photo Johannes Kramer. Scale bar = 2 cm. For a color version of this figure, see www.iste.co.uk/bouquillon/lead.zip*

8.2. Islamic lead silica glass (*mīnā*) from the Near East

Islamic lead silica glass finds are few and far between, but growing scholarly work on the topic in recent years has considerably expanded the corpus of data. This includes studies on the glass finds from Rāya and the Wadi al-Tur on the Sinai Peninsula (Shindo 2006), Caesarea Maritima in Israel (Pollak 2017), the Serçe Limani shipwreck off the coast of Turkey (Barnes et al. 1986), Nishapur (north-eastern Iran) (Wypyski 2015) and a comprehensive review of Islamic lead glass from numerous museum collections and excavations (Krueger 2014). The published examples date from the 9th to 12th century CE and include vessels as well as flat glass (window fragments). Most of the objects are of a deep emerald green color (Figure 8.1) and can be either undecorated or mold blown and/or relief-cut (Barnes et al. 1986; Shindo 2006; Krueger 2014; Wypyski 2015; Pollak 2017). They contain on average about 68% lead oxide (57%–78%), 28% silica and 1% copper oxide as the main coloring agent (Table 8.1). Although a chemical analysis of the glasses could not be carried out for all the bodies, using density measurements, the lead contents of finds from Rāya, the Wadi al-Tur and Caesarea Maritima were estimated to span a similarly wide range of lead contents (Shindo 2006; Pollak 2017). All other elements are at levels that suggest they were part of either the silica source or the lead component. The high lead glasses from the 11th century Serçe Limani shipwreck, a molar flask from Heshbon (Brill 1999), a vessel fragment from Ṣabra al-Manṣūriya (Kairouan, Tunisia; Foy 2020) and a glass chunk from Spain have somewhat higher alumina (~1.5%) and titanium oxide levels (~0.14%) than the other analyzed samples (Figure 8.2). These features indicate the use of a silica source enriched in accessory minerals. Unfortunately, not many trace element data have been published to date. Some unpublished samples from Pechina (Spain), Samarra (Iraq) and Damascus (Syria) have elevated silver (approximately 100–200 ppm), while the two Samarra fragments (Lamm 187 & 270) additionally have elevated arsenic concentrations (As > 2000 ppm) (Lamm 1928; unpublished data).

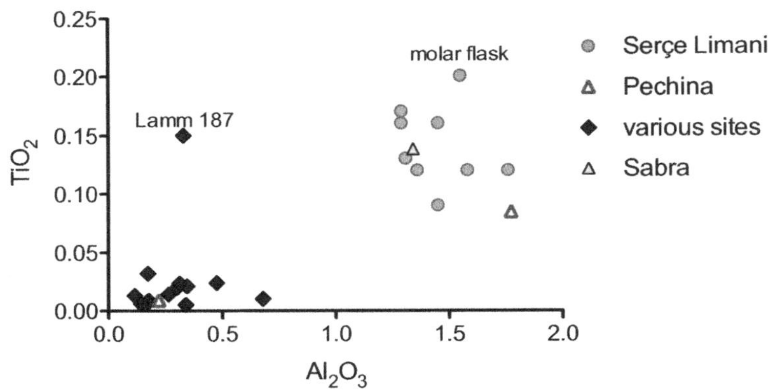

Figure 8.2. *Binary plot of titanium versus aluminum oxide concentrations of the lead silica glass finds from various sites (Serçe Limani, Damascus, Samarra, Ribe, Pechina, Ṣabra al-Manṣūriya), reflecting differences in the silica component. Data sources: (Brill 1999; Brill and Stapleton 2012; Foy 2020, De Juan Ares and Schibille forthcoming; unpublished data). For a color version of this figure, see www.iste.co.uk/bouquillon/lead.zip*

8.2.1. *Geographical and chronological patterns*

In the absence of systematic analyses of lead silica glasses, not much can be said about possible chronological and/or geographical patterns, on the one hand, and the existence of specific compositional signatures of the production sites, on the other hand. It is certain, however, that Islamic lead silica glass is compositionally diverse, pointing to the use of different raw materials and recipes and thus different production locations. The earliest Islamic lead silica glass from a relatively well dated context comes from Samarra (836–892 CE) in Mesopotamia. Two high lead glass fragments from Samarra were analyzed (Figure 8.1): one is a cylindrical cameo cup, where a colorless glass body is overlain by two rectangles of emerald green lead glass patches with linear wheel-cut ornamentation (Lamm 187; Carboni and Whitehouse 2001; No. 98), and the other is an emerald green trail applied to a colorless vessel fragment (Samarra Sam.I.44/Lamm 270; Figure 8.1; Lamm 1928). The glass body of the latter vessel has very low silica related impurities (Samarra group 1), while the cameo glass has slightly higher mineral impurities (Samarra group 2). Both glass types are believed to have been produced in Mesopotamia in or near Samarra itself (Schibille et al. 2018). The lead raw material, however, must have been imported as no lead sources are available in the immediate vicinity of the alluvial plains of Mesopotamia (Yener et al. 1991). This may also explain why lead silica glass is not more common among the Samarra glass finds and only found in the form of decorative applications.

In contrast, numerous fragments of emerald green lead silica glass vessels have been recovered from Fustat (Krueger 2014), Rāya, the Wadi al-Tur (Shindo 2006), and Caesarea (Pollak 2017). Here, the lead silica glass includes molar flasks and small bottles, beakers and jugs that are often mold blown and with linear wheel-cut decorations, as well as some relatively plain, undecorated vessels (Shindo 2006; Krueger 2014; Pollak 2017). Isolated fragments of flat glass from al-Raqqa (Henderson et al. 2004) and the citadel in Damascus (Krueger 2014) are further examples of the different uses of lead silica glass in the Islamic Near East. Further west, lead silica glass has been found at Ṣabra al-Manṣūriya (Tunisia) in 10th century contexts (Foy 2020). This appears to be the westernmost finds of leaded glass objects in the strict sense at present and may be an Egyptian import (Krueger 2014; Foy 2020). In fact, the two samples dated to the 9th or 10th centuries, found in Pechina (Almería, Spain), are problematic: one of them is a glass chunk that may have been used as raw material for the production of translucent green lead glazes (Salinas et al. 2019), while the other sample is an unidentified folded fragment with a lead oxide concentration of almost 80% (de Juan Ares and Schibille forthcoming).

8.2.2. *Potential lead sources*

The relative abundance and distribution of finds suggests that the majority of early Islamic lead silica glasses originated in the eastern Islamic world. This is not to say that all the raw materials were sourced locally. Historically, the main mining districts are located in the mountainous regions of Iran and central Asia (e.g. Khorasan, Transoxiana) as well as the Caucasus and eastern Turkey. These regions can therefore be considered as potential sources of the lead component used for the manufacture of lead silica glass (Al-Hassan et al. 2001; Yener et al. 1991). One of the most promising tools to determine the likely provenance of lead is isotope analysis. Identifying group affiliations and attributing origins on the basis of lead isotope data hinges on the quantity and quality of available data. Only 13 Islamic emerald green lead silica glasses have been analyzed for their lead isotope signatures so far, including four emerald green fragments from the Serçe Limani assemblage, a molar flask from Heshbon and three cameo glasses of unknown provenance (Barnes et al. 1986). The remaining samples are unpublished data from a Viking context in Ribe (Denmark), Damascus (Syria) (both courtesy of B. Gratuze) and a single sample from Samarra (Iraq) (sample Berlin Sam 014 in Schibille et al. 2018). Lead isotopic data from a sgraffito glaze from Samarra (Mason et al. 1991), as well as glazes from the Serçe Limani shipwreck (Barnes et al. 1986; Stos-Gale 2004) and from Fustat were used for comparison (Wolf et al. 2003).

These data reveal different isotopic clusters, representing a considerable diversity of lead sources used in the production of Islamic lead silica glass and glazes (Figure 8.3). As previously reported (Barnes et al. 1986; Stos-Gale 2004), three of the emerald green samples together with the majority of glazes from the Serçe Limani shipwreck and the molar flask from Heshbon form a very tight group in terms of all three lead isotope ratios (Group A). The majority of early glazes (8th/9th century to 1025 CE) from Fustat have also been attributed to the same isotopic group (Wolf et al. 2003). The Samarra lead glass is not far from the main cluster and may belong to the same isotopic field, whereas the sgraffito ceramic glaze from Samarra is clearly distinct (Mason et al. 1991). Unfortunately, only two isotopic ratios are available for this ceramic fragment, but it is safe to assume that it represents a separate isotopic group (Group C) possibly together with a cameo glass of unknown provenance (Barnes et al. 1986). Finally, two cameo glass fragments and an emerald green molar flask from the Serçe Limani shipwreck are isotopically related (Group B), despite an earlier different attribution of one of the cameo glasses (Barnes et al. 1986, cameo glass PB-1030).[2]

2. If the Euclidean distances are calculated for all three artifacts, they are well within the limits of the analytical error.

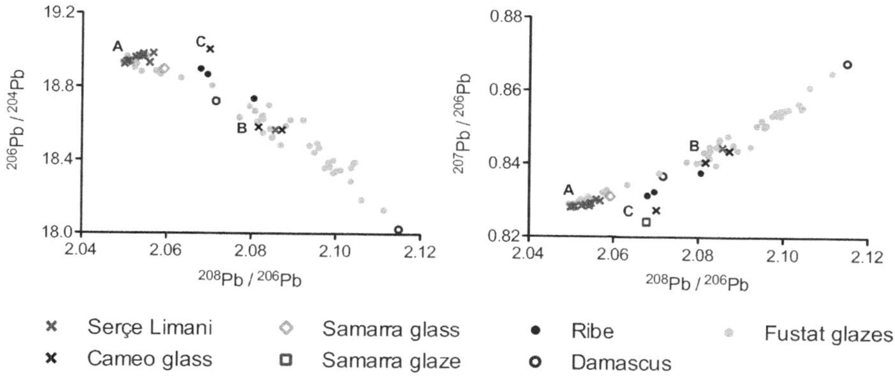

Figure 8.3. *Lead isotope ratios for early Islamic emerald green lead silica glasses and glazes from the Serçe Limani shipwreck and related objects (Barnes et al. 1986; Stos-Gale 2004), Damascus and Ribe (courtesy of B. Gratuze), sample Lamm 270 from Samarra (unpublished) and early Islamic glazes from Fustat (Wolf et al. 2003). For a color version of this figure, see www.iste.co.uk/bouquillon/lead.zip*

A lead source can be considered an isotopic match, if the Euclidean distance between all three lead isotope ratios of a data point and the ore sample are within the limit of the analytical precision estimated at 0.1%. The lead isotope signatures of early Islamic lead silica glasses are thus indicative of different origins of the lead used for their manufacture. None of the lead silica glasses or glazes match the isotopic pattern of lead ores from Egypt that show significantly lower $^{206}Pb/^{204}Pb$ ratios (Figure 8.4). Egyptian lead ores can thus be ruled out as potential sources. They were in fact previously excluded as sources for the production of the glazes from Fustat (Wolf et al. 2003). There are also no conclusive similarities with samples from Syria (Gale et al. 1981) or Turkey, even though the isotopic profiles of some of the Turkish lead ores appear close. One sample from Damascus and two from Ribe may present an exception; they could contain Turkish lead if their Euclidean distance is considered. Overall, however, it can be concluded that Turkish ores probably did not provide the raw material for early Islamic lead silica glasses and glazes.

Ores from north-western Iran in the region of Anguran as well as from Tunisia have been suggested as possible sources of the lead component in the glasses and glazes of group A (Barnes et al. 1986; Wolf et al. 2003; Stos-Gale 2004). The database of Iranian and Tunisian lead ore isotopes has since been significantly expanded, which facilitates the classification of some of the lead silica glasses (Figure 8.5). Iran boasts several major lead deposits, particularly north-west of

Teheran, in the central Iranian zone (Mirnejad et al. 2011; Mirnejad et al. 2015) as well as the Zagros Belt in the south-west (Fazli et al. 2019). There is written and archaeological evidence that many of the lead and silver mines in Iran have been exploited in antiquity (Wertime 1968; Momenzadeh 2004; Nezafati et al. 2008), and it appears that at least some of the lead silica glasses contain lead from Iran. For example, one of the Damascus samples with high $^{207}Pb/^{206}Pb$ and $^{208}Pb/^{206}Pb$ ratios is consistent with the cluster representing lead ores from the Kuh-e-Surmeh reserve in south-west Iran (Fazli et al. 2019)[3]. The glass samples of group B (two cameo vessels and a molar flask from the Serçe Limani) show isotopic similarities with the ores from the historic lead-silver mines of the Nakhlak region in central Iran (Barnes et al. 1986; Stos-Gale 2004).

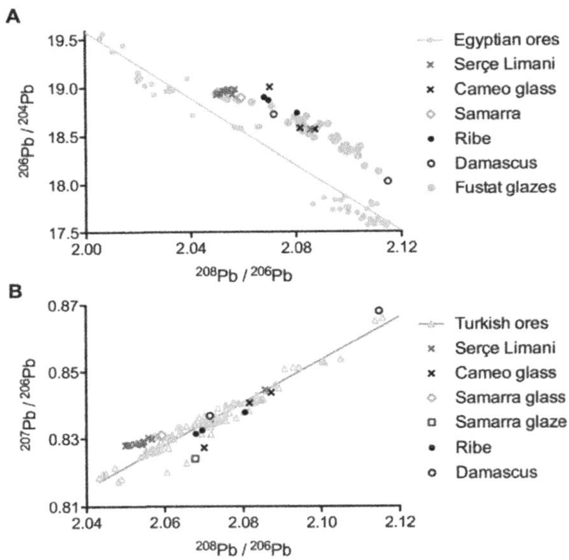

Figure 8.4. *Lead isotope ratios for (a) lead ores from Egypt and (b) lead ores from Turkey compared to early Islamic lead glasses and glazes. Data sources: for glass and glazes see (Barnes et al. 1986; Wolf et al. 2003; Stos-Gale 2004), for Egyptian lead ores (Stacey et al. 1980; Hassan and Hassan 1981; Stós-Gale and Gale 1981; Shortland 2006); for Turkish lead ores (Wagner et al. 1986; Yener et al. 1991)[4]. For a color version of this figure, see www.iste.co.uk/bouquillon/lead.zip*

3. The name of the mining district indicates that the ores have been exploited for the production of *Sormeh* (kohl) since ancient times (Momenzadeh 2004).

4. See also: http://oxalid.arch.ox.ac.uk.

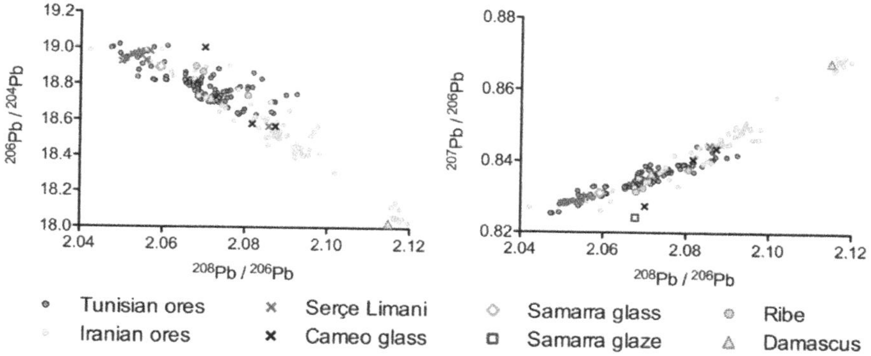

Figure 8.5. *Lead isotope ratios for Islamic lead glasses and glazes in comparison to the isotope signatures of lead ores from Iran and Tunisia. Data sources: for glass and glazes see (Barnes et al. 1986; Stos-Gale 2004), for lead ore data from Iran see (Stos-Gale 2004; Bazargani-Guilani et al. 2011; Mirnejad et al. 2011; Mirnejad et al. 2015; Saeidabadi et al. 2018; Fazli et al. 2019); for Tunisian lead ore data see (Jemmali et al. 2011; Bejaoui et al. 2013; Jemmali et al. 2013; Skaggs et al. 2012; Jemmali and Souissi 2018). For a color version of this figure, see www.iste.co.uk/bouquillon/lead.zip*

The vast majority of the ceramic glazes, the three emerald green glass fragments from the Serçe Limani shipwreck and the molar flask from Heshbon (group A) have ratios consistent with lead ores in either north-western Iran in the Urumieh-Dokhtar zone, more specifically the ores from around Anguran (Wolf et al. 2003; Stos-Gale 2004), or with Tunisian ores from the north-south axis zone around Jebel Ressas, in the hinterland of Tunis (Skaggs et al. 2012). The role of Tunisian ore deposits in the supply of lead during the Roman, medieval and early Islamic period is unclear, but some evidence of ancient mining activities exists (Farquhar and Vitali 1989; Wolf et al. 2003; Skaggs et al. 2012). Euclidean distance calculations show that the Serçe Limani glass and glazes of group A are as close to Tunisian ore sources as they are to the Iranian ones. The lead silica glass fragment from Samarra seems to closely match the lead isotopes of the Oued Maden deposits located within the Nappe zone close to the Algerian-Tunisian border (Jemmali et al. 2013) or with ores from Mecella or Hammam Zriba from the north south axis zone (Skaggs et al. 2012). Again, the overlap between Tunisian and Iranian ore deposits cannot be resolved with certainty.

The provenance of the lead used for the manufacture of lead silica glasses can thus be narrowed down to two likely regions, either Tunisia or Iran. For logistical reasons, one may be inclined to favor the use of lead sources from north-western

Iran around Anguran for the production of glass from Samarra, rather than ores from the more distant Tunisian deposits. At the same time, the glass may very well have been made from an unknown source. Ninth century dirhams from Maden Bajunais in Armenia apparently provide a near perfect isotopic match for the Samarra glass, while the ceramic glaze from Samarra has the same lead isotope ratios as three dirhams analyzed from Samarra (S. Merkel, pers. communication). This underscores the special nature of the production of early Islamic lead silica glasses and suggests that it was a separate industry not directly related to minting or glazing technologies. Either the lead used for the production of early lead silica glass, the raw lead glass or the final product must have traveled some distance. While a correspondence of the lead isotope signatures may point to the origins of the lead raw material, it does not necessarily reveal the location of primary production or where the glass objects were ultimately made.

8.2.3. *The ambiguity of the term mīnā*

Arabic textual sources can shed further light on the possible origins of the emerald green lead silica glass composition. The highly ambiguous old Persian term *mīnā* that in Islamic literature ultimately came to designate (colored) lead glass originally signified among many other things "emerald" (Aga-Oglu 1946; Krueger 2014). A passage in an undated Islamic textual source that is partially preserved in the so-called Karšūnī manuscripts, explicitly differentiates between glass and *mīnā*, which has been variably translated as enamel, glaze or leaded glass (Krueger 2014; Holakooei 2016). While glass according to this source is composed of sand and alkali, the raw materials of *mīnā* are pebbles and burnt lead (Berthelot and Duval 1893; Krueger 2014).

One of the earliest textual references that associates the term *mīnā* with leaded glass is the *Kitāb ad-Durra al-Maknūna* (*The Book of Hidden Pearl*), generally dated from the 8th to 10th century CE. The surviving manuscript is a compilation collected and copied by al-Marrakushı in Baghdad in 1252 CE (Al-Hassan 2009; Krueger 2014). Out of a staggering 51 recipes detailing the production of colored glass, 26 mention different forms of lead compounds such as lead oxide red, lead oxide white, litharge or burnt lead (Al-Hassan 2009). Recipe number 35, for instance, describes the production of white (colorless) *mīnā* from "one part of pulverized white pebbles, one part of lead oxide red, and one part of *natrūn*. Combine and cook and it will come out white" (Al-Hassan 2009, p. 145). In the 11th century, Abu al-Bīrūnī states that *al-mīnā* is a type of glass, but softer and heavier on account of its lead content. It was produced from a mixture of white pebbles and lead oxide red or litharge (Kahle-Bonn 1936; Krueger 2014). Al-Bīrūnī furthermore specifies that the glass owes its color to different metal oxides. A green color, for example, is obtained by adding copper oxide (Kahle-Bonn 1936). The oldest known Persian book of stones, the *Jawâhar-nâma-i Nizâmi*, dating to 1196 CE, concedes that the term *mīnā* has many meanings, linking it to artificially made precious stones (Porter 2004; Fol. 152b). It describes the essence of *mīnā* as a mixture of silica and alkali that was melted so as to produce a crystal-like substance. If the desired color was to be yellow, lead oxide red (*minium* – Pb_3O_4) had to be added to the base material, whereby the amount of lead determined the intensity of the color as well as the hardness and weight of the final product (Porter 2004; Fol. 154a). Instead of lead oxide red, litharge (PbO) could also be used. The treatise contains several recipes for green *mīnā* made from quartz, alkali, lead oxide red and copper, although the amount of copper required to obtain emerald green is drastically exaggerated in the manuscript (Porter 2004; Krueger 2014). About half a century later, Nasīr al-Dīn Tūsī (1201–1274 CE) states that green *mīnā* has the best color, that it is occasionally passed off as emerald, and that many rare and precious things are made of it. Tūsī furthermore remarks that this type of green *mīnā* is particularly abundant in Syria and the Maghreb (Aga-Oglu 1946).

Although the term *mīnā* has many meanings and in principle can refer to all sorts of colored glass, from enamels and gemstone imitations to mosaic tesserae, emerald green *mīnā* seems to have been particularly desirable, perhaps in reference to its original old Persian meaning. All sorts of valuable objects were fabricated from this artificial emerald (Krueger 2014). Judging from the literary sources, the emerald green *mīnā* was known in Persia, Syria and north Africa. It has been claimed that all larger, high lead emerald green luxury vessels originated in Iran (Krueger 2014). However, none of these luxury objects has a secure provenance. The two fragments from Samarra provide the first relatively firm evidence that emerald green lead silica glass was worked in Mesopotamia, because the colorless glass to which the green lead glass was applied can confidently be attributed to a local glassmaking technology. The fact that some of the largest assemblages of emerald green lead glass have been found in Egypt and the Levant, including relatively simple and undecorated wares, suggests a more widespread availability compared to Mesopotamia and Iran, where the lead raw material was predominantly used for the production of objects of high value and prestige. They are nonetheless unlikely to represent independent developments, since there is a clear material and technological link between the finds from the different regions. The etymology of the term *mīnā* and its Persian roots suggests that the lead silica glassmaking technology originated in central Asia and spread along the Silk Roads to the shores of the eastern Mediterranean.

8.3. Lead slag glass from Šaqunda (Córdoba)

At the other end of the Mediterranean, the situation is rather different. Here, new types of high lead glass appear between the 8th and 10th century CE. Conquered by Muslim forces in 711 CE, al-Andalus was integrated into a vast Umayyad Empire and formed the western boundary of the Islamic world until the fall of Granada in 1492 CE. Córdoba, seat of the local government since 716 CE, became the capital of the independent Umayyad Emirate in 756 CE (Casal García 2018). As a consequence, the urban area of Córdoba was significantly enlarged and a new suburb (Rabad of Šaqunda) was founded on the southern shore of the Guadalquivir. In the wake of a popular uprising, the Rabad of Šaqunda was completely obliterated in 818 CE (Casal García 2018). The finds from Šaqunda thus represent one of the earliest and best dated glass assemblages from the early Islamic period in the Iberian Peninsula. One of the most remarkable findings in Šaqunda is the presence of a unique type of lead silica glass that makes up about a quarter of the glass assemblage, and represents one of the earliest examples of lead glass both in Europe as well as in the Islamic world (Schibille et al. 2020).

When comparing the lead glass from Šaqunda to the Islamic lead silica glass from the eastern Mediterranean and Mesopotamia, they are clearly distinct. The Šaqunda samples have lower lead (mean ~52%) and somewhat higher alkali and alkaline earth elements, as well as considerably higher alumina relative to the silica and high lime concentrations (Table 8.1, Figure 8.6). Due to its high content of mineral impurities, the glass is usually yellowish green or amber in color (Figure 8.7). Hence, the high lead glass from Šaqunda cannot be a simple mixture of lead oxide and a quartz rich silica source similar to the finds from the Islamic Near East. Instead, the Šaqunda samples are compositionally reminiscent of some Carolingian lead smoothers from Melle (France) that were made from lead slag (Gratuze et al. 2003, 2014, 2017; Chapter 6): their high alumina (4.7%), iron and barium contents, and the presence of low silver and bismuth concentrations (Figure 8.9) similarly suggest the use of lead slag (Schibille et al. 2020). No other glass finds are known with a comparable composition. There are, however, similarities with lead glazes from Córdoba that date to the second half of the 9th century CE, highlighting the close relationship between the two technologies in the Iberian Peninsula during the early Islamic period (Salinas et al. 2021).

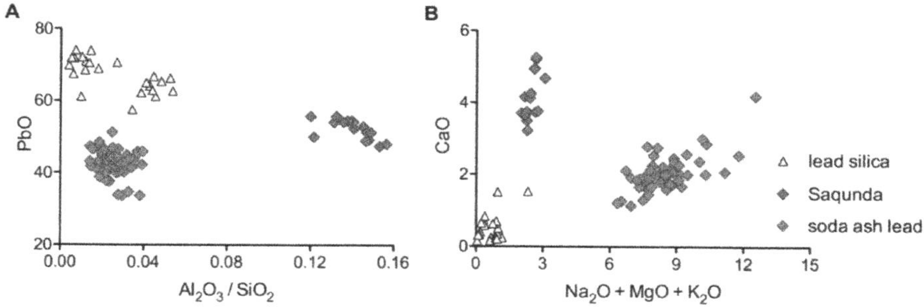

Figure 8.6. *Comparison of lead silica glass from the eastern Islamic lands with the compositional data from Šaqunda and soda ash lead silica glasses from al-Andalus. (a) Lead versus alumina to silica ratios reflect differences in the silica raw material; (b) lime compared to the sum of soda, magnesia and potash indicate different recipes. Data sources: for Islamic lead silica glass (Brill 1999; Brill and Stapleton 2012; unpublished data); for Andalusian soda ash lead glass (Duckworth et al. 2015; De Juan Ares and Schibille 2017a; de Juan Ares et al. 2021); for Šaqunda (Schibille et al. 2020). For a color version of this figure, see www.iste.co.uk/bouquillon/lead.zip*

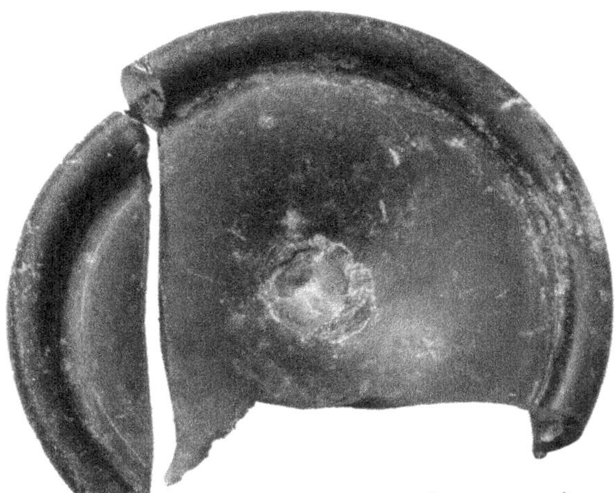

Figure 8.7. *Fragment of vessel base (MIR 040) from the lead slag glass from Šaqunda (courtesy of J. de Juan Ares; Schibille et al. 2020). Scale bar = 1 cm*

Lead isotope data confirm a local origin of the lead from the nearby mining districts north of Córdoba (Schibille et al. 2020). The Šaqunda assemblage thus marks the beginnings of a local Iberian glassmaking industry that may have been triggered by shortages in glass supplies from Egypt and the Levant. It may also have been inspired by a proliferation of silver mining in the area, which in turn was probably motivated by the introduction of a mono-metallic silver based monetary system and the establishment of a mint in Córdoba in the early 8th century CE (Barceló 1975). Šaqunda's 8th- to early 9th-century high lead glass was relatively short lived and was in fact the first instalment in a longer process of developing an independent Iberian glassmaking industry.

8.4. Soda ash lead glass from al-Andalus

About a century after the glassmaking innovations at Šaqunda, soda ash lead glass became the major glass type in and around Córdoba. Almost half of the glass finds in the archaeological record of Córdoba dating to the 10th and 11th centuries are of the soda ash lead glass (Duckworth et al. 2015; De Juan Ares et al. 2021). At the palace city of Madīnat al-Zahrā', about 8 km west of Córdoba, founded in 936 CE, soda ash lead glass makes up as much as 70% of the vitreous material (Figure 8.8; unpublished data).

Excavations of 15 early Islamic domestic dwellings in a western suburb of Córdoba (so-called *Piscinas Municipales de Poniente*) in 2005 yielded an interesting array of glass finds, about half of which are soda ash lead glasses (De Juan Ares et al. 2021). They are mostly colorless, occasionally with a greenish or yellowish tinge, and only one small purple flask that is intentionally colored. All the soda ash lead glasses from Piscinas Municipales are mold blown with or without embossed surface decorations (De Juan Ares et al. 2021). The glasses are compositionally clearly distinct from the previously discussed lead silica and lead slag glass types in that they contain appreciable amounts of alkali and alkaline earths in addition to lead and silica as the main constituents (Table 8.1; Figure 8.6). They are made from a ternary mixture of lead, silica and a soda-rich plant ash with high soda, moderate magnesia and potash and relatively low lime concentrations (Figure 8.6). Their composition is consistent with that of more or less contemporary glass from other Iberian sites from the 10th and 11th centuries CE (Duckworth et al. 2015; De Juan Ares and Schibille 2017a). The silica component of these Iberian soda ash lead glasses is different from that of the Šaqunda glasses but bears some resemblance with the Near Eastern lead silica group. The lead silica glasses from the eastern Mediterranean and the soda ash lead glasses from Iberia have considerably lower alumina to silica ratios compared to the Šaqunda slag glass, indicative of the use of a quartz rich raw material poor in accessory minerals (Figure 8.6). The use of a clean silica source recalls Arab textual sources, all of which describe the use of (white) pebbles, possibly quartz pebbles, for the production of leaded glass (*mīnā*).

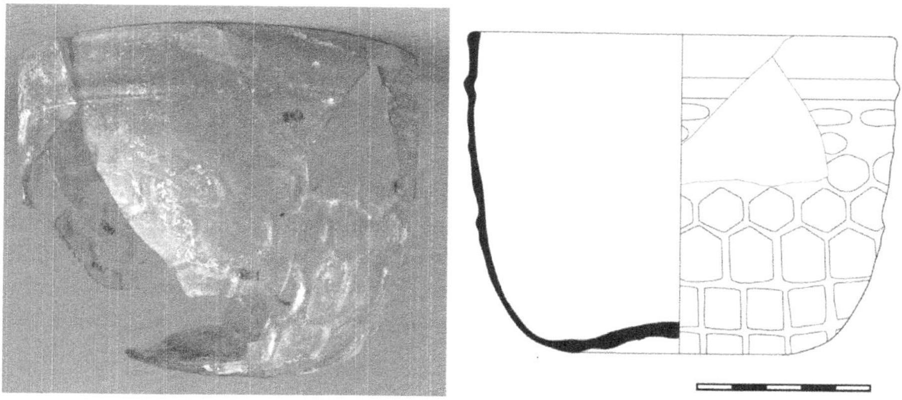

Figure 8.8. *Mold blown beaker from Madīnat al-Zahrā' made from typical Iberian soda ash lead glass (photo & drawing by A. Zamorano). Ana Zamorano*

Perhaps the most peculiar feature of the soda ash lead glasses from the Iberian Peninsula is their high chlorine concentration with an average of 1.8%, reaching up to 2.5% (Table 8.1), which is about twice the usual chlorine content in conventional natron-type and plant ash glass. A correlation matrix shows that chlorine correlates positively with lead, providing strong evidence that the lead raw material was accompanied by chlorine in one way or other. Two hypotheses can be advanced for the elevated chlorine concentrations of Iberian soda ash lead glasses: either chlorine was added separately, or it is in itself a by-product of yet another metallurgical process. In antiquity, sodium chloride (common salt) was used in the cementation process to purify gold. During this process, the silver contained in the gold was converted to silver chloride (Ramage et al. 2000, p. 11). In the early 12th century, Theophilus recommends in his *De diversis artibus* the addition of urine and powdered brick to the salt mix (Brepohl 2013, book 3, ch. 33). If the silver is recovered from the silver chloride through cupellation, chlorine would be incorporated in the glassy waste-product of the cupellation process. Alternatively, as described in Theophilus' second book that is devoted to glass working and coloring, the production of glass for rings could be achieved by mixing ash, salt (sodium chloride), powdered copper and lead (*cineres, sal, pulverem cupri et plumbum*; (Brepohl 2013, book 2, ch. 31).

Arabic alchemical and technological literature on the coloring of glass and the production of *mīnā* likewise recommend adding *natrūn* to the mix. As mentioned above, recipe 35 in the *Kitāb ad-Durra al-Maknūna* gives explicit instructions on how to prepare a mixture of white pebbles, lead oxide read and *natrūn* for the purpose of producing white *mīnā* (Al-Hassan 2009, p. 145). For producing emerald green glass, the author recommends irrigating the pulverized glass mixture with a *natrūn* solution (recipe 18), while the making of ruby requires washing the raw materials in salt water and an additional "one hundred dirhams of *natrūn*" during the initial stages of the production process (recipe 46) (Al-Hassan 2009). The addition of sodium chloride may explain the high chlorine content in Andalusian soda ash lead glasses. None of the other high lead glasses tend to have as much chlorine. Hence, high chlorine contents appear to be characteristic of a specifically Iberian manufacturing process of soda ash lead glass, which was produced between the 9th and 11th century CE.

With regard to the lead component, it has previously been proposed that lead had been added in a relatively pure form of calcined lead (Duckworth et al. 2015). However, lead used for the production of the soda ash lead glass was accompanied by trace elements such as silver, tin and bismuth. This trace element pattern is in stark contrast to the composition of the Šaqunda glass that has lower silver, tin and bismuth, but considerably higher barium and uranium (Figure 8.9). Lead is one of the most important by-products of silver production and some experimental results have shown that antimony, tin, bismuth and silver are oxidized to litharge at different stages of the cupellation process (L'Héritier et al. 2015). Tin and antimony tend to form a solid phase at the beginning of cupellation (Shortland et al. 2000; Mass et al. 2002; L'Héritier et al. 2015). A small proportion of both elements is retained in the initial part of the litharge. Bismuth is only partially oxidized during the early stages of the cupellation process and it is only enriched in the final fractions alongside silver (L'Héritier et al. 2015). Here, the fact that tin, antimony, silver and bismuth are found confirms that litharge from earlier stages of the cupellation process rather than slag was used. Slag was used for the production of the Šaqunda glass, reflected in the comparatively high barium and uranium concentrations. Barium and uranium are associated with the galena ore and are usually dissolved in the slag during the initial smelting process and not retained during the silver cupellation (Goldenberg et al. 1996).

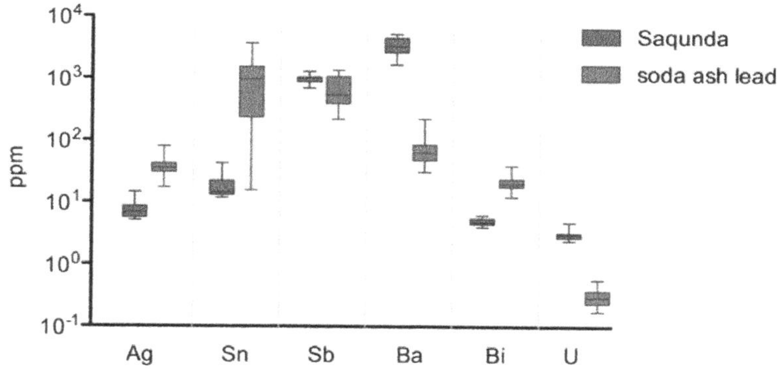

Figure 8.9. *Trace elements associated with the lead component in the Šaqunda high lead glasses and the soda ash lead glasses from various Iberian sites. Data sources: Duckworth et al. 2015; De Juan Ares and Schibille 2017a; Schibille et al. 2020; De Juan Ares et al. 2021. For a color version of this figure, see www.iste.co.uk/ bouquillon/lead.zip*

Iberian lead sources were exploited from the Phoenician through to the Roman and Islamic periods and traded widely in the form of either metal or galena. Lead from Spain and/or Sardinia was, for example, supplied to Islamic potters for the production of some of the later lead glazes from Fustat in Egypt (Wolf et al. 2003; Marzo et al. 2009). Lead isotope analysis of soda ash lead glasses from the Iberian Peninsula and the high lead glasses from Šaqunda (Schibille et al. 2020) show that most of the lead-soda glasses originate from the two mining districts of Linares La Carolina and Los Pedroches, located less than 100 km north and northeast of Córdoba. Only two soda ash lead glasses from Pechina are clearly different. No unambiguous match was found among the published isotopic data of Iberian ore deposits, but the glasses from Pechina appear to be related to the isotopic field defined by samples from the Roman mines of Cartagena (Trincherini et al. 2009), and closer still to some ore deposits from the Province of Almería, specifically the eastern Betic Cordillera (Murillo-Barroso et al. 2019). At least one of the Pechina samples lies within the Euclidean space of ores from Cerro Minado (Sierra de Almagro), located about halfway between Almería and Cartagena, an important mining district during prehistoric times due to opencast mining (Murillo-Barroso et al. 2019). The fact that the lead source of the Pechina samples differs from the other Andalusian high lead and soda ash lead glasses strongly supports the hypothesis of some local small-scale productions of these types of glasses, using more or less locally accessible plumbiferous material.

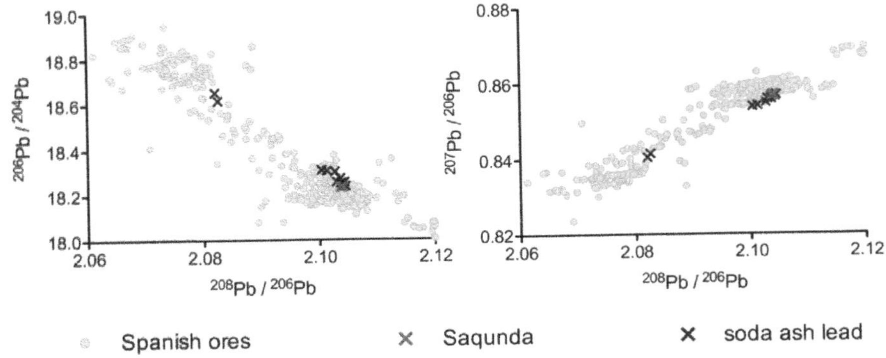

Figure 8.10. *Lead isotope ratios for the high lead glasses from Šaqunda (Schibille et al. 2020) and some soda ash lead glasses from Vascos, Albalat, Gauzon, Silves and Pechina (unpublished data). For a color version of this figure, see www.iste.co.uk/bouquillon/lead.zip*

Most of the lead used for the production of soda ash lead glass in al-Andalus appears to derive from a localized source in the Province of Córdoba, where about 50% of 10th- to 11th-century glass assemblages from domestic contexts correspond to soda ash lead glass (Duckworth et al. 2015; De Juan Ares et al. 2021). Soda ash lead glass with the same isotopic signatures have been identified among assemblages from Albalat (Cáceres), Vascos near Toledo (De Juan Ares and Schibille 2017a), Silves in Portugal as well as Gauzón in the Christian kingdom of Asturias and León to the north (De Juan Ares et al. 2018). Pechina (Almería) appears to be an exception, inasmuch as different lead sources were undoubtedly used. With increasing distance from Córdoba, soda ash lead glass becomes more sporadic. Only a single sample has thus far been found outside the Iberian Peninsula, in 10th- to 12th-century Sicily (unpublished data). The prevalence of this type of glass in the immediate vicinity of Córdoba reinforces the argument for a local primary glassmaking center, specializing in the manufacture of soda ash lead glasses in the caliphal capital. These glasses may be a direct descendant of the high lead glass recovered from the suburb of Šaqunda, and they may have been the result of some experimentation with waste products from lead and silver mining in the region.

The development of a local glassmaking tradition based on the use of lead may have been the direct result of dwindling glass imports from the eastern Mediterranean, from where the Iberian Peninsula had drawn its supplies throughout the Roman and late antique periods (De Juan Ares and Schibille 2017b; De Juan Ares et al. 2019). The re-introduction of glazed ceramics to Iberia in the second half of the 9th century (Molera et al. 2018) is probably related to this new glassmaking technology in terms of both the raw materials used as well as the production technologies (Salinas et al. 2019, 2021). Archaeological evidence from the 9th- and 10th-century ceramic workshops at Pechina (Almería) and San Nicolás (Murcia) supports the close link between the manufacture of glass and glazes (Molera et al. 2009, 2018; Salinas et al. 2019, 2021). There is no conclusive evidence for any technological transfer from the eastern Mediterranean to al-Andalus. On the contrary, the technological progress appears to have been related to the increasing independence of the Umayyad emirate, and the establishment of the Umayyad caliphate of al-Andalus in 929 CE.

8.5. Concluding remarks

The eastern Mediterranean lead silica glass is predominantly of an emerald green color and belongs without a doubt to the category of specialized glass reserved for more valuable items, even though it was also used for undecorated, more mundane objects. In 10th-century al-Andalus, soda ash lead glass becomes the main glass type used in domestic and palatial contexts alike. It was accordingly used for the entire range of vitreous objects from luxury items to lamps and window panes, and it is typically colorless or only weakly tinted (Figure 8.8).

Compositional and isotopic data provide ample evidence for the diversity of early Islamic high lead glasses, reflecting the use of different raw materials, different purification processes and different production locations. Unfortunately, the lead isotope signatures of the eastern lead silica glasses have not yet been unambiguously assigned to a geological lead source, making it difficult to be sure where exactly such glasses were made. Nevertheless, it is very likely that at least some lead silica glasses originated in Mesopotamia, and the etymology of the term *mīnā* suggests that the technique may have been transferred from east to west. In contrast, the lead glass in al-Andalus can be considered an independent innovation that evolved from related metallurgical technologies, in particular lead and silver mining processes in which a glassy slag is one of the waste products. There is also a close material and technological link between lead glass and translucent lead glazes in al-Andalus. The earliest exemplar is the glass from Šaqunda that may simply represent re-melted and shaped lead slag. It marks the incipient stage of experimentation and emancipation of a local glass industry in Islamic Spain. Subsequently, slag was replaced by litharge, and soda-rich plant ash was added to perhaps improve the workability of the vitreous material. Temperatures needed for the manufacture of high lead glass are significantly lower than for the production of traditional alkali silica glass, and it may have been for this reason that lead glass was the first primary glass produced in Iberia.

As to the silica and lead raw materials, interesting variations between the high lead glass groups are apparent. The eastern lead silica glasses were made with a mixture of quartz rich silica, in some cases possibly even quartz pebbles, and lead oxide with elevated silver. The potential use of quartz is interesting in light of Arabic alchemical and technological sources that all specify the use of pebbles instead of sand as raw material for the manufacture of lead silica glass (*mīnā).* It must nonetheless be stressed that in spite of these written sources some of the eastern lead silica glasses as well as Andalusian soda ash lead glasses were certainly produced from a quartz-rich sand, reflecting different glassmaking recipes.

8.6. References

Aga-Oglu, M. (1946). The origin of the term mīnā and its meanings. *J. Near East. Stud.*, 5, 241–256.

Al-Hassan, A.Y., Ahmed, M., Iskandar, A.Z. (eds) (2001). *The Different Aspects of Islamic Culture: Science and Technology in Islam. Vol. 4 – Science and Technology in Islam.* Unesco Publishing, Beirut.

Barceló, M. (1975). El hiato en las acuñaciones de oro en al-Andalus, 127-316/744 (5)-936 (7). *Moneda y crédito*, 33(132), 33.

Barnes, I.L., Brill, R.H., Deal, E., Piercy, G. (1986). Lead isotope studies of some of the finds from the Serce Liman shipwreck. In *Proceedings of the 24th International Archaeometry Symposium*, Olin, J.S., Blackman, M.J. (eds). Smithsonian Institution Press, Washington.

Bazargani-Guilani, K., Nekouvaght Tak, M., Faramarzi, M. (2011). Pb–Zn deposits in Cretaceous carbonate host rocks, northeast Shahmirzad, central Alborz, Iran. *Aust. J. Earth Sci.,* 58, 297–307.

Bejaoui, J., Bouhlel, S., Cardellach, E., Canals, À., Perona, J., Piqué, À. (2013). Mineralization and fluid inclusion studies of the Aptian carbonate-hosted PbZnBa ore deposits at Jebel Hamra, Central Tunisia. *J. Geochem. Explor.*, 128, 136–146.

Berthelot, M. and Duval, R. (1893). *La chimie au Moyen Age.* Imprimerie Nationale, Paris.

Brepohl, E. (2013). *Theophilus Presbyter und das mittelalterliche Kunsthandwerk.* Bölau Verlag, Cologne, Weimar, Vienna.

Brill, R.H. (1989). Thoughts on the glass of central Asia with analyses of some glasses from Afghanistan. In *Proceedings of the XV International Congress on Glass, Leningrad, 1989, Archaeometry*, Mazurin, O.V. (ed.). Rotoprint VNIIZS, Moscow.

Brill, R.H. (1999). *Chemical Analyses of Early Glasses.* Corning Museum of Glass, Corning.

Brill, R.H. and Stapleton, C.P. (2012). *Chemical Analyses of Early Glasses, Volume 3.* Corning Museum of Glass, Corning.

Casal García, M.T. (2018). The Rabad of Saqunda in Umayyad Córdoba (750–818 AD). In *Entre civitas y madīna: el mundo de las ciudades en la Península Ibérica y en el norte de África (siglos IV–IX),* Callegarin, L., Panzram, S. (eds). Casa de Velázquez, Madrid.

De Juan Ares, J. and Schibille, N. (2017a). Glass import and production in Hispania during the early medieval period: The glass from Ciudad de Vascos (Toledo). *PLoS One,* 12, e0182129.

De Juan Ares, J. and Schibille, N. (2017b). La Hispania antigua y medieval a través del vidrio: La aportación de la arqueometría. *Bol. Soc. Esp. Ceram. V.*, 56, 195–204.

De Juan Ares, J. and Schibille, N. (forthcoming). El taller de Pechina y el inicio de la producción de vidrio en Al-Andalus. *El alfar de cerámica vidriada de Pechina (siglo IX).* Universida de Jaén.

De Juan Ares, J., Fernández Calderón, N., Muñiz López, I., García Álvarez-Busto, A., Schibille, N. (2018). Islamic soda-ash glasses in the Christian kingdoms of Asturias and León (Spain). *J. Archaeol. Sci.: Reports*, 22, 257–263.

De Juan Ares, J., Guirado, A.V.-E., Gutiérrez, Y.C., Schibille, N. (2019). Changes in the supply of eastern Mediterranean glasses to Visigothic Spain. *J. Archaeol. Sci.*, 107, 23–31.

De Juan Ares, J., Cáceres Gutiérrez, Y., Moreno Almenarar, M., Schibille, N. (2021). Composition and origins of decorated glass from Umayyad Cordoba (Spain). *Heritage Science*, 9.

Duckworth, C.N., Córdoba De La Llave, R., Faber, E.W., Govantes Edwards, D.J., Henderson, J. (2015). Electron microprobe analysis of 9th–12th century Islamic glass from Córdoba, Spain. *Archaeometry*, 57, 27–50.

Farquhar, R.M. and Vitali, V. (1989). Lead isotope measurements and their application to Roman lead and bronze artifacts from Carthage. *MASCA Research Papers in Science and Archaeology*, 6, 39–45.

Fazli, S., Taghipour, B., Moore, F., Lentz, D.R. (2019). Fluid inclusions, S isotopes, and Pb isotopes characteristics of the Kuh-e-Surmeh carbonate-hosted Zn–Pb deposit in the Zagros Fold Belt, southwest Iran: Implications for the source of metals and sulfur and MVT genetic model. *Ore Geology Reviews*, 109, 615–629.

Foy, D. (2020). *Le verre de Sabra al-Mansuriya (Kairouan, Tunisie) milieu Xe-milieu XIe siècle : Production et consommation vaisselle – contenants – vitrages. Archaeology of the Maghreb Vol. 1.* Archaeopress Publishing Ltd., Oxford.

Gale, N., Spooner, E., Potts, P. (1981). The lead and strontium isotope geochemistry of metalliferous sediments associated with Upper Cretaceous ophiolitic rocks in Cyprus, Syria, and the Sultanate of Oman. *Can. J. Earth Sci.*, 18, 1290–1302.

Gan, F. (2009). Origin and evolution of ancient Chinese glass. In *Ancient Glass Research Along the Silk Road*, Gan, F., Brill, R.H., Tian, S. (eds). World Scientific Publishing, Singapore.

Goldenberg, G., Otto, J., Steuer, H. (1996). *Archäometallurgische Untersuchungen zum Metallhüttenwesen im Schwarzwald.* Thorbecke Jan Verlag, Sigmaringen.

Gratuze, B., Foy, D., Lancelot, J., Téreygeol, F. (2003). Les "lissoirs" carolingiens en verre au plomb : mise en évidence de la valorisation des scories issues du traitement des galènes argentifères de Melle (Deux-Sèvres). In *Échanges et commerce du verre dans le monde antique (Actes du colloque de l'AFAV, Aix-en-Provence et Marseille, 7–9 juin 2001)*, Foy, D., Nenna, M.-D. (eds). Éditions Monique Mergoil, Montagnac.

Gratuze, B., Guerrot, C., Foy, D., Bayley, J., Arles, A., Téreygeol, F. (2014). Melle : mise en évidence de l'utilisation des scories vitreuses issues de la chaîne opératoire de production de l'argent comme matière première de l'industrie verrière. In *Du monde franc aux califats omeyyade et abbasside : extraction et produits des mines d'argent de Melle et de Jabali.* Deutschen Bergbau-Museums, Bochum.

Gratuze, B., Guerrot, C., Foy, D., Arles, A., Téreygeol, F. (2017). Les galets de verre au plomb carolingiens issus des scories de Melle : élaboration et distribution. In *Mine, métal, monnaie, Melle – Les voies de la quantification de l'histoire monétaire du haut Moyen Age.* Droz, Paris.

Hassan, A. and Hassan, F. (1981). Source of galena in predynastic Egypt at Nagada. *Archaeometry*, 23, 77–82.

Henderson, J., McLoughlin, S.D., McPhail, D.S. (2004). Radical changes in Islamic glass technology: Evidence for conservatism and experimentation with new glass recipes from early and middle Islamic Raqqa, Syria. *Archaeometry*, 46, 439–468.

Henderson, J., An, J., Ma, H. (2018). The archaeometry and archaeology of ancient Chinese glass: A review. *Archaeometry*, 60(1), 88–104.

Holakooei, P. (2016). A Mediaeval Persian treatise on coloured and enamelled glass: Bayan Al-Sana'at. *Iran*, 54, 95–106.

Jemmali, N. and Souissi, F. (2018). Lead isotopes as tracers of metal sources and timing of the carbonate-hosted Pb-Zn deposits in the Nappes Zone, Northern Tunisia. In *Contributions to Mineralization*, Al-Juboury, A.I. (ed.). IntechOpen, London.

Jemmali, N., Souissi, F., Villa, I.M., Vennemann, T.W. (2011). Ore genesis of Pb–Zn deposits in the Nappe zone of Northern Tunisia: Constraints from Pb–S–C–O isotopic systems. *Ore Geology Reviews*, 40, 41–53.

Jemmali, N., Souissi, F., Carranza, E.J.M., Bouabdellah, M. (2013). Lead and sulfur isotope constraints on the genesis of the polymetallic mineralization at Oued Maden, Jebel Hallouf and Fedj Hassene carbonate-hosted Pb–Zn (As–Cu–Hg–Sb) deposits, Northern Tunisia. *J. Geochem. Explor.*, 132, 6–14.

Kahle-Bonn, P. (1936). Bergkristall, Glas und Glasflüsse nach dem Steinbuch von el-Bērūnī. *Zeitschrift der deutschen morgenländischen Gesellschaft*, 90, 322–356.

Krueger, I. (2014). Islamisches Bleiglas und mīnā. *J. Glass Stud.*, 56, 61–84.

L'Héritier, M., Baron, S., Cassayre, L., Téreygeol, F. (2015). Bismuth behaviour during ancient processes of silver–lead production. *J. Archaeol. Sci.*, 57, 56–68.

Lamm, C.J. (1928). *Das Glas von Samarra.* Verlag von Dietrich Reimer/Ernst Vohsen, Berlin.

Liu, S., Li, Q., Gan, F., Zhang, P., Lankton, J. (2012). Silk Road glass in Xinjiang, China: Chemical compositional analysis and interpretation using a high-resolution portable XRF spectrometer. *J. Archaeol. Sci.*, 39, 2128–2142.

Marzo, P., Laborda, F., Pérezarantegui, J. (2009). Medieval and postmedieval Hispano-Moresque glazed ceramics: New possibilities of characterization by means of lead isotope ratio determination by Quadrupole ICP-MS. In *Isotopes in Vitreous Materials*, Degryse, P., Henderson, J., Hodgins, G. (eds). Leuven University Press.

Mason, R.B., Farquhar, R.M., Smith, P.E. (1991). Lead-isotope analysis of Islamic glazes: An exploratory study. *Muqarnas Online*, 9, 67–71.

Mass, J.L., Wypyski, M.T., Stone, R.E. (2002). Malkata and Lisht glassmaking technologies: Towards a specific link between second millennium BC metallurgists and glassmakers. *Archaeometry*, 44, 67–82.

Mirnejad, H., Simonetti, A., Molasalehi, F. (2011). Pb isotopic compositions of some Zn–Pb deposits and occurrences from Urumieh–Dokhtar and Sanandaj–Sirjan zones in Iran. *Ore Geology Reviews*, 39, 181–187.

Mirnejad, H., Simonetti, A., Molasalehi, F. (2015). Origin and formational history of some Pb-Zn deposits from Alborz and Central Iran: Pb isotope constraints. *International Geology Review*, 57, 463–471.

Molera, J., Pradell, T., Salvadó, N., Vęndrell-Saz, M. (2009). Lead frits in Islamic and Hispano-Moresque glazed productions. In *From Mine to Microscope, Advances in the Study of Ancient Technology*, Shortland, A., Freestone, I.C., Rehren, T. (eds). Oxbow Books, Oxford.

Molera, J., López, J.C.C., Molina, G., Pradell, T. (2018). Glazes, colourants and decorations in early Islamic glazed ceramics from the Vega of Granada (9th to 12th centuries CE). *J. Archaeol. Sci.: Reports*, 21, 1141–1151.

Momenzadeh, M. (2004). Metallic mineral resources of Iran, mined in ancient times: A brief review. In *Persian Antiques Splendor, Mining Crafts and Metalurgy in Ancient Iran*, Stollner, T., Slotta, R., Vatandoust, A. (eds). Deutsches Bergbau-Museum Bochum.

Murillo-Barroso, M., Montero-Ruiz, I., Nieto, J.M., Massieu, M.C., Socas, D.M., Martinón-Torres, M. (2019). Trace elements and lead isotopic composition of copper deposits from the eastern part of the Internal Zone of the Betic Cordillera (SE Iberia): Application to provenance of archaeological materials. *J. Iber. Geol.*, 45, 1–24.

Nezafati, N., Momenzadeh, M., Pernicka, E. (2008). New insights into the ancient mining and metallurgical researches in Iran. In *Ancient Mining in Turkey and the Eastern Mediterranean*, Yalçin, Ü., Özbal, H., Paşamehmetoğlu, A.G. (eds). Atılım University, Ankara.

Pollak, R. (2017). The early Islamic green lead glass from the excavations at Caesarea Maritima, Israel. In *Annales Fribourg/Romont 7–11 septembre 2015 du 20e Congrès de l'Association Internationale pour l'Histoire du Verre*, 288–292.

Porter, Y. (2004). Le quatrième chapitre du Javaher-Name-Ye Nezami. In *Sciences, techniques et instruments dans le monde iranien (Xe–XIXe siècle)*, Pourjavady, N., Vesel, Z. (eds). Presse Universitaires d'Iran, Institut Français de Recherche en Iran, Tehran.

Ramage, A., Craddock, P.T., Cowell, M.R. (2000). *King Croesus' Gold: Excavations at Sardis and the History of Gold Refining*. British Museum Press, London.

Saeidabadi, Z.K., Karimpour, M.H., Shafaroudi, A.M., Rahimi, B., Corfu, F. (2018). Origin of the Kaviro lead deposit in the Neyganan area, Lut Block, Eastern Iran: Constraints from geology, fluid inclusions, and isotope geochemistry. *J. Geochem. Explor.*, 192, 85–102.

Salinas, E., Pradell, T., Molera, J. (2019). Glaze production at an early Islamic workshop in al-Andalus. *Archaeol. Anthropol. Sci.*, 11, 2201–2213.

Salinas, E., De Juan Ares, J., Piñero, J.M., Casal, M.T., Schibille, N., Pradell, T. (2021). From glass to glaze in al-Andalus: Local invention and technological transfer. *European Journal of Archaeology*, 25, 22–41.

Sayre, E.V. and Smith, R.W. (1961). Compositional categories of ancient glass. *Science*, 133, 1824–1826.

Schibille, N., Meek, A., Wypyski, M.T., Kröger, J., Rosser-Owen, M., Haddon, R.W. (2018). The glass walls of Samarra (Iraq): Ninth-century Abbasid glass production and imports. *PLoS One*, 13, e0201749.

Schibille, N., De Juan Ares, J., Casal García, M.T., Guerrot C. (2020). Ex novo development of lead glassmaking in early Umayyad Spain, *Proc. Natl. Acad. Sci.*, 117(28), 16243–16249.

Shindo, Y. (2006). Lead glass found in the Räya and the monastery of Wâdï al-Tur sites. In *Archaeological Survey of the Räya/al-Tür Area on the Sinai Peninsula, Egypt, 2005 and 2006. Islamic Archaeology and Culture, Vol. 1.* Kawatoki, M. (ed.). Koweït.

Shortland, A.J. (2002). The use and origin of antimonate colorants in early Egyptian glass. *Archaeometry*, 44, 517–530.

Shortland, A.J. (2006). Application of lead isotope analysis to a wide range of Late Bronze Age Egyptian materials. *Archaeometry*, 48, 657–669.

Shortland, A., Nicholson, P., Jackson, C. (2000). Lead isotopic analysis of Eighteenth-Dynasty Egyptian eyepaints and lead antimonate colourants. *Archaeometry*, 42, 153–157.

Skaggs, S., Norman, N., Garrison, E., Coleman, D., Bouhlel, S. (2012). Local mining or lead importation in the Roman province of Africa Proconsularis? Lead isotope analysis of curse tablets from Roman Carthage, Tunisia. *J. Archaeol. Sci.*, 39, 970–983.

Stacey, J.S., Doe, B.R., Roberts, R.J., Delevaux, M.H., Gramlich, J.W. (1980). A lead isotope study of mineralization in the Saudi Arabian Shield. *Contributions to Mineralogy and Petrology*, 74, 175–188.

Stos-Gale, S. (2004). Lead-isotope analyses of glass, glazes, and some metal artifacts. In *Serçe Limani: An Eleventh-Century Shipwreck. Volume I: The Ship and Its Anchorage, Crew, and Passengers*, Bass, G.F., Matthews, S.D., Steffy, J.R., Van Doorninck, F.H. (eds.). Texas University Press, College Station.

Stos-Gale, Z.A. and Gale, N.H. (1981). Sources of galena, lead and silver in Predynastic Egypt. *ArchéoSciences-revue d'Archéométrie*, 1, 285–296.

Trincherini, P.R., Domergue, C., Manteca, I., Nesta, A., Quarati, P. (2009). The identification of lead ingots from the Roman mines of Cartagena: The role of lead isotope analysis. *Journal of Roman Archaeology*, 22, 123–145.

Turner, W.E.S. and Rooksby, H.P. (1959). A study of the opalising agents in ancient opal glasses throughout three thousand four hundred years. *Glastechnische Berichte, Sonderband*, 32K, 17–28.

Wagner, G., Pernicka, E., Seeliger, T.C., Lorenz, I., Begemann, F. (1986). Geochemische und isotopische charakteristika früher rohstoffquellen für kupfer, blei, silber und gold in der Türkei. *Jahrbuch des Römisch-Germanischen Zentralmuseums Mainz*, 33, 723–752.

Wertime, T.A. (1968). A metallurgical expedition through the Persian desert. *Science*, 159, 927–935.

Wolf, S., Stos, S., Mason, R., Tite, M.S. (2003). Lead isotope analyses of Islamic pottery glazes from Fustat, Egypt. *Archaeometry*, 45, 405–420.

Wypyski, M.T. (2015). Chemical analysis of early Islamic glass from Nishapur. *J. Glass Stud.*, 57, 121–136.

Yener, K.A., Sayre, E., Joel, E., Özbal, H., Barnes, I., Brill, R. (1991). Stable lead isotope studies of Central Taurus ore sources and related artifacts from Eastern Mediterranean Chalcolithic and Bronze Age sites. *J. Archaeol. Sci.*, 18, 541–577.

9

Lead in the Enamels of the Middle Ages and Renaissance

Isabelle BIRON[1] and Marco VERITÀ[2]
[1]*C2RMF, IRCP, Paris, France*
[2]*LAMA-Iuav, University of Venice, Italy*

9.1. Limoges champlevé enamels on copper from the Middle Ages

The invention of the new technique of "champlevé enamels on copper" marked an exceptional flourishing throughout Roman Europe of the 12th century (Durand 2005). Cavities are carved in a copper plate by burin and chisel and delimited by partitions or cloisons. The cavities are filled with wet colored glass powders and the partitions form the drawing. Once the powders dry, they are molten in an enamel furnace. The surface of the objects is finally polished and the metallic parts are gilded. This technique appears nearly simultaneously in the first half of the 12th century, in Central and South-Western France, the valleys of Meuse and Rhine, Northern Spain (Monastery of Silos) and in Lower Saxony with the workshops of Brunswick and Hildesheim (Taburet-Delahaye and Boehm 1995, 1996; Durand 2005, p. 182).

The origin of Southern enameling essentially overlaps with that of the treasure of Conques dating from the end of the 11th century to the beginning of the 12th century (Taburet-Delahaye 1995, pp. 33–39). In terms of number of works, Limoges champlevé enamels on copper represent a very significant production, which lasted from this period until the mid-14th century.

Lead in Glassy Materials in Cultural Heritage,
coordinated by Anne BOUQUILLON and Patrice LEHUÉDÉ.

The production of Limoges enamels is tremendously creative and varied, offering for each period a surprising technical diversity for the metal work, which accompanies a stylistic evolution presented in detail in the catalogue of the exhibition[1] of these works which took place in Paris, and then in New York (Taburet-Delahaye and Boehm 1995, 1996). Two examples are presented in Figures 9.1 and 9.2.

Most of these objects of various sizes (from 8–10 cm to over 100 cm in height) and various forms are decorated with religious patterns, while some of them are decorated with fantastic animals or secular subjects. These are medallions, plaques, cross plaques, book covers decorating objects (coffret, book, altar, etc.), chasses, tabernacles, pyxes, incense boats, Eucharistic doves, reliquaries, chrismatories, candlesticks and tomb effigies.

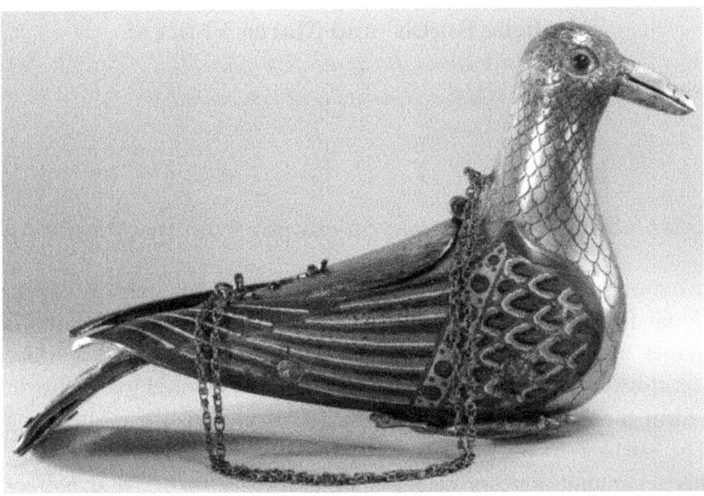

Figure 9.1. *Eucharistic dove, Limoges, 1215–1235, preserved at the Louvre museum (H: 14.7; L: 19.8; thickness: 6.9 cm) (inv. OA8104) (cat. 107). Formed, champlevé, engraved, enameled and gilded copper (© C2RMF/A.Maigret). For a color version of this figure, see www.iste.co.uk/bouquillon/lead.zip*

For further details on the history, techniques and materials employed for these enamels, please refer to the catalogue of the exhibition (Taburet-Delahaye and

1. Two versions of the catalogue have been published, one in French in 1995 and the other one in English in 1996.

Boehm 1995, 1996; Biron et al. 1995, 1996) and the recently published and most complete synthesis (Biron et al. 2015).

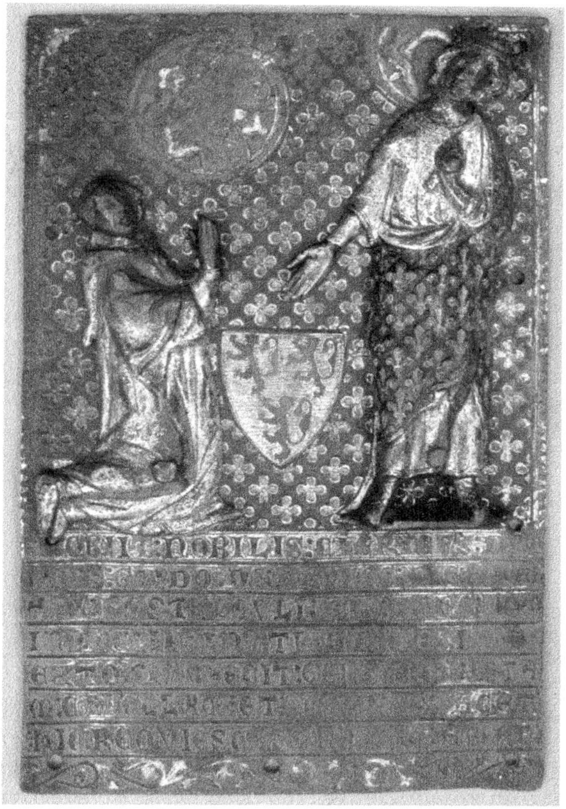

Figure 9.2. *Funerary plaque of Guy de Meyos, Limoges, 1307, preserved at the Louvre museum (H: 32; L: 20 cm) (inv. OA7495) (cat. 153). Champlevé, engraved, enameled, gilded and silvered copper (© C2RMF/D.Bagault). For a color version of this figure, see www.iste.co.uk/bouquillon/lead.zip*

9.1.1. *Enameling techniques*

Our results on the enameling techniques and on the chemical analysis of glasses employed to make these objects have shown a high consistency throughout the period of production from the late 11th century/early 12th century to the middle of the 14th century. While the exact location and organization of Limoges enamel workshops are not known (Taburet-Delahaye 1995, p. 36), our results support the idea of centralized and controlled production in large centers or possibly in small structures brought together by know-how, supply sources and common glassmaking practices.

9.1.2. Composition of enamels and lead content

Limoges enamellers mixed powders of opaque and translucent glasses conferring to enamels a very specific heterogeneous texture (see Figure 9.3).

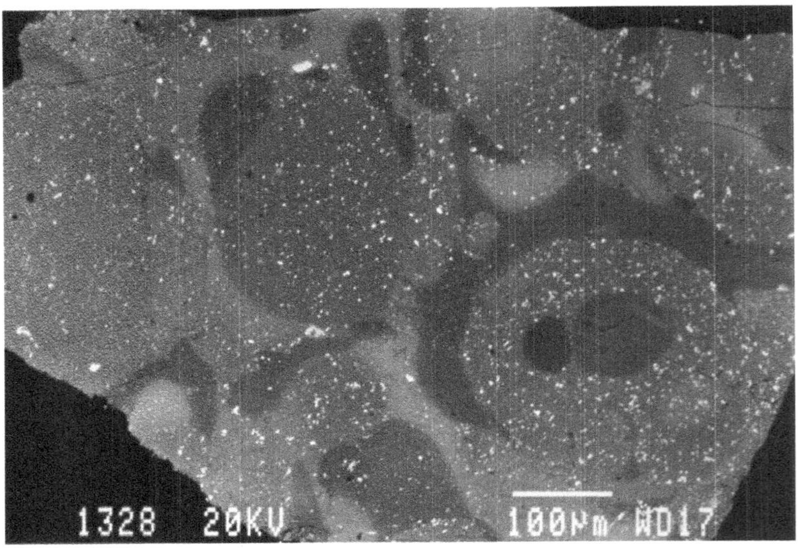

Figure 9.3. *Scanning electron micrograph in backscattered electrons (BSE) of a Limoges champlevé enamel on copper. The opaque glass grains containing opacifier crystals (white contrast) appear light gray. The dark gray areas correspond to translucent glass grains – sample of a Limoges champlevé enamel on copper (© C2RMF/I.Biron)*

They employed two types of main compositions of soda-lime-silica glass[2] that feature a chronological evolution in relation to one another (see Table 9.1):

– *Type 1 composition* in use from the end of the 11th century/beginning of the 12th century until the end of the 12th century/beginning of the 13th century. The

2. Four other compositions of potassium silicates, potash-lime-silicates, soda-lime-silicates and mixed alkali silicates have also been used in various periods, but they seem marginal, representing only 1–2% of our analyses.

sodium sources are of mineral origin and opacifiers are antimony based. Lead contents are very low for white and turquoise glasses (PbO < 1%) and low for red glass (<2%) and blue glass (<3%) (see Table 9.2). The presence of lead in these glasses is very likely involuntary but rather associated with raw materials, as the sources of cobalt for certain blue colors (Biron et al. 2015, pp. 214–215). On the other hand, the higher lead contents of yellow glasses (6–12%) and green glasses (1–7%) correspond to a deliberate addition by means of opacifiers and coloring particles of lead antimonates (see Table 9.2). White, blue, turquoise and partly green glasses are opacified with calcium antimonates. Their fabrication technology was the object of detailed studies (Moretti and Hreglich 1984; Lahlil et al. 2011).

– *Type 2 compositions* used from the end of the 12th century/beginning of the 13th century to the middle of the 14th century. The sodium sources are vegetal soda ash and opacifiers are tin based. These enamels contain higher lead levels that may reach 34% (see Table 9.2). The highest lead contents can be found, roughly at the same level, in white glass (PbO 13.6–33.7%), blue glass (mainly 10–28% and 1–3% for four enamels) and turquoise (12–27.4%). All these colors are with the highest opacifier content (maximum for the white enamels). Tin oxide crystals were used to color and opacify white glass, as well as to opacify blue, turquoise and green glasses, while lightening their shades. The presence of lead in glass is strictly related to the methods used to synthesize the tin oxide crystals (white opaque glass) and the lead stannate crystals (yellow opaque glass).

	Type 1	Type 2
Na_2O	10–25	7–20
MgO	0.2–1.0*	1.2–5.0
Al_2O_3	1.5–3.0	0.5–4.0
P_2O_5	<0.3	>0.3
SiO_2	55–70	39–70
K_2O	0.6–2.5	1.1–2.5
CaO	5–11**	3–8.6
SnO_2	<0.6	5.2–12
Sb_2O_3	0.5–8	<0.5
PbO	0.5–11	5–22§

*Several rare cases 1–1.5%; **several rare cases 13%, §several rare cases 1–5%.

Table 9.1. *Results of the analyses of type 1 and 2 compositions of Limoges enamels of the Middle Ages, summarized as ranges in oxide mass percentage (results of C2RMF laboratories – analyses by ion beam using PIXE-PIGE methods – and of MET – analyses by SEM-EDX) (C2RMF, I. Biron)*

Red glasses are colored and opacified by cuprite crystals (Cu_2O) or metallic copper. Their composition varies widely and their lead concentration indicates manufacturing methods or copper sources that differ between type 1 glasses (PbO 0.24–1.9%) and type 2 glasses (0.1–13.2%) (see Table 9.2).

Color	Type 1 PbO %	Type 2 PbO %
White	0–0.8	13.6–33.7
Blue	0.1–2.6	1.0–28
Turquoise	0.3–0.8	12–27.4
Yellow	6.3–11.9	5.7
Green	0-6.8	2–18.4
Red	0.2–1.9	0.1–13.2

Table 9.2. *Minimum and maximum contents of lead (oxide mass percentage) in Limoges champlevés enamels on copper from the Middle Ages analyzed at C2RMF by color and for the two main types of compositions (C2RMF, I. Biron)*

9.1.3. *Recipes employed for type 2 enamels*

The high lead content in the opaque glasses of type 2 (see Table 9.2) proves to be particularly useful, as it facilitates the dispersion of tin oxide crystals in the vitreous matrix by reducing the viscosity.

9.2. The so-called Venetian enameled coppers of the Italian Renaissance

Enameled coppers of the Italian Renaissance are a limited but very sophisticated class of products, whose origin is traditionally attributed to Venice. While the art historians agree that this technique originates in Northern Italy, they diverge with respect to the exact location. The production sites as well as the organization of work (copperware, goldsmith's objects and glassmaking) are in fact difficult to identify given the lack of precise notes on enameled coppers in the ancient texts (Barbe et al. 2018a, pp. 10–15).

These objects are part of a dinnerware composed of bowls, which are sometimes covered, plates, dishes, saltcellars, ewers and flasks (see Figures 9.4 and 9.5). There are also other types of objects, with caskets, torches, candlesticks, peace kisses, oil cans, reliquaries or monstrances.

For further details on the history of these objects, their manufacturing techniques (copper substrate, enameling and gilding), materials employed (copper, glasses and gold leaves) and more generally on the research conducted on these productions, please refer to some previous studies (Barbe et al. 2018a; Cagnini et al. 2018; Verità et al. 2018).

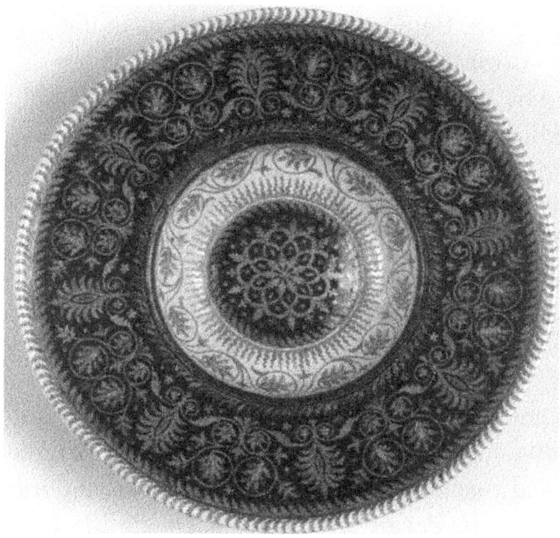

Figure 9.4. *Deep plate, Paris, Louvre museum, inv. R283 (corpus 6 in Barbe et al. (2018b)) (© C2RMF/A.Maigret)*

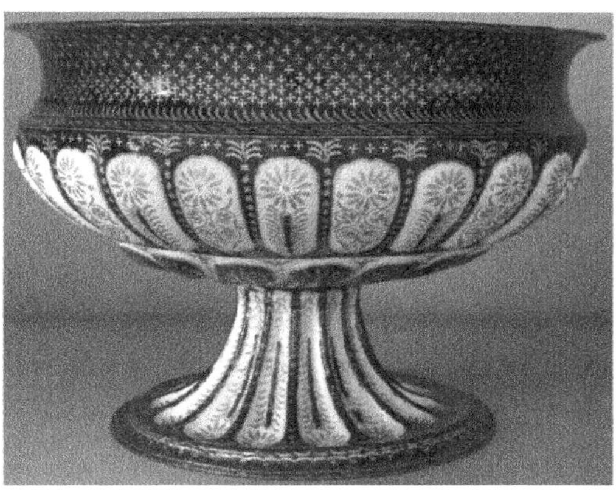

Figure 9.5. *Deep bowl with high foot. Paris, Louvre museum, inv. N1221 (corpus 164 in Barbe et al. (2018b)) (© C2RMF/A.Maigret)*

9.2.1. *Manufacturing techniques*

In parallel with a typological and stylistic study, an in-depth study of metal and enamels technology was conducted by the detailed examination of objects (with the naked eye and under high magnification) and X-ray imaging. These works highlight the technical specificities in the manufacturing and set-up of various copper elements (Biron et al. 2018; Beillard 2018a), as well as in the practices of workshops related to enamel art (Biron 2018; Cagnini et al. 2018). These objects are made from copper plates formed by hammering, then mechanically assembled by "tubular ties", setups sewn by metallic wires and specific flaps.

As for the enameling art, one of the specificities of this production is the use of a preparation layer of translucent colorless glass containing a high density of air bubbles, as well as grains of opaque white glass in variable amounts. This layer is directly deposited on the copper substrate; it is then covered by various layers of blue, or green, or purple translucent glasses and opaque white glasses forming the pattern (Biron 2018, p. 196). This preparation layer limits the mechanical stresses between the copper and the various layers of glass by increasing the total thickness of the deposited glass; furthermore, it protects the blue or green translucent glass layers against chemical reactions with the copper substrate during firing and finally contributes to the visual rendering of colors. Rich gilding made with gold leaves deposited with iron tools differentiate this production from other enameled coppers (Biron 2018; Beillard 2018b).

9.2.2. *Chemical composition of the base glass used for these enamels*

The base glass[3] of colored enamels is soda-lime-silicates, whose sodium sources are soda plant ashes. Four compositional groups (A, B, C and D) have been identified depending on the nature of silica and sodium sources. The base glass of group A comply with the Venetian glass composition of the Renaissance, while those of type B, C and D show great similarities with compositional groups of Tuscan glasses of the Renaissance (Cagno et al. 2008, 2010, 2012; Fenzi et al. 2009; Verità et al. 2018).

The use of mainly Tuscan glasses for all these objects and of Venetian glasses for only a part of them proves the non-Venetian character of this production –

3. The base glass is made by melting a silica source and alkaline sources, to which the glassmaker adds coloring agents and opacifiers, if needed. It is therefore calculated by subtracting from the total composition the content of colorants and opacifiers and then normalizing to 100%.

Venice being at that time the largest glassmaking center manufacturing all sorts of glass of various qualities.

9.2.3. *Lead glasses*

Among the enamels decorating the surface of these objects, only the opaque white and the turquoise enamels contain lead in concentrations above 15% weight for about half of them. These are soda–lime–silica glasses containing high but variable amounts of lead (white: PbO 3–30% – turquoise: 8–29%) and tin (white: 8–26% – turquoise: 9–14%) (see Table 9.3). Blue, green, purple and red translucent enamels have very low lead contents (<0.7%, and between 1 and 1.5% for three of them).

White glasses are a significant proportion of the enamels employed to decorate these objects, while the use of turquoise glasses is limited to ornamental touches applied on the top of gold leaves (see Figures 9.4–9.6).

White glasses are colored and opacified by cassiterite crystals dispersed in a colorless glass matrix. The turquoise enamels are opacified by cassiterite crystals dispersed in a glass matrix colored with copper (CuO 2–3%).

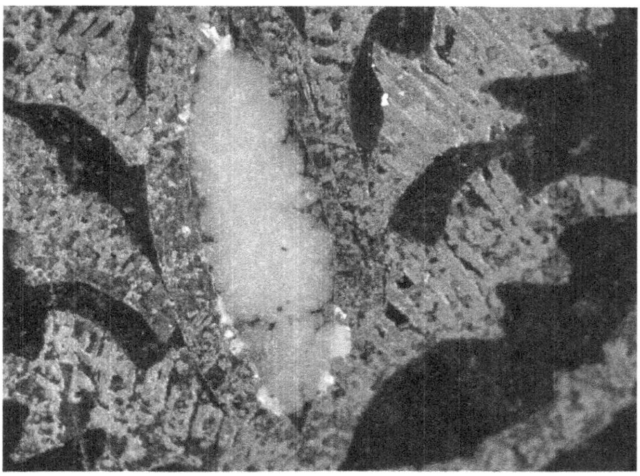

Figure 9.6. *Detail of an opaque turquoise enamel of the deep plate, inv. R283 (Figure 9.4), deposited on gold leaf, which is applied on translucent blue glass. Clear dots represent the distribution of white crystals of tin oxide in the turquoise glass (© C2RMF/I.Biron). For a color version of this figure, see www.iste.co.uk/bouquillon/lead.zip*

Number	C	M	G	Period	Na_2O	MgO	Al_2O_3	SiO_2	P_2O_5	Cl	K_2O	CaO	TiO_2	MnO	Fe_2O_3	CuO	BaO	PbO	SnO_2	CoO
R 254	B	L	D	XVI	12.8	3.31	4.21	51.4	0.15		5.30	5.62	0.11	0.26	0.35	0.04		2.71	13.7	
N 1221	B	L	B	XV	14.2	1.85	3.10	49.0	0.36		3.24	4.86	0.11	0.13	0.38	0.05	0.099	6.71	15.8	
MR 2448	B	L	D	??	14.8	1.77	2.55	35.8	0.18		1.05	8.20	0.12	0.80	0.47	0.02	0.065	9.5	24.5	
4470	B	CP	B	XV	12.5	2.	2.70	51.3	0.15	0.80	4.70	5.90	0.10	0.80	0.55	0.10		10.0	8.40	
R 252	B	C	D	XVI	13.7	1.48	2.90	42.2	0.13		3.30	3.90	0.09	0.51	0.58	0.11	0.045	11.0	20.0	
OA 9458	B	L	B	XV	14.4	1.53	3.59	46.0	0.16		2.09	4.20	0.09	0.36	0.53	0.83	0.055	11.5	14.5	
FGC 40053	B	V	A	XV	15.7	0.80	0.70	51.7		0.80	1.95	2.90		0.37	0.38			12.4	12.3	
FGC 40058	B2	V	D	XV	14.3	2.20	4.40	46.0		0.30	3.15	6.41	0.10	0.29	0.68	0.20		12.5	9.5	
FGC 40058	B1	V	D	XV	14.8	1.60	3.70	44.6		1.10	2.40	4.00	0.07	0.43	0.49			13.4	13.4	
655	B	M	D	XV	11.8	1.76	2.59	41.2	0.42	1.19	2.40	7.00		0.54	0.45	0.63		14.8	15.2	
	B	A	B	XV	10.4	1.60	3.40	46.0	0.53	0.85	2.50	6.00	0.	0.38	0.35			15.5	12.3	
OA 2631	B	L	B	XV	13.0	2.10	3.10	44.2	0.19		3.28	4.56	0.07	0.27	0.31	0.020	0.034	15.7	13.1	
R 258	B	L	B	XV	16.5	1.70	3.50	39.3	0.50		1.95	4.32	0.11	0.57	0.45	0.038	0.053	16.5	14.5	
N 1223	B	C	B	XV	15.5	1.70	2.85	39.3	0.35		2.50	5.80	0.08	0.32	0.50	0.05	0.033	16.5	14.5	
R 248	B	L	D	XVI	13.2	1.52	1.36	27.8	0.31			8.80	0.06	0.37	0.55	0.59	0.049	19.1	26.3	
FGC 40051	B	V	C	XVI	11.0	1.63	0.70	39.4		0.35	2.2	3.13	0.10	0.10	0.37			22.5	18.5	
2G	B	F	D	XVI	12.7	2.25	2.80	35.0	0.10	0.67	3.00	5.30		0.80	0.50			25.0	11.9	
1184	B	F	B	XV	8.2	1.35	0.56	37.9		0.56	1.66	2.81		0.10	0.25			30.0	16.7	
R 283	T	L	A	XV	17.7	0.31	0.92	51.9	0.07		2.34	1.58	0.02	0.01	0.28	2.09		8.31	14.3	
	T	A	A	XV	15.0	0.30	0.75	53.0	0.08	0.55	2.60	1.70			0.23	2.40		9.00	14.0	
CL 22603	T	C	B	XV	11.5	0.91	2.76	38.0	0.15		1.58	2.74	0.13	0.12	0.67	2.98	0.021	29.3	8.99	0.011

Table 9.3. *Results of analyses (in weight percent of the oxides) of opaque white (B) and turquoise (T) enamels (column C) of enamelled coppers from French and Italian museums (column M). L = Louvre, CP = Civici Pesaro, C = Cluny, V = Palazzo Cini, Venice, M = Bagatti Valsecchi, Milan, A = Amalfi cathedral, F = Bargello, Florence. The symbols (column G) correspond to the compositional groups A, B, C and D. XV and XVI correspond, respectively, to the manufacturing period before 1520-1530 and afterwards (M.Verità and I.Biron)*

9.2.4. *Lead origin: manufacturing a lead-tin calx powder*

During the elaboration of opaque white and turquoise glasses, lead is not added directly to the batch as main raw material, but it is a component of the opacifier. Among the three modes of opacification known to this day (*in situ* crystallization, addition of crystals and addition of *corpo*) (Lahlil 2008), the opacifier crystals of these enamels were independently synthesized and then added to the batch or to the molten glass.

The presence of micrometric tin oxide crystals dispersed in a lead-rich matrix (see Figure 9.7) suggests the use of *lead-tin-calx* for white and turquoise enamels, whose fabrication is presented in detail in the Venetian recipes (see Chapter 5). These recall the medieval recipes (see the same chapter), but describe the raw materials and the process in far greater detail. Certain recipes recommend the use of the highest quality lead, particularly copper-free, in order to avoid any contamination of the final color. A recipe can be also found in a Tuscan treatise from the end of the 14th century (recipe 24 of the first Trattatello: *per fare il lattimo bianco*; lead to tin ratio 1:2) (Milanesi 1968). According to Luigi Zecchin, it is one of the most ancient recipes known to this day for making opaque white glass (Zecchin 1987, pp. 108–112).

The lead-tin calx powder was prepared in the same glassmaking workshops by melting in open air a mixture of metallic lead and tin at above 600°C. Laboratory experiments were performed to gain a better understanding of the mechanisms involved in these processes (Matin 2019). The Pb/Sn ratio of the mixture controls the reactions occurring during calcination:

a) $2Pb + O_2 \rightarrow 2PbO$;

b) $Sn + O_2 \rightarrow SnO_2$;

c) $SnO_2 + 2PbO \rightarrow Pb_2SnO_4$.

The color of the resulting powder varies with the Pb/Sn ratio. If the ratio is below 3.5 (stoichiometry required for the formation of Pb_2SnO_4), only reactions (a) and (b) occur (white powder), while if it is above 3.5, reactions (a)–(c) occur (yellow powder whose color is more or less intense depending on lead concentration) (Matin 2019).

The white powder separated at the surface (crystallized mixture of tin and lead oxides) was then added to the batch of a transparent soda–lime–silica glass. During melting, lead oxide dissolves in the glass, leaving tin oxide crystals (cassiterite) dispersed in the vitreous matrix. Pb/Sn ratios vary between 1/3 and 3/1 (Zecchin

1989). These variations were probably related to the enamel viscosity control. For example, a less viscous white glass containing more lead and less tin allowed the deposition of a thin enamel layer on the substrate. A more viscous white glass with less lead and more tin was better adapted for depositing at equal temperature the enamel in a thicker layer (in the form of droplets, for example, as on the Venetian enameled glasses of the Renaissance).

Lead and tin contents of these white and turquoise enamels prove to be heterogeneous. At the scale of the small surfaces analyzed (20 × 20 μm^2 – SEM), Pb/Sn ratios vary from 0.5 to about 1.5 and get close to 1 for larger surfaces (at least 500 × 500 μm^2 – PIXE-PIGE). These variations are produced by a heterogeneous distribution of micrometric crystals both isolated and agglomerated in clusters that can reach 50 μm (see Figure 9.7).

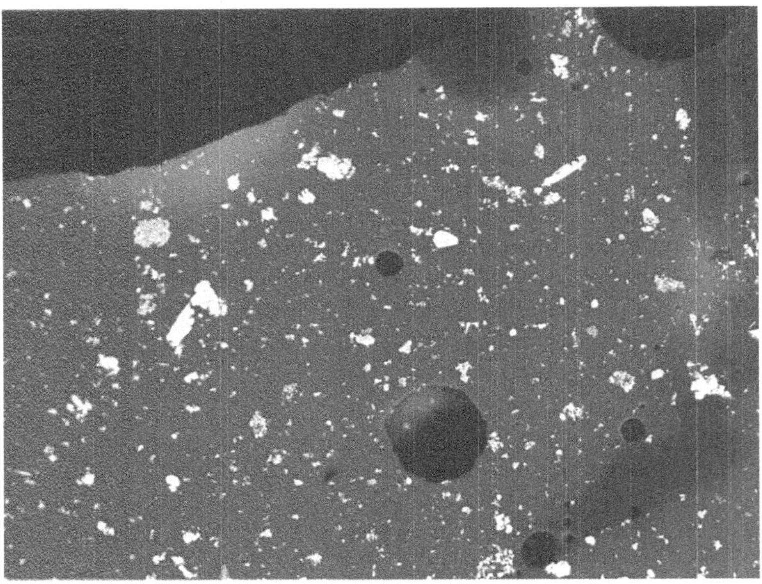

Figure 9.7. *Scanning electron micrograph in backscattered electrons (BSE) of the polished surface of a sample of opaque white enamel taken from the object, inv. B2G museum of Bargello in Florence. The white color particles are cassiterite crystals and crystal clusters (© M.Verità)*

As a conclusion, the two examples of enamels on metal of the objects presented in this chapter match well the information provided in the books of recipes. In most of the applications of the Middle Ages and of the Renaissance, lead is not added to the glass batch as a main component, but as a component of the manufactured opacifier (lead-tin calx) as described in the recipes for the opaque white glass. Furthermore, the lead to tin ratios reported in the books of recipes are almost identical with those observed in the analyses of opaque white enamels and glasses of those times in general.

9.3. References

Barbe, F., Biron, I., Caselli, L. (eds) (2018a). *Les cuivres émaillés dits vénitiens de la Renaissance italienne – Actes du colloque international d'études, volume I.* Silvana Editoriale, Milan.

Barbe, F., Caselli, L., Dantan, M.-E. (eds) (2018b). *Les cuivres émaillés dits vénitiens de la Renaissance italienne – Corpus des oeuvres en collections publiques et privées, volume II.* Silvana Editoriale, Milan.

Beillard, B. (2018a). Restauration du chandelier CL.22603 du musée de Cluny. In *Les cuivres émaillés dits vénitiens de la Renaissance italienne – Actes du colloque international d'études, volume I*, Barbe, F., Biron, I., Caselli, L. (eds). Silvana Editoriale, Milan.

Beillard, B. (2018b). Le décor à la feuille d'or des cuivres émaillés dits vénitiens : premier classement. In *Les cuivres émaillés dits vénitiens de la Renaissance italienne – Corpus des oeuvres en collections publiques et privées, volume II*, Barbe, F., Caselli, L., Dantan, M.-E. (eds). Silvana Editoriale, Milan.

Biron, I. (2015). Les émaux cloisonnés sur or byzantins du IXe au XIIe siècle. In *Les émaux sur métal d'Occident, du IXe au XIXe siècle, histoire, technique et matériaux*, Biron, I. (ed.). Faton, Dijon.

Biron, I. (2018). Les cuivres émaillés dits vénitiens : l'art de l'émaillerie et les dorures. In *Les cuivres émaillés dits vénitiens de la Renaissance italienne. Actes du colloque international d'études, volume I*, Barbe, F., Biron, I., Caselli, L. (eds). Silvana Editoriale, Milan.

Biron, I., Dandridge, P., Wypyski. M.T. (1995). Le cuivre et l'émail : technique et matériaux. In *L'œuvre de Limoges, émaux limousins du Moyen Age*, Taburet-Delahaye, E., Boehm, B. (eds). RMN, Paris.

Biron, I., Dandridge, P., Wypyski. M.T. (1996). Technique and materials in Limoges enamels. In *Enamels of Limoges 1100–1350*, Boehm, B., Taburet-Delahaye, E. (eds). The Metropolitan Museum of Art, New York.

Biron, I., Lambert, E., Marsac, J., Maigret, A. (2018). Les cuivres émaillés dits vénitiens : le travail du métal et l'assemblage des pièces. In *Les cuivres émaillés dits vénitiens de la Renaissance italienne, Actes du colloque international d'études, volume I*, Barbe, F., Biron, I., Caselli, L. (eds). Silvana Editoriale, Milan.

Cagnini, A., Porcinai, S., Verità, M., Biron, I. (2018). Il contributo delle indagini scientifiche alla conoscenza e alla conservazione. In *Les cuivres émaillés dits vénitiens de la Renaissance italienne, Actes du colloque international d'études, volume I*, Barbe, F., Biron, I., Caselli, L. (eds). Silvana Editoriale, Milan.

Cagno, S., Janssens, K., Mendera, M. (2008). Compositional analysis of Tuscan glass samples: In search of raw material fingerprints. *Anal. Bioanal. Chem.*, 391, 1389–1395.

Cagno, S., Mendera, M., Jeffries, T., Janssens, K. (2010). Raw materials for medieval to post-medieval Tuscan glassmaking: New insight from LA-ICP-MS analyses. *J. Archaeol. Sci.*, 37, 3030–3036.

Cagno, S., Favaretto, L., Mendera, M., Izmer, A., Vanhaecke, F., Janssens, K. (2012). Evidence of early medieval soda ash glass in the archaeological site of San Genesio (Tuscany). *J. Archaeol. Sci.*, 39, 1540–1552.

Durand, J. (2005). L'apogée de l'art roman, l'art roman dans le Saint-Empire. In *L'art roman au Louvre*, Gaborit, J.-R. (ed.). Fayard/Musée du Louvre, Paris.

Fenzi, F., Mendera, M., Vigato, P.A (2009). Technology of Mediaeval and Renaissance glass in Tuscany (Italy): An interdisciplinary approach to the study of production indicators. In *Annales du 17e Congrès de l'Association internationale pour l'histoire du verre (Antwerp)*, Janssens, K., Degryse, P., Cosyns, P., Caen, J., Van't Dack, L. (eds). University Press Antwerp, Antwerp.

Lahlil, S., Cotte, M., Biron, I., Szlachetko, J., Menguy, N., Susini, J. (2011). Synthesizing lead antimonate in ancient and modern opaque glass. *J. Anal. At. Spectrom.*, 26(5), 1040–1050.

Matin, M. (2019). Tin-based opacifiers in archaeological glass and ceramic glazes: A review and new perspectives. *Archaeol. Anthropol. Sci.*, 11, 1155–1167.

Milanesi, G. (1968). *Dell'arte del vetro per mosaico. Tre trattatelli dei secoli XIV e XV, ora per la prima volta pubblicati*. Romagnoli, Bologna.

Moretti, C. and Hreglich, S. (1984). Opacification and coloring of glass by the use of "anime". *Glass Technol.*, 25, 277–282.

Taburet-Delahaye, E. and Boehm, B. (1995). *L'œuvre de Limoges, émaux limousins du Moyen Âge*. RMN/Musée du Louvre, Paris.

Taburet-Delahaye, E. and Boehm, B. (1996). *Enamels of Limoges 1100–1350*. The Metropolitan Museum of Art, New York.

Verità, M., Biron, I., Cagnini, A., Porcinai, M., Barbe, F. (2018). The so-called Venetian enamelled copper artworks of the Italian Renaissance: The technology and provenance of the enamels – An analytical approach. *Archaeometry*, 60(2), 233–254.

Zecchin, L. (1987). *Vetro e vetrai di Murano*, volume 1. Arsenale, Venice.

Zecchin, L. (1989). *Vetro e vetrai di Murano*, volume 2. Arsenale, Venice.

PART 4

History, Implementation and Evolution of Lead Glazes

10

History of Lead in Ancient Ceramic Materials

Anne BOUQUILLON
C2RMF, IRCP, Paris, France

10.1. Introduction

The first glazes applied to small objects or to quartz or steatite beads were found in archeological contexts dating from the fifth millennium, in Egypt, Mesopotamia and in the Indus sites (Tite et al. 2008a, p. 23 and associated references). In most cases, it was a blue-green glaze resulting from a mix of sand and alkaline flux (natron or *Reh* in the Gangetic plains) and colored by copper oxides.

This know-how persisted for a very long time in various civilizations, even after the appearance of "archeological faïence" composed of a silica body consisting of quartz grains incorporated in a more or less developed glassy phase decorated by an alkaline or calc-alkaline glaze. The colorway diversified including green, yellow, dark blue, white, turquoise blue because of the addition of coloring metallic oxides, copper, iron, cobalt, manganese, etc. (Tite et al. 2008b). Lead was sometimes detected in variable amounts, but it was always related to the specific presence of lead antimonate microcrystals ($Pb_2Sb_2O_7$), colorants and yellow opacifiers that had been widely used since Antiquity (Kaczmarczyk and Hedges 1983).

Over one millennium would pass (the Bronze Age, toward the 14th century BC) until the potters managed to durably stick glaze onto fired clay. In the current state of knowledge, it was in the Mediterranean Basin, in Mesopotamia, particularly at Nuzi and a little while later in Syria at Ougarit that were discovered the first proofs

Lead in Glassy Materials in Cultural Heritage,
coordinated by Anne BOUQUILLON and Patrice LEHUÉDÉ.

of a calc-alkaline glaze colored in blue-green by copper oxides on a clay object (Matoïan and Bouquillon 1999; Matoïan and Bouquillon 2003; Tite and Shortland 2008, p. 189). Most of the time, this glaze was very weathered and, similar to what happened to the vases discovered at Ougarit, it only lasted in the form of small infra-centimetric flakes, often almost colorless. The appearance of these first glazed clay ceramics is an essential stage in the ceramic art, which was probably facilitated by the experience acquired while working with the Egyptian glassmakers of the new Empire who created hollow objects by depositing fine glass melt threads on a clay core that they removed when the material cooled (Nicholson 1993, p. 57). The technical constraints to be addressed were significant (Chapter 11). Glazed clay ceramic was actually developed only by the Mesopotamians, especially by large-scale use of decorated bricks in the first millennium BC. It was only in the first century BC that this technique was introduced in Egypt (Tite et al. 2008c, p. 188). By contrast, in China, since 1500–1050 BC, under the Shang dynasty, thanks to the use of specific clays and to the deep knowledge of materials, on the one hand, and to the perfect mastery of kilns and high temperature firing, on the other hand, calcium-rich glazes obtained from wood ashes covered stoneware and proto-porcelain objects (Wood 1999).

The first lead-rich glazes appeared only in the first century BC, in China and in the Roman world. In the history of ceramic technology, this is considered an essential innovation. It is worth considering why.

10.2. Properties and implementation of lead glazes

According to the definition provided by the technical vocabulary of ceramics, the glaze is a "vitrified coating whose composition is designed to adhere, when molten, to the surface of the shard [ceramic object] conferring impermeability to the porous paste and/or serving as decoration" (Blondel 2000, p. 51). A proper glaze should meet several requirements: no crazing, homogeneous coating, no flaking, an optimized state of the surface without bubbles or pinholes, and a pleasant aesthetic appearance (brightness and colors).

All these considerations involve very specific properties: (1) sufficiently low surface tension of the molten glaze to avoid shrinkage, (2) optimized viscosity of about 10^4–10^5 poise toward 900°C–1,000°C (Tite et al. 1998; Roisine 2018), sufficiently low to avoid bubble trapping, and quite high to avoid glaze flow, (3) compatible thermal contraction of the two materials (paste and glaze): the optimum contraction coefficient of the glaze is estimated to be 5–10% lower than that of the clay body (Tite et al. 1998).

In a glaze, lead addition increases the fusibility of the mixture, decreases thermal expansion coefficients, and reduces viscosity and surface tension. Brightness and elasticity also increase. This powerful flux allows the potter to obtain a bright, transparent glaze, without secondary crystallizations in the glassy phase, and perfectly coating (Tite et al. 1998), while remaining at quite low firing temperatures, thus saving time and fuel. It is easy to use, its maturation time[1] is long and regular, which makes it possible to obtain bright and shiny colors, but ceramists must be extremely cautious with the atmosphere of the furnace, as lead oxides are very easily reduced, conferring a dirty gray color to the material.

There are several manners to introduce lead oxide in the glazing mixture: in the form of cerussite ($PbCO_3$), ceruse ($2PbCO_3.Pb(OH)_2$), galena (PbS) (Pradell and Molera 2020), litharge (PbO) or minium (Pb_3O_4) – lead oxides have the advantage of not being soluble in water – or fritted with other elements (silica, alumina, etc.) to reduce its toxicity, particularly by limiting its volatilization in the furnaces during firing (Hamer and Hamer 1997, pp. 198–199).

But lead is dangerous for the health of ceramists and to a lesser extent for the health of those who use lead-glazed dinnerware. The modification of compositions by adding alumina for example, or calcium in the formulation of the recipe, reduces the leaching of lead contained in the glazes and renders them more resistant to alteration (Lehman 2002) (see Chapter 13).

In order to identify the implemented recipes, many authors propose to recalculate the composition of the glaze after having withdrawn the contributions of lead and colorants or opacifiers. Comparing these data with the composition of the underlying paste, it is possible to determine whether the glaze was obtained by uniquely depositing a lead component (litharge or minium) or a more complex silica-lead mixture. In the first case, the compositions of the recalculated glaze and the paste are identical; in the second case, the comparison shows an excess of silica in the recalculated glaze. A similar calculation is applied with alumina in order to understand if clay was used in the glazing mixture or it only came by diffusion from the paste to the glaze (Walton 2004; Walton and Tite 2010; Matin 2019). These calculation techniques are appealing and rapid, but they do not take into account a certain number of parameters that will be discussed in Chapter 11.

1. Period during which the temperature is sufficient for the fully molten mixture to homogeneously coat the shard.

10.3. The first lead glazes

10.3.1. *In China*

It is under the Han dynasty (206 BC to 220 AD) that lead-glazed ceramics experienced for the first time in the history of Chinese ceramics a notable development; more specifically, it should be noted that this production was preceded by the creation of several rarer pieces that were excavated in funerary contexts dating from the Warring States period (475–221 BC). But these were sporadic discoveries concerning several rare ceramics with polychrome glazes or beads constituted of an alkali-rich silica paste (Needham 2004, p. 478) whose glaze, colored in red, white, yellow or blue-grey, has high lead contents (above 20% PbO) and barium contents (over 9% BaO)[2].

	Na_2O	MgO	Al_2O_3	SiO_2	K_2O	CaO	TiO_2	MnO	Fe_2O_3	CuO	BaO	SnO_2	PbO
Blue-green bead	5.9	0.6	0.9	41.4	0.2	1.4			0.3	3.7	9.7		37.4
Red glazed bead	1.1	0.4	2	42.3	0.7	2.2	< 0.1	<0.1	17.8	<0.1	8.9		24.2
White glazed bead	2.6	0.4	4.8	58.4	1	0.7	0.2	<0.1	0.5	<0.1	11.2		19.1
Yellow glazed bead	0.8	0.7	8.6	27	0.7	1.5	0.4	<0.1	4.7	<0.1	20.1		35
Han glazes	0.4	0.7	3.9	33.4	0.5	2	0.6		2	3	7.7	1.2	43.5

Table 10.1. *Comparison of chemical compositions of Pb-Ba glasses dating from the period of Warring States and of Han glazes (weight % oxides). Data from Wood (1999) and Needham (2004)*

A large majority of Han ceramics with lead glazes were found in funerary contexts and their typology clearly imitates that of bronze objects. The brown-red clay of their paste, typical for the loess of Northern China (Wood 1999; Needham 2004, p. 147), similar to that implemented for the painted or not decorated ceramics of that period, is rich in silica and alumina and their iron contents are high. This

2. Until present, these Pb-Ba glazes seemed only present in the ancient Chinese productions, but they were revealed by very recent discoveries in Oman, in contexts dating from the late Islamic period (17th–19th century AD) (Živković et al. 2019).

orange red paste after firing modifies and spoils by transparence the final aspect of the glazes whose basic coloration is green, brown or yellow. Since the origin of the technique, the chemical compositions of glazes lead to differentiate two main recipes, one (Table 10.1) drawing heavily on the glass compositions from the Warring States period (Cui et al. 2011), containing both lead and barium, the second, simpler, rich in silica, lead and alumina (Table 10.2). The coloring metallic oxides are traditionally copper for green, iron for honey colors, manganese for browner shades.

	Na_2O	MgO	Al_2O_3	SiO_2	K_2O	CaO	TiO_2	Fe_2O_3	CuO	BaO	SnO_2	PbO
Han	0.2	0.5	3.7	29.5	0.9	1.2		1.3	1.2	0.2	0.2	59.7
Tang	3	0.5	5.8	33.6		1			2			55.6
Tang	0.2	0.7	6.7	32.4	0.4	1.2		0.3	0.6			53.6
Liao		1.4	3.4	33.3		0.9		1.5	2.1			57.2
Liao		0.3	4.1	33.2	1	0.9		0.5	1.9			56.7
Lohan Green glaze*	0.8	0	0.7	33.3	0	0.6	0	0.3	2.9			61.3
Brown*	1.4	0.5	6.1	38.5	0.9	1.1	0.4	6	0			46.1
Yellow-brown*	0.8	0	4.5	37.1	0.5	0.6		1.7	0			55.9

Table 10.2. *Chemical compositions of Han, Tang and Liao lead glazes (weight % oxides). Data from Wood (1999) and for the elements marked by * (Feugère et al. 2002)*

Research studies conducted to identify lead and barium sources show the abundance of galena and baryte deposits in Yangtze valley. A map illustrating both the distribution of geological deposits and the location of the discoveries of Pb-Ba glasses shows that the manufacturing workshops were probably near the lead mines (Gan et al. 2006, p. 709). The copper used to color in green certain glazes seems associated with tin traces, which might be explained by the recycling of bronze scraps. Are clays also added to the initial recipe? The presence of alumina detected in the composition and also the alkali contents would suggest it (Needham 2004, p. 483, p. 116), but it is very difficult to draw a conclusion, as the processes of diffusion from the paste to the glaze during firing are complex and may sometimes influence the final concentrations (Roisine et al. 2017).

The glazing technique is revealed both by the persistence of dark areas in the paste[3], characteristic for incompletely oxidized zones, and by the study of the SiO_2/PbO relation between paste and glaze (see Chapter 11): everything stands as a proof that for these first Chinese lead-glazed ceramics, the craftsmen deposited a silica/lead oxide mix directly on the raw paste (Needham 2004).

10.3.2. *In Asia Minor and in the Roman world*

The first lead glazed ceramics in the Roman world and in Asia Minor date from the first century BC. A wide majority of the shapes are inspired from those of silverware (skyphos, cantharus, chalices) (Hochuli-Gysel 2002, p. 308) and, for certain bowls, from typologies known for terra sigillata ware[4]. A short time later (in the second century AD), particularly at Mytilene, Tanagra figurines can be found (Caubet and Pierrat 2005).

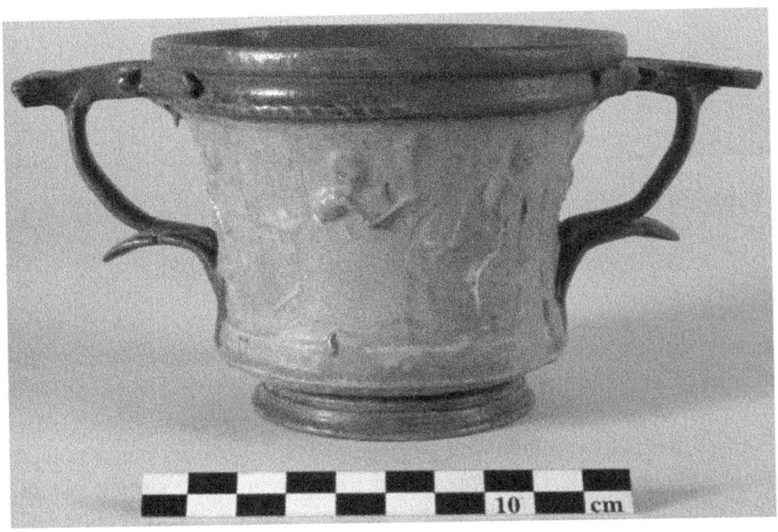

Figure 10.1. *Skyphos with skeletons, Smyrna, fourth quarter of the first century BC, Louvre Museum – inv. CA 1486. ©C2RMF/photo A. Chauvet*

3. In certain clays, organic matter or Fe^{2+}-rich components give a dark color to the paste; they can still persist if oxidation at the firing moment is not sufficient.

4. Terra Sigillata is a term applied to tableware ceramics from the Roman period, molded, with standardized shapes covered with a red slip. It derives its name from the stamps or seals used to create the decoration pattern.

In Anatolia, in Asia Minor, since the first century BC, many remains of workshops, misfires, molds, firing stands and various objects, glazed or not, have been discovered around Perga, Mytilene or Smyrna (Hatcher et al. 1994; Hochuli-Gysel 2002; Goslier 2018). According to these archeological data, it seems that the technique appeared earlier in Anatolia and spread to the Roman world during the first century AD until the fourth century. The archaeologists thus determine several large manufacturing centers, each identified by the composition of clays constituting the clay bodies: first Smyrna and Clazomenae between the first century BC and the first century AD, then Lyon and its region during the Augustan period, then in Central Gaul, Allier; all these centers are characterized by the use of more or less alumina and silica-rich clay pastes; on the other hand, Tarsus (first century BC to first century AD) and the Italian workshops in Latium and Campania can be distinguished by the calcium-rich clays (over 15% CaO). However, this division seems to be a bit too general in the light of recent studies, more manufacturing sites being now identified, such as Canosa in Apulia in Southern Italy (De Benedetto et al. 2004).

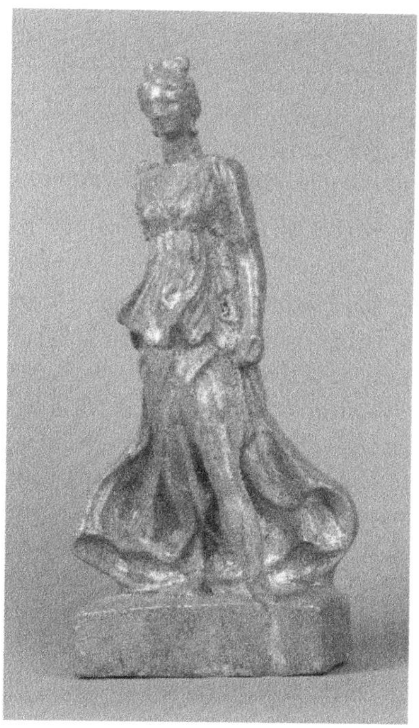

Figure 10.2. *Victory, Kymé, second half of the first century BC, signed Phanitès. Louvre Museum – Myr 677, 18 cm. C2RMF/photo D. Bagault*

Despite the significant number of workshop remains and discoveries of objects throughout the Roman world, there are still quite few actually exhaustive studies on the material characteristics of these various productions. However, the founding works of Hatcher et al. (1994) and Perez-Arantegui et al. (1996), then of Walton and Tite (2010) in the years 2000–2010, and finally the recent synthesis of Goslier (2018), associated with more focused studies, offer a quite precise view on the materials employed and the techniques developed. As a general rule, the characteristics of glazes, all very lead-rich, are homogenous (Table 10.3) irrespective of the type of underlying paste and the geographical origin of the objects.

	Na_2O	MgO	Al_2O_3	SiO_2	K_2O	CaO	Fe_2O_3	CuO	PbO	TiO_2	Paste
MNC304 yellow	0.5	0.2	2.8	28.4	0.0	1.0	1.9		66.7		Si 70, Al 10
MNC304 green	0.0	0.0	0.9	31.1	0.0	1.5	0.0	4.6	61.9		
CA1486 brown	0.1	0.6	9.5	26.0	0.4	1.7	2.2	.	60.5	0.5	Si 55, Al 23
CA1353 green	0.2	0.4	3.3	30.2	0.2	2.0	0.8	1.0	63.7		Si 51, Al 17, Ca 15

Table 10.3. *Several examples of compositions of glazes from objects dating from the Hellenistic and Roman periods conserved at the Louvre (Caubet and Pierrat 2005, pp. 191–197). The column on the right provides information on the nature of the paste (in oxide weight %)*

They typically present lead oxide contents near 50–60%, silica contents from about 30–40% SiO_2 and alumina contents from 2 to 10%. Alkali and calcium have a small contribution (generally less than 3%). This chemical composition is close to that of eutectic point (31.7% SiO_2, 61.2% PbO and 7.1% Al_2O_3) (Chapter 3). Similar to China, the colors of these glazes are green, brown and a variation from yellow to honey, creating sometimes highly sophisticated patterns due to their finesse and play of colors: green background and yellow-orange patterns, or variation of honey yellow, for example (Figure 10.1). Straw yellow coloration is obtained from the impurities naturally contained in lead compounds (see Chapter 4), and for the other colors by the addition of traditional metallic oxides, copper for green, iron for honey shades and manganese for brown shades (Table 10.3). Some authors mention red glazes, but no precise information was found on this coloration until now.

Many researchers focused on the origin of the lead employed. In 2010, Walton and Tite (2010) proposed, relying on isotopic analyses, the Rhodopes in the Balkans for the objects discovered in Serbia and Romania, the Massif central or the British

Isles for the others, either originating in central Gaul or in Italy. In 2020, the latest research studies published by Medeghini et al. (2020) bring clarifications and promote the hypothesis of lead coming from the British Isles for the vases found in Rome. All these authors highlight the difficulty of drawing a conclusion on the unique basis of isotopic ratios, associated or not with the research of trace elements, the overall characteristics remaining close and sometimes overlapping from one source to another (see Chapter 2). The recycling possibility can never be entirely excluded.

The various stages of the operating chain leading to these ceramics are deduced from the excavation results, the analyses of the compositions of materials and the observations on the interfaces between paste and glaze. These objects, most of which are molded (Atik 1995; Hochuli-Gysel 2002), are biscuit-fired first at temperatures near 900–1,000°C. Then, depending on the type of clay, two different glazing techniques can be used: on silica and alumina-rich pastes, lead is sprinkled directly on the surface; on calcium-rich pastes, a silica-lead mixture, perhaps associated with a bit of clay, is deposited; the temperature of the second firing is estimated at around 800°C. The addition of clays to the glazing mixture is often mentioned to explain aluminum contents of nearly 8% Al_2O_3. Even though these clay minerals can obviously be already present as impurities in the sand or they can serve as media to facilitate glazing, caution is needed in the interpretation: reconstitutions of lead glazes have shown that during firing, the diffusion of aluminum from the paste to the glaze may by itself also lead to such concentrations (Godet et al. 2019).

10.3.3. *Reasons for an emergence*

Why and how does a new know-how or a new technique develop? Except for serendipity, in general there must be an idea, a new technique, a need, a market or at least a strong interest for this new material and a demand that justifies the investment. Then the technique must be disseminated and other groups, other cultures, must take ownership of and adapt it to their own needs and further develop it.

Greene (2007) proposes a very thorough analysis of the possible reasons for the “sudden” emergence of lead-glazed ceramics both in the Roman world and in China. He relies on historical, sociological and technological considerations. It can thus be seen that in both cultures, the typologies of metallic objects were resumed by the ceramists, which led many authors to propose the hypothesis of a cheap imitation of the bright appearance of metallic surfaces by yellow or green lead glazes.

The glassmakers' experience was a source of inspiration for the ceramists; indeed, in the Near-East and Western world, toward the 6th century BC, red glass beads, colored by nanoparticles of metallic copper, have high lead contents, very different from those detected in most of the glasses of that period (Brun and Pernot 1992). The same (imported?) beads can be found on contemporary Chinese sites. More significant, still in China, during the Warring States period, the very first glazing tests based on a recipe containing PbO and BaO definitely recall the compositions of contemporary glass beads (see Chapter 6).

The interaction between potters and glassmakers in an active context of stimulation in the field of arts of fire around the second century BC may be at the origin of this technical evolution. Where does the emergence originate? In China? In the Roman world? Is it a simultaneous emergence? The discussion is complicated and of narrow interest. It is known from texts that the Romans were present in China toward 166 BC (Cabouat and Ogaki 2016). It is also known that the two empires had very active exchanges starting with the Augustan period by the intermediary of many trading posts in India. Furthermore, in these two regions of the world, lead was heavily used, particularly in relation with the significant increase of silver coinage, lead being a by-product of silver working. In the Roman world, it was moreover used in various daily life activities (make-up, construction and even the food industry) to such an extent that the term "lead civilization" was justifiably coined, and many researchers advanced the hypothesis that it might have partly caused the end of the Roman civilization due to the appearance of massive lead pollution and diseases it causes (Delile 2014). In China, it was also part of the pharmacopoeia.

Therefore, in both civilizations, all the parameters are met for the development of a new ceramic technology. While the emergence is quasi-contemporary, this know-how had a very different posterity in China and in the Western world.

10.4. Dissemination and evolution of the lead glazing technique in the high Antiquity

10.4.1. *In the Near-Eastern and Western world*

Goslier (2018, p. 150, fig. 102), using both the data in the literature and her own research, proposed a very complete synthesis of the dissemination of this know-how between the first and fourth century AD: starting with the main workshops in Asia Minor (Pergamon, Tarsus, Perga, Smyrna, Mytilene, etc.) in the first century BC, the technique would have been transmitted in workshops located in Northern Italy. Very rapidly, craftsmen from these workshops arrived at two sites of Gaul, Lyon and Saint-Romain-en-Gal. The manufacturing, probably reserved for funerary practices, remained limited and local. Starting with the first century AD, the imports of lead-glazed objects increased particularly in Southern Gaul sites. Then thanks to the arrival of craftsmen from Lyon or Asia Minor, major manufacturing workshops developed both in Central Italy and in Central Gaul (Lezoux). Little by little, particularly starting with the second century AD, the workshops of Latium started exporting their ceramics up to the borders of the Roman Empire (Great Britain, Austria, Hungary, Croatia, Poland), while the workshops of Central Gaul rather exported their productions to Northern Gaul, to Raetia, Germany and Brittany. At the same moment, the number of these ceramic objects appears to decline. However, in the Roman provinces in the center of the Balkans, discoveries of many glazed objects, often for practical purposes, reappear with varied typologies and often sophisticated patterns. It could be shown that these objects were all manufactured in contexts entirely related to the Roman troops stationed in the region (Cvjeticanin 1997, 2000). Apart from this Danube region, in the rest of the Roman world, lead-glazed ceramics fell from grace and their manufacturing drastically declined during the fourth and fifth centuries.

	Na_2O	MgO	Al_2O_3	SiO_2	K_2O	CaO	TiO_2	Fe_2O_3	PbO
Vichy (6)*	0.2	0.6	8.5	18.7	0.6	0.9	0.5	1.4	69.2
Lezoux (2)*	0.1	0.5	7.4	23.4	0.7	0.6	0.5	1.7	65.6
Latium (2)*	0.3	0.6	4.3	31.3	0.6	3.5	0.3	2.2	56.6
Byzantium (5)**	0	0.2	8.2	28.7	0.3	0.8	0.4	3.1	58.3

Table 10.4. *Chemical compositions of glazes in the Roman and Byzantine worlds. MEB and PIXE data expressed in weight % oxide. *Novel data (Mainguéné 2002); **(Waksman et al. 2007). In brackets, number of studied glazes*

There are ongoing research projects to understand what happened between this moment and the seventh century, and how and why this technique was revived in Byzantium and from there would conquer the entire Western world constantly maintaining its attractiveness until the 19th century. It is in fact starting with the seventh century, at Constantinople, capital of the Eastern Roman Empire, that high technical quality tableware appeared, referred to as "glazed white wares", characterized by a very white kaolinitic paste covered by a lead-rich glaze of green, brown and orange color (Waksman et al. 2007, 2008). Their chemical composition, very close to those of the initial Roman glazes (Table 10.4), is overwhelmingly dominated by the three common oxides of lead, silica and alumina, while alkali represent maximum 5%. The colorants, copper, iron or manganese oxides, provide the same colored ranges: green, yellow, brown. The technology transfer from the

Roman world is obvious. There is an important dissemination of this production between the seventh and ninth century in Asia Minor, Greece and along the Western coast of the Black Sea. This tradition of lead-rich glazes continued throughout the Byzantine world, until the beginning of the 12th century, with evolutions in the typologies and the variety of patterns and colors employed (e.g. cobalt blue appeared in the 10th century). Byzantine potters deposited the glazes directly on the paste without intermediate slip. At least at the beginning of these manufacturing activities, the glazing technique is the simplest: direct deposition of lead on the paste. Afterward, it gained in complexity.

The production of very lead-rich glazes would go on without interruption, though in France at least it appears that it actually developed only starting with the Carolingian period in the Northern regions (10th–11th century) and from the 13th century in the Mediterranean South (cat. The art of glazed clay 1999, p. 16). This subject is extensively explored by Brongniart, in his *Traité des Arts céramiques* (Treatise on Ceramic Arts or Pottery). They are referred to as *poteries plombées* in the 13th century and *poteries plommées* in the 14 and 15th centuries. It is only since the end of the 16th century, in the dictionary of Furetière, that they would be referred to as lead-glazed clays or potteries ("céramiques vernissées", a name that is still employed nowadays (Blondel 2001)). These glazes are used on all sorts of productions: architectural objects (floor tiles or roof tiles), common objects (tableware, pots, etc.), technical objects (stove tiles, for example). There are small variations of the technique, the two types of glazing mixtures are found, lead compounds applied directly on a biscuit or on an unfired ceramic object or a mixture of lead + silica or lead + silica + clay. The treatises of the 19th century contain "recipes", as illustrated by the example extracted from the work of Garnier (1893):

> One of the most ancient and strangest glazing processes, which was preserved for several centuries and which, until recently, was employed in several countries, particularly in Brittany, consisted of mixing lead scraps with cow manure or with blackwheat flour, so that it formed a more or less thick paste that was spread on the parts to be glazed and was fired while lead was oxidized and started to melt, combined with the silica in the paste, and gave to the objects a very pure glassy varnish, which was well spread and had a beautiful dark green color.

It is known that on the American continent, lead glazed ceramics (60–85% PbO) were discovered in Peru or Mexico in the 16th century, but the technique, brought by the Spanish settlers, does not appear to have ever been mastered in that period by the natives outside the workshops near Mexico's capital (Lister and Lister 1982,

p. 81; Van Valkenburgh et al. 2015, 2017). On the North American continent, the potters in the Colorado region of Rio Grande had used glazes containing lead as paint since the 6th century AD and more widely starting with the 13th century in the Pueblos. The appearance of these first lead vitrified paints is, according to Shepard (1956), very sporadic and had no major impact on the art of ceramics. They are highlighted only in a very narrow geographical area. The presence of lead in the material does not seem to be voluntary, but rather resulted from the use of a black manganese pigment, naturally containing significant lead amounts that allowed reaching a vitrification effect during firing; The quite disappointing aesthetic result might explain why the material was very rapidly abandoned.

Starting with the 13th century, the same region of Colorado was the development center of a culture characterized, among others, by very high quality ceramics, whose polychrome pattern was marked by the use of lead-rich black vitrified paints. The analysis of paints and more precisely the study of lead isotopes show the use of local raw materials and sometimes suggest a mix of supply sources (Huntley et al. 2007; Van Keuren et al. 2013). This technique, originating in the higher regions of Little Colorado, in the region of Western Pueblo, spread along Rio Grande (Habicht-Mauche et al. 2002; Thibodeau et al. 2013).

10.4.2. *In China*

The destiny of lead glazing was very different in China. After a period of significant production of lead glazed ceramics under the Han dynasty, the number of excavated objects discovered seems to drastically decrease during the fourth and fifth centuries. It is only in the Tang period (618–907 AD) that the interest for these productions was revived. They are then much brighter than those of Han, simply due to the use of a very white underlying paste that enhances the colors of the transparent glaze. Due to the three typical shades, white, green and yellow, the first ceramics of this type discovered in the tombs in the 19th century would be called "sancaï", which means three colors in Chinese. But this name, which is now part of the common language, has become erroneous given the more significant diversity of the colored palette used in that period: blue, black, brown, turquoise, etc.

Either ornamental, architectural or funerary (Figure 10.3), these ceramics are in general of recognized quality and while they have as a primary source of inspiration Chinese goldsmiths and metal objects, they also draw on influences of Islamic and Western art, particularly appreciated by the court of Tang. In return, "sancaï" were exported to Iraq, and more broadly to the Middle East, where they also incited research on glazes, particularly in the Abbassid world (see section 10.6).

The chemical compositions of Tang glazes have such a degree of similarity to those of Han (Table 10.2) that they cannot be reasonably used as a single way to differentiate between the two productions: here, too, SiO_2, Al_2O_3 and PbO contents are very close to eutectic point. Only colorant oxides are diversified, as cobalt appeared. The firing temperatures of these glazes remain, as previously, around 900–950°C, but the use of refractory white clays as material of the object requires high-temperature firing (1,100–1,200°C); the ceramists had to choose double firing, as confirmed by the discoveries in several kilns of unglazed ceramics: high temperature for the paste and, in a second stage, lower temperature for the silica-lead glazing mixture (Wood 1999, p. 205).

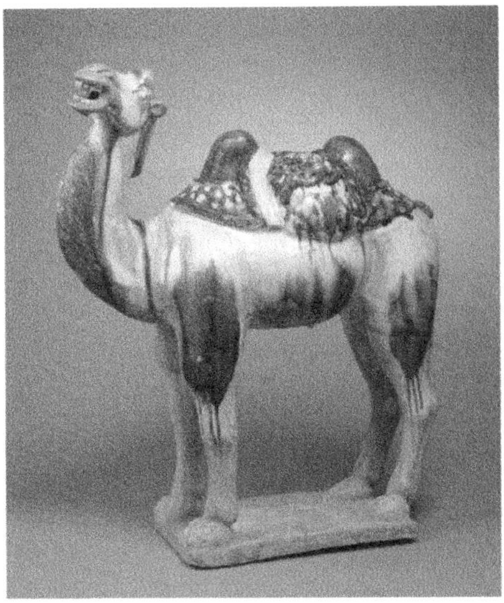

Figure 10.3. *Camel, "sancai" ceramics, Tang period (seventh to eighth century) – inv 49.2383 – Walters Art Museum*[5]

The fall of the Tang empire in the 10th century marked the end of manufacturing these ceramics, but the tradition of lead glazes was resumed by other kingdoms at the north of the Great Wall, in the province of Hebei: lead-glazed Liao ceramics (Table 10.2), common ceramics, were manufactured starting in the 10th century; the techniques are close to those of Tang and most of the authors consider that there has been a technology transfer by the intermediary of potters in the regions of Hebei or

5. Wikimedia Commons (CC BY-SA 3.0: https://creativecommons.org/licenses/by-sa/3.0/deed.fr) – licences Creative Commons Attribution-Share Alike 3.0.

Henan that had migrated or had been taken prisoners. The only change is that of the compositions of coloring metallic oxides: for blue, cobalt is associated with nickel (it is rather associated with manganese for Tang), green is obtained by a mix of copper and iron, while that of Tang presented high copper contents, probably witnessing different sources of coloring raw materials.

There are also other innovations worth being mentioned for Liao ceramics: (1) the use of a white slip under the glaze, which allows masking a too annoying color of the paste and thus regaining the brightness of "sancaï" glazes. This technique was broadly implemented in the Islamic world; (2) the use of another flux that replaced lead in the glaze formulations: boron (possibly as borax?). This discovery marked a new stage in the Chinese ceramic art, but it remained highly confidential and was rapidly abandoned.

The tradition of "sancaï" lasted until Song (960–1279 AD) and Jin Cizhou (1115–1234 AD) with the same technical characteristics and a diversification of colors (turquoise blue, iron red and iron-manganese violet).

Wood (1999) notes that lead-glazed ceramics specific to Northern Chinese provinces are also present in Southern China (Jizhou, in the province of Jaingxi, Qionlai in Sichuan) with the same technological characteristics.

But anyway, all over China, stoneware or porcelains, materials fired at very high temperature, more resistant and with remarkable aesthetic qualities, are the essential part of the productions. Lead glazes, requiring low firing temperatures, would disappear or be rather used as overglaze enamels painted onto already fired porcelain or stoneware glazes for decoration. This turning point in the 12th century would last until the present day (see section 10.9).

10.5. Hybridization of lead glazes with other ceramic traditions

A new stage in the art of ceramics is the use of lead not as single flux, but in addition to more complex glazing mixtures, either alkaline or calco-alkaline in the Near-Eastern world, or in calcium-rich glazes in the Chinese culture.

It is first in Han China that can be followed the tests conducted by Southern Chinese potters in order to try and acquire the new know-how of the Northern potters that mastered the lead glazes. Proto-porcelains had been developed since the middle of the Warring States period by using "porcelain rocks", quartz and sericite. A lime glaze, very thin but a little dim, is deposited on this white paste. Shards found in one of the kilns in Shangyu, analyzed by Wang et al. (2019), reveal the attempts to progressively introduce lead in this initial recipe, probably in order to obtain a brighter and more aesthetic glaze. The results do not always meet the expectations: bubbling, inhomogeneities, etc. All this explains why these attempts remained isolated and the craftsmen from this part of China rather turned to the manufacturing of celadons.

In the Mediterranean world, and more specifically in that of the Abbassids, starting with the ninth century, the potters progressively achieved a "hybridization" of ancient Near-East traditions of alkaline or calco-alkaline glazes with Roman traditions. It can be seen that lead gradually entered in the composition of glazes leading to alkaline-lead and then lead-alkaline glazes (Table 10.5) that characterize a large part of ceramic productions first in the Islamic world and then beyond it.

		Na_2O	MgO	Al_2O_3	SiO_2	K_2O	CaO	MnO	Fe_2O_3	SnO_2	PbO
Suse	MAX	8.3	3.6	3.8	73.3	5.5	7.8	0.7	1.07	12.8	8.7
	MIN	0.6	1.2	1.9	61.5	2.3	2.3	0.06	0.3	3.2	1.4
Samarra	MAX	3.9	3.1	2.1	76.3	6.3	5.3	0.4	0.9	15.3	7.7
	MIN	2.2	1.8	1.1	61.5	3.2	3.5	0.1	0.55	2.0	1.4
Fustat	MAX	4.9	3.14	2.03	75.2	6.4	6.7	0.7	1.0	9.2	9.4
	MIN	1.0	2.3	1.4	61.5	3.6	3.3	0.1	0.6	2.8	1.7

Table 10.5. *Examples of ranges of compositions of Abbassid glazes (Suse and Samarra) and Fatimid glazes (Fustat). PIXE data expressed in weight % oxides obtained on a body of ceramics from the collections of the Department of Islamic Arts at the Louvre Museum ((Bouquillon 2012), unpublished data)*

The addition of lead in an alkaline glaze improves several properties due to which the alkaline mixtures are difficult to implement: the glazing mix has a lower tendency to flocculate, the adherence to the clay object, irrespective of its composition, is better, the crazing tendency decreases (Tite et al. 1998). On the other hand, the palette of possible shades is enhanced; finally, when alkaline elements are added to a lead glaze, there is lower risk of lead reduction during heating and therefore lower risk of glaze blackening or bubbling, particularly when a lustre decoration is applied[6].

6. Lustre decoration is a thin layer containing copper and/or silver nanoparticles dispersed in a glassy matrix (Mason 2004; Sciau 2012).

10.6. The importance of lead-rich glassy materials in the race for porcelain

What J.P. Desroches refers to as "ceramic routes" between China and the Western world opened between the 9th and 15th centuries. They are fully sea routes and follow the ancient jade routes and then the silk routes (second century BC and ninth century AD), which are essentially land routes. It is in this period simultaneously marked by strong trade flows between China and the Islamic world, but also by an extraordinary flourishing of arts and by a renewal of sciences and techniques, that ceramics and glassy materials in general (see Chapter 8) would experience a golden age. Besides the tea and silk imports, starting with the ninth century AD and on a wide scale, there will be imports of Chinese ceramics, more particularly porcelains and celadon stoneware, especially starting with the 14th century, white-blue porcelains. This trade would soon be amplified by the exchanges with the trading posts of the East India Company under the auspices of the Portuguese, English and Dutch, and then later with the French (Nicolas 2020). The Western and Middle Eastern civilizations would thus discover these extraordinary productions and would relentlessly try to reproduce them. But in that period, potters did not have the kilns allowing them to reach the temperatures required for porcelain firing, and they had not yet understood it was essential to use highly refractory clay, the kaolin, as basic ingredient for the paste. Between the ninth and 18th centuries (the first actual porcelain was manufactured at Meissen in 1709, and then at Sèvres starting with 1767), various types of materials would be imagined in order to obtain a very white paste, a white and bright surface and a solid product, with fine and translucent walls that resisted thermal chocks. The glassy materials containing lead would be abundantly used not only for glazes, but also as components of the paste or of the slip.

10.7. Innovations brought by Islamic potters of the eighth to ninth century: the place of lead glassy materials

The fact that the fascination for Chinese porcelains and the relentless search for whiteness, purity, solidity and translucency played a driving role in this diversity of new approaches is undeniable and recognized by all the historians of techniques, but it is undoubtedly too restrictive to reduce all innovations to the simple quest for porcelain, to imitation. Among all the inventions of Islamic potters that durably nourished the art of ceramics in its entirety, I have chosen to focus on two of them: the voluntary addition of milled glass to the slips and pastes and the discovery of lead-tin glazes. In both cases, after a detailed presentation of the originality of new materials and their main characteristics, their posterity will be shown.

10.7.1. *Addition of milled lead glass in the paste or in the slip*

To obtain a bright and "pure" colored effect, a white or at least neutral background is needed, and this is not always compatible with the types of clays available and/or chosen by the ceramists. When the clay is dark, a solution, used for a long time particularly by the Liao potters (see section 10.4.2), is to deposit an intermediate layer, a clay or silica slip, whose color after firing, the clearest possible, hides that of the underlying paste. Besides this aesthetic function, certain slips, particularly those rich in quartz, play also a more technical role facilitating the adherence of an alkaline glaze on a clay body.

10.7.1.1. *Stonepastes and slips*

Within this work, the pastes and slips of interest are only a very small part of them, those containing lead that is voluntarily added to the initial recipe, most often through the addition of fragments of lead glasses. These are referred to as stonepastes. The oldest mention of a written recipe is attributed to an Arab ceramist, Abul'Qasim in 1301 (Allan 1973), but the very first testimonies of crushed glass in a clay paste date from the Abbassid period in Baghdad (Mason and Tite 1994). At that time, clay was still very widely dominant in the mix, while quartz and crushed glass were a minority. It was essentially in the 10th century in Egypt, at Fustat (Tite 2011), that the technique developed thanks to the arrival of Abbassid potters; it then disseminated to Syria and Iran starting with the 11th century (Rugiadi 2016). The evolution of paste recipes is clearly followed throughout time: the proportions of clay progressively decrease to the benefit of quartz and glass fragments. The latter, of highly variable compositions initially, containing alkali or lead, within the same object, become increasingly homogeneous. As for the slips with fragments of lead glasses, this practice seems less frequent, according to the literature referenced here: there are such examples in certain Ottoman Iznik objects or Miletus ware (Burlot et al. 2020). It is important to note that each time the compositions of these fragments of lead glass differ from those of the lead glaze covering the object. Particularly in the productions from Central Asia to Kazakhstan (Klesner et al. 2021), or in Uzbekistan (Molera et al. 2020), lead (14–25% PbO in the slip) seems directly added to the clay slip in order to modify its mechanical properties (better adherence of the glaze, better paste/glaze compatibility). The most accomplished objects are those created by the Ottomans at Iznik, and then by the Safavids.

10.7.1.2. *"Faïences fines"*

Toward the middle of the 18th century, new productions referred to as "*faïences fines*" appeared first in England, then in France and elsewhere in Europe

(Switzerland, Italy, Central Europe) (Maggetti 2018). They differ from the traditional faïence (see next section) by the very white pastes obtained from highly aluminous white clays in combination with milled flint or feldspar (in England), or with grog (in France). Other manufacturers chose calcareous clays with chalk, particularly in Lorraine (Lorraine, Bois d'Epense), where the technique of adding fragments of glass frit containing lead is practiced in order to have a very white and quite translucent paste.

In 1540, Bernard Palissy (1510–1590), a major artist of the Renaissance, invented *rustic figulines*. He developed a molding technique on living creatures in order to conceive ceramic objects celebrating nature and creation. Besides the precision of the molds and the extraordinary complexity of his compositions, Palissy pushed mimicry to the extreme. For the artist, the aim was not beauty, but authenticity. Each color, as close as possible to the reality, is the result of patient dosing, of a research that is often quite difficult to follow by the analysis of the glassy matter. All the more so as in this precise domain, the ceramist shares none of his secrets, except for a quite common list of components (Bouquillon et al. 2017). The presence in the excavations of technical material and particularly of albarello that he coated with various coloring mixtures in order to test their fire resistance and the effect after firing are already valuable pieces of evidence.

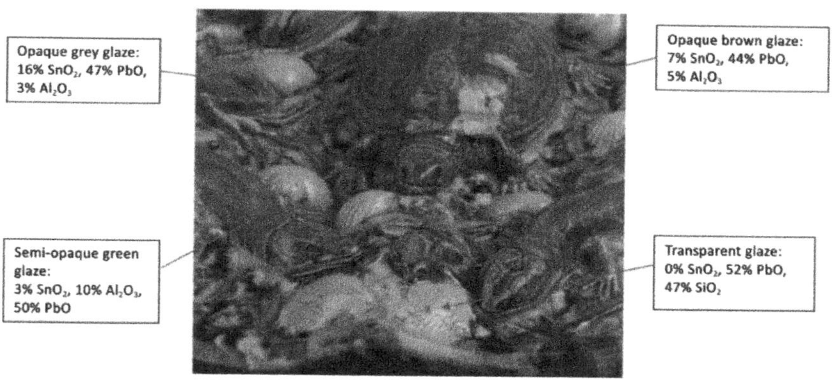

Figure 10.4. *Detail of the rustic figulines platter (Louvre Museum, inv MR 3395) with some information on the glazes (PIXE data in % of oxides). For the complete composition of glazes, see Bouquillon et al. (2017, 2018). ©C2RMF/photo D. Bagault. For a color version of this figure, see www.iste.co.uk/bouquillon/lead.zip*

The non-destructive study of several dishes conserved in the French collections and the chemical and microstructural analysis of the glazes of a dozen fragments of frogs, snakes and shells show that the mixes are highly variable (Figure 10.4); similarly, the measure on stratigraphic cross-sections or, in a non-destructive manner, by confocal interference microscopy of the thickness highlights the remarkable mastery of the artist in order to find the right value that does not erase the lizard skin for example or that renders perfectly the slightly flabby and sticky appearance of a frog's skin. But the most surprising in these investigations of color is the multiplication on the same dish of compositions: transparent glazes with 60% lead oxides are in the vicinity of others containing over 10% of alumina, with others opacified by over 15% tin oxide. However, there is no paint run, no major defect (there is just slight crazing), the coating is perfect and the colors do not mix.

Palissy overcame all these difficulties. He finally excelled in representing everything by clay and glass. The ceramists that tried to imitate him did not reach such subtlety, probably with the exception of Avisseau or Pull in the 19th century. Most of the time they were satisfied with adding different coloring oxides to a single base of glaze (Bouquillon et al. 2018).

Box 10.1. *Bernard Palissy: rustic figuline and the art of ceramics*

All these "*faïences fines*" are covered by bright and transparent glazes, which are very rich in lead (50–60% PbO), silica (35–40% SiO_2) and sometimes alumina and potash. They are very brittle and soft, and have a tendency toward rapid crazing and alteration: various solutions were found to address these defects by the addition of cassiterite (4–7% SnO_2) in the factories of Lorraine (Maggetti 2018) or boron (3–7% B_2O_3), in the factories of Bordeaux, for example (Beauvoit et al. 2021). Though successful on the market and in terms of their large-scale production, these productions that flooded the market in the 18th–19th centuries were progressively partially abandoned because of their too high brittleness and also for economic reasons.

Certain authors associated these "*faïences fines*" with very specific ceramics (Box 10.1), created during the Renaissance by an extraordinary artist, Bernard Palissy, who used for his shaped objects very white kaolinitic clays covered by very bright glazes (Heimann and Maggetti 2014, p. 271).

10.7.2. *Tin-opacified alkali-lead glazes: emergence of faïences*

While the previous innovation concerned the stoneware pastes appreciated for their whiteness, the one concerning the white opaque glaze had a significant posterity. The Abbassid ceramists developed a solution to obtain a white and bright surface resembling certain porcelains: the use of lead glazes opacified with tin oxide micro-crystals, cassiterite (SnO_2). The quality of these lead-tin glazes is far higher than the one of the more ancient glazes opacified by the presence of crystallized phases (such as calcium or lead antimonates), by the presence of unmolten crystals (quartz) or recrystallized phases (diopsides), or even the presence of bubbles. The direct addition of tin calcined at high temperature seems possible but rare (Matin 2019), even if certain authors question it (Kingery and Vandiver 1986). In fact, the common technique involves the use of calx, a mixture of calcined lead and tin; this calx containing PbO, SnO_2 and Pb_2SnO_4 in the form of powder is then incorporated in a mixture of silica and alkali, pre-fritted or not. Lead addition allows a better distribution of cassiterite microcrystals (SnO_2) in the final glass.

Once again, the Islamic potters relied on a glassmaking know-how; indeed, tin oxide-opacified glasses were discovered in the first century AD on several sites of Roman-controlled regions, then in the Mediterranean and Levant starting with the fourth century (Chapter 6) (Matin 2019 and associated references), and later on in Byzantium. There was therefore an already well-established tradition and thereby a good mastery of the fabrication of these pigments when the ceramists from Egypt and Levant got interested in it in the eighth century at Fustat, Tod or Al-Mina (Tite et al. 2015; Matin et al. 2018; Matin 2019) in order to obtain yellow glazes (lead stannates) or opaque white glazes. The know-how was disseminated toward Mesopotamia by the potters migrating to Baghdad, capital of the Abbassid caliphate. The ancient descriptions of these recipes first of all in the treatise dating from 1301 of Abul Qasim (Allan 1973) then in treatises of the Renaissance (Picolpasso 1557), from the 19th century (Brongniart 1877, p. 25, volume II), in the treatise of Esfahani written in 1888 at Sir Robert Murdoch Smith's request (Matin 2019), as well as the recent experiments conducted by Tite et al. (2008d) and Matin et al. (2018) show the variability of these recipes depending on the chronology and highlight the essential role of various parameters in obtaining the sought-for yellow or white shade: the Pb/Sn ratio, the firing temperature and the alkali content of the glass frit to which the calx is integrated. These data are thoroughly presented in Chapter 9. Even though there is a certain variability according to the periods and the geographical areas, main trends emerge: the most ancient yellow glazes (8th, 9th and 10th centuries) at the same time in Egypt, Levant and Mesopotamia, and later in Iran and central Asia, are characterized by quite high lead contents (sometimes above 60% PbO), and low contents of tin and alkali. Most of the time, white glazes contain less lead and more alkali (up to about 13%) and the Pb/Sn ratio widely exceeds 3.5. The only ones that actually stand out are the Italian archaic majolicas, the Islamic archaic objects and certain Persian ceramics, Mina'ï and Haft Rang (Matin 2019, p. 1161).

Thanks to the dynamism of the commercial exchanges and the rapid dissemination of know-how in the empire of Abbassids and then of Fatimids, these tin opacified glazes had an extraordinary posterity first throughout the Islamic world, both to the East (Iran, central Asia, East Asia) and even greater to the West: Egypt, Tunisia (Salinas et al. 2020) then Al-Andalus via Cordoba. In all these regions, the arrival of this new technique stimulated the inventiveness of potters who adapted their local ceramic traditions and renewed their aesthetics, as shown by the recent researches conducted on the sites of Kazakhstan or Azerbaijan (Holakooei et al. 2019; Klesner et al. 2019; Klesner et al. 2021). The beauty of these ceramics and the increasing demand by the populations may also draw the interest of the craftsmen and allow the development of local glazes productions in regions where they were not known previously, for example in Al-Andalus at the end of the ninth century under the influence of the Califate (Salinas et al. 2018; Salinas and Pradell 2020).

In the 15th century, at Florence in Italy, Luca della Robbia (1399–1480) founded a family workshop in which he adapted the faïence technique to sculpture and architecture (Gaborit and Bormand 2002). The objective of this invention was twofold: (1) obtain a very bright polychromy that lasted in external environments, (2) allow all the modeling freedoms for a large size sculpture with aesthetic performances rivaling the most beautiful marble. For this purpose, he used a clay mix most of which was composed of a calcium-rich clay (22% CaO) to which a small part of silica-aluminous red Impruneta clay is added. The color of the clay after the first firing is pale. The glaze is very specific: if the base is close to that of faïences, marzacotto (fritted silica-potash glass) + calx + quartz, the coloration is not obtained here by surface painting, but by the addition of coloring oxides in the initial mix so that the color is distributed throughout the thickness of the glaze (about 200 μm) and thus the intensity of the shade is maximal.

To achieve a perfect rendering, Luca della Robbia conducted many tests that can be followed by studying the early productions (Zucchiatti and Bouquillon 2008); he obtained an optimized recipe in which lead contents are very high ($PbO/SnO_2 < 2$) and those of alkali are of about 5–7% ($Na_2O + K_2O$). This way he reached a compromise between a very good distribution of cassiterite grains, the absence of traces of yellow lead stannates and a higher viscosity that allowed juxtaposing various colors without mixing them. However, this required higher firing temperatures than for traditional majolicas.

Once the recipe was developed, several generations implemented it without major changes (Figure 10.5). But when Girolamo della Robbia and Luca the younger came to France upon request of Francis first to decorate the front of the Château de Madrid, they were not able to adapt the *terracotta invetriata* technique to the materials they found in France (Meunier and Bouquillon 2004).

Box 10.2. *Della Robbia: adaptation of faïence techniques to sculpture and architecture*

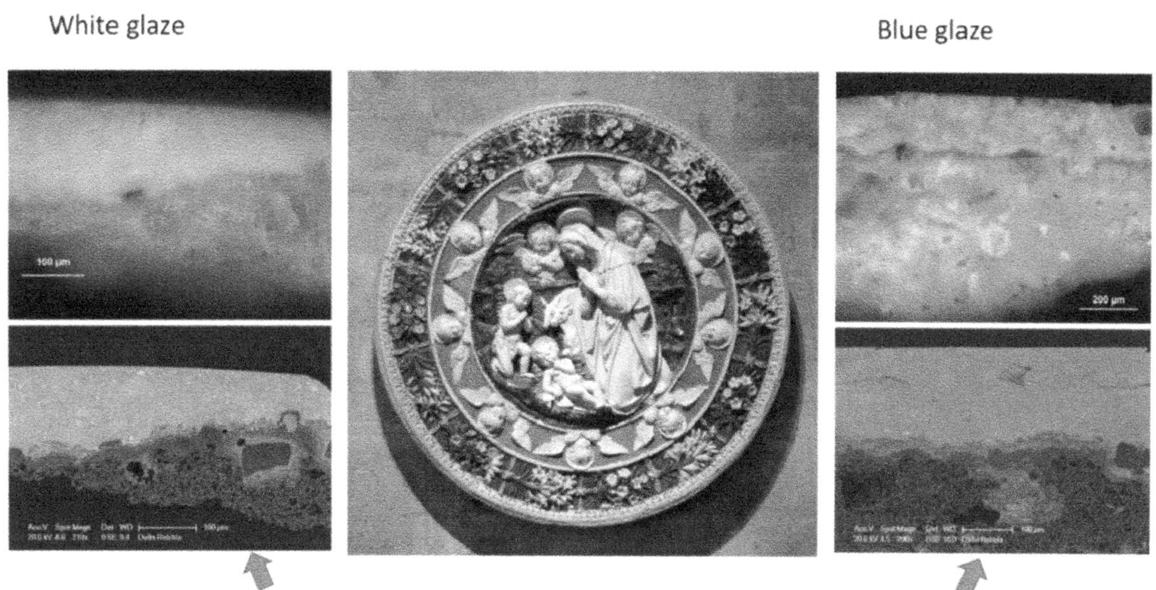

Na2O	MgO	Al2O3	SiO2	SO3	K2O	CaO	TiO2	Fe2O3	SnO2	PbO
2.1	0.4	1.5	41.2	0.2	3.8	1.9	0.2	0.7	20.8	27.3

Na2O	MgO	Al2O3	SiO2	SO3	K2O	CaO	TiO2	Fe2O3	CoO	NiO	ZnO	As2O3	SnO2	PbO
1.7	0.3	1.8	52.8	0.3	6.1	2.7	0.1	1.2	0.5	0.5	0.7	0.07	13.6	17.8

Figure 10.5. *Examples of white and blue Della Robbia glazes: The Virgin adoring the child in the presence of the infant Saint John the Baptist and two cherubs, workshop of Andrea and Giovanni della Robbia (inv LP 3410). Louvre museum/C2RMF/A. Chauvet. Optical microscope and SEM images of microstructures. Chemical compositions of glazes (in % of oxides). ©C2RMF/C. Doublet. For a color version of this figure, see www.iste.co.uk/bouquillon/lead.zip*

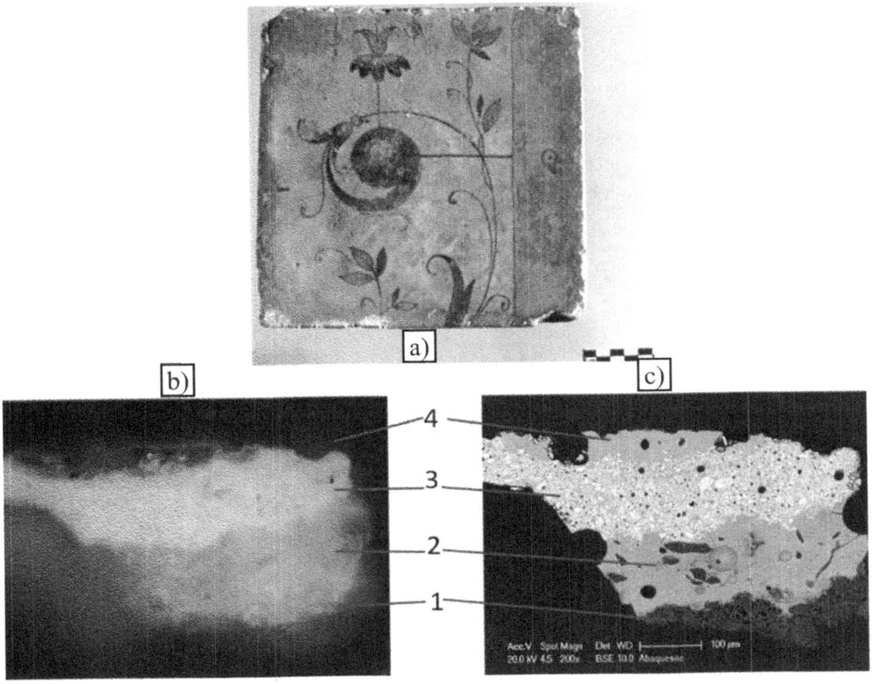

Figure 10.6. *Stratigraphic section of a sample of yellow glaze taken from floor tile of Ecouen pavement. Masséot Abaquesne, ECI 21014. (a) tile; (b) optical microscope image of a cross-section; (c) scanning electron microscope image of the section. (1) Clay body, (2) tin opacified ground glaze, (3) yellow paint layer (here lead antimonate crystals + cassiterite) and (4) coperta. (C2RMF/photos C. Doublet). For a color version of this figure, see www.iste.co.uk/bouquillon/lead.zip*

The techniques of opacification by cassiterite would reach the rest of Europe via Spain in the 11th century: first Italy, France starting with the 13th century in Marseille (Rosen 2018). The beginnings in France are hesitant, but facilitated by the arrival of the "Sarazins", Spanish craftsmen (Amouric et al. 1995, p. 185); they concern, especially at the beginning, Provençal and Languedocian pavements dating from the 13th century and the several fewer discoveries of floor tiles in royal or rich residences (Vincennes, Chartreuse de Champmol (cat. Images of power 2000)). However this is a period when the Italian influence can also be perceived, though it is especially very clear starting with the 15th century (Rosen 2018, p. 47). These are the premises of a major development of the faïence[7]. It is not our aim to retrace here the history of faïence in Europe, this being dealt with in many existing syntheses (Barbe 2016; Rosen 2018). It suffices to say that under the impetus and the extraordinary vitality of Italian artists and craftsmen of the Renaissance (Box 10.2) prestigious factories will develop in France (Lyon, 1494, Rouen with the workshop of Masséot Abaquesne), in the Netherlands (Anvers with the Andries family, Delft), but also in Switzerland (Maggetti 2020).

The recipes of this period are known and thoroughly detailed in *Li tre libre dell'arte del vasaio* by C. Picolpasso (v. 1577), but also in a more ancient treatise (1540) of V. Béringuccio, *Pirotechnia*. The glaze stratigraphy can be described as a "multilayer" (figure 10.6); indeed, there is most often a layer of thick white glaze obtained by mixing fritted and milled glass, marzacotto (sand + wine lees) with calx (lead/tin) and quartz grains. Above this white background and before firing, a layer containing coloring pigments and lead fluxes is painted. In the case of majolicas s.str., a third layer, known as "coperta" in Italy, a perfectly transparent, fine glaze, consisting of a lead-silica mixture, covers everything and provides protection to the underlying colored scene. This perfectly white and homogeneous background will be widely used for cobalt-based blue patterns that for quite a time would be the best imitation of Chinese "white-blue" porcelains of the Ming period, highly valued in the 17th century.

10.8. Soft-paste porcelains: lead glaze on translucent paste

The French and Belgian ceramists of the 18th century would explore another direction in order to reach the appearance of Chinese porcelains, partly inspired from the pottery know-how. Soft-paste porcelains are characterized by a very white and translucent body, achieved with a paste that Brongniart qualified as "artificial" given its high complexity (Brongniart 1877, p. 444, volume II). Following the experiments conducted by Louis Poterat at Rouen before 1696, several renowned ceramists developed a paste containing a frit composed of sand and soda first, or of sand, soda, alum[8], saltpeter, sea salt and a marly white clay (rich in calcium) to which over 13% chalk was added in the later pastes of Sèvres and Tournai. Very subtle colors magnify and characterize the decoration of these soft porcelains: dark blue, green or pink. These colors are applied either on the biscuit, or directly on the raw substrate before firing. The lead-rich transparent glaze is deposited after this firing. Depending on the manufacturers, the glaze recipes change (Heimann and

7. A faïence is defined as terra cotta of variable texture that is dissimulated by a white or sometimes colored opaque tin glaze that may itself have a pattern (according to (Blondel 2001, p. 72)).

8. Alum is a double potassium and hydrated aluminum sulfate.

Maggetti 2014, p. 310 and the following), but they all correspond to what Brongniart qualifies as flint glass, a mix of silica, lead and alkali: they are lead-alkaline for Chantilly soft-paste porcelains (1725-1800), slightly enriched in tin (2% SnO_2) at the beginning of the productions in order to mask the yellowish color of the paste; their lead content is higher for Sèvres (for example 40% PbO, 40% SiO_2, 6% alkaline).

Either way, these soft porcelains have low resistance to fire, as their coating is brittle and they can be easily scratched, but given the diversity of the colorway and the finesse of the patterns, they were in high demand by a wide range of customers.

10.9. Lead-rich vitrified paints

With the rise of porcelain and stoneware, lead can no longer be part of the composition of glazes. Indeed, its use is incompatible with the high temperatures required for firing these ceramics, as for most of the common pigments, unstable at high temperature. Therefore, the artists cannot obtain a subtle and diversified polychromed pattern unless they use vitrifiable paints with low melting point that they apply on the already fired base glaze. Brongniart (1877, p. 542) provides their main properties: a good paint should correctly adhere to the substrate, it should be sufficiently fusible to avoid modifying the underlying glaze and it should definitely not be subjected to crazing after firing. From an aesthetic perspective, it should be possible to mix it with others in order to augment the colorway and to preserve its coloring properties and its purity irrespective of the thickness of the colored layer.

The analyses of Chinese paints used starting with the 12th century (Needham 2004, p. 616) show that most of the time the recipes resemble those of traditional lead glazes, very rich in silica and lead with very low alkali content, particularly for the yellow and green ones in which the lead content is sometimes over 60% PbO (Vandiver et al. 1997; Wood 1999). The mix, different for each color, is fritted, then finely milled, mixed with a medium allowing the painting. Other colors, especially the blue ones, require rather a lead-alkaline basis, in which potassium, probably from a mineral source (saltpeter?) may sometimes reach 8% K_2O. For these enamels on glaze, it is important to note an obvious relation with the “enamels” on metal of the same period (Biron and Quette 1997). Hematite-based red paints are more difficult to obtain, as oxide crystals must be directly added to the lead-silica frit; this technique of adding coloring oxides to a pre-fritted base would predominate in more recent period and would be resumed in the West (d’Albis 2003). In the Persian world, too, starting with the 12th century, the paint on glaze technique is widely used for *Mina’i* and *Haftrang* ceramics. There are two types of recipes for the Persian *Haftrang* vitrified paints on glaze: blue and white are rather lead-alkali glazes, while brown, yellow and green correspond to lead-silica glazes (Holakooei 2019). This similarity between Persian and Chinese traditions and the contemporary emergence of these materials raise the question of the respective influences of the Persians and Chinese on the ceramic art of these two worlds.

Generally speaking, in the more ancient periods, the analyses of these enamels on glazes reveal that each color was prepared separately, a standardization actually appearing in the more recent periods (Ming), in which the preparation of several colors was simplified by using a single base; only the coloring agents differ. The analyses also show the constant experimentation conducted by the ceramists in order to expand the diversity of the colorway and thus address the fluctuations of fashion, local or global taste and meet the export demands. It is also possible to follow the transfers of technology from one region to another and show how the innovations developed by folk potters were resumed by the imperial potters and vice versa (Vandiver et al. 1997).

The same types of compositions were implemented in the large European factories. Several coloring agents used for the porcelains, particularly at Sèvres (Brongniart 1877), contain always a lead base (ceruse, minium) for certain cream-colored shades, yellow, white or mauve (lead chromates). It is all the more valid for soft pastes (pink, ruby, yellow (lead uranates)). Generally speaking, paint contains one or more colorants dispersed in one or more fluxes. It is important to note that at Sèvres, all the 13 fluxes listed for paintings on porcelain contain variable lead proportions (d'Albis 2003, p. 215). This element is therefore still widely present in these prestigious productions for both porcelains and soft pastes.

10.10. Conclusion

This history of lead glazes is just touched upon in these several pages. Given its abundance and multitude of shapes, choices had to be made and many of its aspects were not developed. We wanted to focus our attention on the emergence of know-how, on the pivotal periods, and we therefore had to leave out many significant productions or major artists. We also voluntarily stopped the history presentation at the dawn of the 19th century, as it was the beginning of a new world, which benefitted from the discoveries of modern chemistry, but also tried to limit the use of lead, which was considered too dangerous for health. In his book of reference, Alexandre Brongniart mentions the research works conducted at that time by the Manufacture de Sèvres particularly for finding substitutes for lead and tin, which were considered too costly for ordinary productions. Several years later, the preface of *L'Enquête des préfets* (Survey of prefects) (1805–1810), conducted at the request of the same Brongniart, mentioned that the removal of lead from glaze formulations was a public health measure (Dubus and Pannequin 1989). It is true that nowadays lead content was significantly reduced, and even fully banned in many formulations. Even though certain great artists continued to use it, most of them have often used stonewares and their high temperature alkaline glazes (Chaplet, Carriès, Gauguin, etc.). Even at Vallauris, the *alquifou* (name given to galena), a symbolic major component of ceramics with typical yellow or green glazes, was banned in the 1950s. It is now the time for writing another history of lead in the ceramics, namely that of its gradual replacement.

10.11. References

d'Albis, A. (2003). *Traité de la porcelaine de Sèvres*. Faton, Dijon.

Allan, J.W. (1973). Abü'l-Qäsim's treatise on ceramics. *Iran*, 11, 111–120.

Amouric, H., Démians d'Archimbaud, G., Vallauri, L. (1995). De Marseille au Langudoc et au Comtat Venaissin : les chemins du vert et du brun. In *Le Vert et le Brun – De Kairouan à Avignon, céramiques du Xe au XVe siècle*, Musées de Marseille (ed.). RMN, Paris.

Atik, N. (1995). *Die Keramik aus den Südthermen von Perge*. Ernst Wasmuch, Tübingen.

Barbe, F. (2016). *Majolique – L'âge d'or de la faïence italienne au XVIe siècle*. Citadelles et Mazenod, Paris.

Beauvoit, E., Ben Amara, A., Tessier-Doyen, N., Frugier, C., Lemasson, Q., Moignard, B., Pacheco, C., Pichon, L., Chapoulie, R., Gratuze, B. (2021). Chemical and mechanical characterisation of white earthenware glazes from the Johnston-Vieillard manufactory (France, 19th Century). *Archaeometry*. doi: 10.1111/arcm.12656.

Biron, I. and Quette, B. (1997). Les premiers émaux chinois. *Technè*, 6, 35–40.

Blondel, N. (2001). *Céramique – Vocabulaire technique. Monum*. Patrimoine Centre des monuments nationaux, Paris.

Bouquillon, A. (2012). Connaissance des verres et glaçures du patrimoine : une approche analytique par spectroscopies et faisceaux d'ions. HDR Thesis, Université de Lille, Lille.

Bouquillon, A., Barthelemy de Saizieu, B., Duval, A. (1995). Glazed steatite beads from Mehrgarh and Nausharo (Indus valley). In *Materials issues in Art and Archaeology*, Druzig, J., Vandiver, P. (eds). Cambridge University Press, Cambridge.

Bouquillon, A., Aucouturier, M., Chabanne, D. (2008). Analyses des céramiques lustrées au C2RMF. In *Reflets d'or d'orient en occident, la céramique lustrée – IX–XVe siècle*, Dectot, X., Makariou, S., Miroudot, D. (eds). RMN, Paris.

Bouquillon, A., Bormand, M., Zucchiatti, A. (2011). *Della Robbia. Dieci anni di Studi – Dix ans d'études*. Sagep, Genoa.

Bouquillon, A., Castaing, J., Barbe, F., Paine, S.R., Christman, B. (2017). Lead-glazed rustiques figulines (rustic ceramics) of Bernard Palissy (1510–1590) and his followers. *Archaeometry*, 59(1), 69–83.

Bouquillon, A., Castaing, J., Barbe, F., Crepin-Leblond, T., Tilliard, L., Paine, S.R., Christman, B., Heuer, A.H. (2018). French decorative ceramics mass-produced during and after the 17th century: Chemical analyses of the glazes. *Archaeometry*, 60(5), 946–965.

Brongniart, A. (1877). *Traité des Arts céramiques ou des poteries*. Hachette, Paris.

Brun, N. and Pernot, M. (1992). The opaque red glass of celtic enamels from continental Europe. *Archaeometry*, 34(2), 235–252.

Burlot, J., Waksman, S.Y., Bellot-Gurlet, L., Simsek Franci, G. (2020). The glaze production technology of an early Ottoman pottery (mid-14th(?)-16th century): The case of "Miletus Ware". *J. Archaeol. Sci.: Reports*, 29, 102073.

Cabouat, P. and Ogaki, A. (2004). *Des romains au cœur de la Chine*. Documentary, NHK, France 5, Point du Jour [Online]. Available at: https://www.youtube.com/watch?v=5fBc-ou_GT0 [Accessed 11 August 2020].

Caubet, A. and Pierrat, G. (eds) (2005). *Faïences de l'antiquité, de l'Egypte à l'Iran*. Cinq Continents/Musée du Louvre, Paris.

Cui, J., Wu, X., Huang, B. (2011). Chemical and lead isotope analysis of some lead-barium glass wares from the Warring States Period, unearthed from Chu tombs in Changde City, Hunan Province, China. *J. Archaeol. Sci.*, 38(7), 1671–1679.

Cvjeticanin, T. (1997). Late Roman glazed pottery as a military commodity. *Rei cretariae romanae fautorum acta*, 35, 17–25.

Cvjeticanin, T. (2000). Eastern Mediterranean elements in technology and design of Upper Moesian pottery: The case of glazed pottery. *Rei cretariae romanae fautorum acta*, 36, 319–327.

De Benedetto, G.E., Acquafredda, F., Masieri, M., Quarta, G., Sabbatini, L., Zambonin, P.G., Tite, M.S., Walton, M. (2004). Investigation on Roman lead glaze from Canosa: Results of chemical analyses. *Archaeometry*, 46(4), 615–624.

Delile, H. (2014). Les paléo-pollutions au plomb, témoins des conditions socio-économiques de la Rome antique. *Med. Sci.*, 30(10), 831–833.

Dubus, M. and Pannequin, B. (1999). La céramique française sous l'Empire à travers l'enquête des préfets (1805–1810). Notes et Documents des Musées de France/RMN, Paris.

Feugère, L., Bouquillon, A., Beillard, B., Bailly, M. (2002). Le Lohan en terre vernissée du Musée Guimet. *Techne*, 16, 20–25.

Gaborit, J.-R. and Bormand, M. (2002). *Les Della Robbia, sculptures en terre cuite émaillée de la Renaissance italienne*. RMN, Paris.

Gan, F., Cheng, H., Li, Q. (2006). Origin of Chinese ancient glasses – Study on the earliest Chinese ancient glasses. *Science in China, Series E: Technological Sciences*, 49(6), 701–713.

Garnier, E. (1893). *Dictionnaire de la céramique : faïences, grès, poteries*. Librairie de l'art, Paris.

Godet, M., Roisine, G., Beauvoit, E., Caurant, D., Majérus, O., Menguy, N., Dargaud, O., Bouquillon, A., Cormier, L. (2019). Multi-scale investigation of body-glaze interface in ancient ceramics. *Heritage*, 2(3), 2480–2494.

Goslier, P. (2018). *Les céramiques à glaçure plombifère antiques en Gaule méridionale et dans la vallée du Rhône (Ier s. av. J.-C.–IIIe s. apr. J.-C.)*. Mergoil, Drémil-Lafage.

Greene, K. (2007). Late Hellenistic and Early Roman invention and innovation: The case of lead-glazed pottery. *Am. J. Archaeol.*, 111(4), 653–671.

Habicht-Mauche, J.A., Glenn, S.T., Schmidt, M.P., Franks, R., Milford, H., Russell Flegal, A. (2002). Stable lead isotope analysis of Rio Grande Glaze paints and ores using ICP-MS: A comparison of acid dissolution and laser ablation techniques. *J. Archaeol. Sci.*, 29(9), 1043–1053.

Hamer, F. and Hamer, J. (1997). *The Potter's Dictionary of Materials and Techniques*. A and C, London.

Hatcher, H., Kaczmarczyk, A., Scherer, A., Symonds, R.P. (1994). Chemical classification and provenance of some Roman glazed ceramics. *Am. J. Archaeol.*, 98(3), 431.

Heimann, R.B. and Maggetti, M. (2014). *Ancient and Historical Ceramics – Materials, Technology, Art, and Culinary Traditions*. Schweizerbart Science Publishers, Stuttgart.

Hochuli-Gysel, A. (2002). La céramique à glaçure plombifère d'Asie Mineure et du bassin méditerranéen oriental (du Ier s. av. J.-C. au Ier s. ap. J.-C.). In *Céramiques hellénistiques et romaines, productions et diffusion en Méditerranée orientale (Chypre, Égypte et côte syro- palestinienne)*, Blondé, F., Ballet, P. (eds). Maison de l'Orient méditerranéen, Lyon.

Holakooei, P., de Lapérouse, J.-F., Carò, F., Röhrs, S., Franke, U., Müller-Wiener, M., Reiche, I. (2019). Non-invasive scientific studies on the prkilnance and technology of early Islamic ceramics from Afrasiyab and Nishapur. *J. Archaeol. Sci.: Reports*, 24, 759–772.

Huntley, D.L., Spielmann, K.A., Habicht-Mauche, J.A., Herhahn, C.L., Russell Flegal, A. (2007). Local recipes or distant commodities? Lead isotope and chemical compositional analysis of glaze paints from the Salinas pueblos, New Mexico. *J. Archaeol. Sci.*, 34(7), 1135–1147.

Kaczmarczyk, A. and Hedges, R. (1983). *Ancient Egyptian Faience: An Analytical Survey of Egyptian Faience from Predynastic to Roman Times*. Aris & Phillips Ltd, Warminster.

Kingery, W.D. and Vandiver, P.B. (1986). *Ceramic Masterpieces: Art, Structure and Technology*. The Free Press, New York.

Klesner, C.E., Akymbek, Y., Vandiver, P.B. (2019). Local production and long-distance trade of Islamic glazed ceramics in Central Asia: A compositional analyses of ceramics from Southern Kazakhstan by NAA and LA-ICP-MS. *J. Archaeol. Sci.: Reports*, 26, 101905.

Klesner, C.E., Akymbek, Y., Vandiver, P.B. (2021). Lead-glazing technology from Medieval Central Asia: A case study from Aktobe, Kazakhstan. *J. Archaeol. Sci.: Reports*, 36, 102825.

Lehman, R.L. (2002). *Lead Glazes for Ceramic Foodware*. The International Lead Management Center, Research Triangle Park.

Lister, F.C. and Lister, R.H. (1982). *Sixteenth Century Maiolica Pottery in the Valley of Mexico*. University of Arizona Press, Tucson.

Maggetti, M. (2018). Archaeometric analyses of European 18th–20th century white earthenware. A review. *Minerals*, 8(7), 269.

Maggetti, M. (2020). The introduction of the tin opacified glaze technique in Switzerland at the end of the 15th century CE – A SEM study of stove tiles from the canton Bern. *J. Archaeol. Sci.: Reports*, 34, 102601.

Mason, R.B. (2004). *Shine like the Sun: Lustre-painted and Associated Pottery from the Medieval Middle East*. Mazda Press, Toronto.

Mason, R.B. and Tite, M.S. (1994). The beginnings of Islamic stonepaste technology. *Archaeometry*, 36(1), 77–91.

Matin, M. (2019). Tin-based opacifiers in archaeological glass and ceramic glazes: A review and new perspectives. *Archaeol. Anthropol. Sci.*, 11, 1155–1167.

Matin, M., Tite, M.S., Watson, O. (2018). On the origins of tin-opacified ceramic glazes: New evidence from early Islamic Egypt, the Levant, Mesopotamia, Iran, and Central Asia. *J. Archaeol. Sci.*, 97, 42–66.

Matoïan, V. and Bouquillon, A. (1999). La céramique argileuse à glaçure de Ras-Shamra–Ougarit. *Syria*, 76, 57–82.

Matoïan, V. and Bouquillon, A. (2003). Vitreous materials in Ugarit: New data. In *Culture through Objects: Near Eastern, Studies in Honour of P.R.S. Moorey*, Potts, T., Roaf, M., Stein, D. (eds). Griffith Institute, Oxford.

Medeghini, L., Fayek, M., Mignardi, S., Coletti, F., Contino, A., De Vito, C. (2020). A prkilnance study of Roman lead-glazed ceramics using lead isotopes and secondary ion mass spectrometry (SIMS). *Microchem. J.*, 154, 104519.

Molera, J., Cristóbal, J., López, C., Molina, G., Pradell, T. (2018). Glazes, colourants and decorations in early Islamic glazed ceramics from the Vega of Granada (9th to 12th centuries CE). *J. Archaeol. Sci.: Reports*, 21, 1141–1151.

Molera, J., Martínez Ferreras, V., Fusaro, A., Gurt Esparraguera, J.M., Gaudenzi, M., Pidaev, S.R., Pradell, T. (2020). Islamic glazed wares from ancient Termez (southern Uzbekistan). Raw materials and techniques. *J. Archaeol. Sci.: Reports*, 29, 102169.

Needham, J. (2004). Ceramic technology. In *Science and Civilisation in China: Volume 5, Chemistry and Chemical Technology, Part 12*, Kerr, R., Wood, N. (eds). Cambridge University Press, Cambridge.

Nicholson, P.T. (1993). *Egyptian Faience and Glass. Shire Egyptology Series, 19*. Shire Publication, Buckinghamshire.

Nicolas, B. (2020). La formidable épopée des Compagnies des Indes. *L'objet d'Art*, 570, 56–71.

Palissy, B. (ed.) (1580). Discours Admirables de la nature des eaux et fontaines tant naturelles qu'artificielles, des sels et salines, des pierres, des terres et du feu et des esmaux. In *Oeuvres de Bernard Palissy*. Charavay Frères, Paris.

Pérez-Arantegui, J., Uruñuela, M.I., Castillo, J.R. (1996). Roman glazed ceramics in the Western Mediterranean: Chemical characterization by inductively coupled plasma atomic emission spectrometry of ceramic bodies. *J. Archaeol. Sci.*, 23(6), 903–914.

Piccolpasso, C. (1559). *Li tre libri dell'arte del vasaio*. Edizioni del Giglio, Florence.

Pradell, T. and Molera, J. (2020). Ceramic technology. How to characterise ceramic glazes. *Archaeol. Anthropol. Sci.*, 12(189) [Online]. Available at: https://doi.org/10.1007/s12520-020-01136-9.

Rosen, J. (2018). *La faïence en France du XIIIe au XIXe siècle : technique et histoire*. Errance, Paris.

Roisine, G. (2018). Céramiques glaçurées de Bernard Palissy : à la recherche des secrets d'un maître de la Renaissance. PhD Thesis, PSL Université, Paris.

Roisine, G., Capobianco, N., Caurant, D., Wallez, G., Bouquillon, A., Majerus, O., Cormier, L., Gilette, S., Gerbier, A. (2017). The art of Bernard Palissy (1510–1590): Influence of firing conditions on the microstructure of iron-coloured high-lead glazes. *Appl. Phys. A*, 123(8), 1–9.

Rugiadi, M. (2016). Stonepaste technology in Syria and Iran. In *Court and Cosmos: The Great Age of the Seljuqs*, Canby, S.R., Beyazit, D., Rugiadi, M., Peacock, A.C.S. (eds). The Metropolitan Museum of Art/Yale University Press, New York/New Haven.

Salinas, E. and Pradell, T. (2018). The transition from lead transparent to tin-opacified glaze productions in the western Islamic lands: al-Andalus, c. 875–929 CE. *J. Archaeol. Sci.*, 94, 1–11.

Salinas, E. and Pradell, T. (2020). The introduction of the glaze in al-Andalus: Technological waves and oriental influences. *Libyan Studies*, 51, 87–98.

Salinas, E., Pradell, T., Matin, M., Tite, M.S. (2019). From tin- to antimony-based yellow opacifiers in the early Islamic Egyptian glazes: Regional influences and ruling dynasties. *J. Archaeol. Sci.: Reports*, 26, 101923.

Salinas, E., Reynolds, P., Tite, M.S., Pradell, T. (2020). Polychrome glazed ware production in Tunisia during the Fatimid-Zirid period: New data on the question of the introduction of tin glazes in western Islamic lands. *J. Archaeol. Sci.: Reports*, 34, 102632.

Sciau, P. (2012). Nanoparticles in ancient materials: The metallic lustre decorations of medieval ceramics. In *The Delivery of Nanoparticles*, Hashim, A.A. (ed.). IntechOpen, London.

Shepard, A.O. (1956). *Ceramics for the Archaeologist*. Carnegie Institution of Washington, Washington.

Thibodeau, A.M., Habicht-Mauche, J.A., Huntley, D.L., Chesley, J.T., Ruiz J. (2013). High precision isotopic analyses of lead ores from New Mexico by MC-ICP-MS: Implications for tracing the production and exchange of Pueblo {IV} glaze-decorated pottery. *J. Archaeol. Sci.*, 40(7), 3067–3075.

Tite, M.S. (2011). The technology of glazed Islamic ceramics using data collected by the late Alexander Kaczmarczyk. *Archaeometry*, 53(2), 329–339.

Tite, M.S. and Shortland, A.J. (2008). *Production Technology of Faience and Related Early Vitreous Materials*. Oxford University School of Archaeology, Oxford.

Tite, M.S., Freestone, I., Mason, R., Molera, J., Vendrell-Saz, M. (1998). Lead glazes in antiquity methods of production and reasons for use. *Archaeometry*, 40, 241–260.

Tite, M.S., Shortland, A.J., Bouquillon, A. (2008a). Glazed steatite. In *Production Technology of Faience and Related Early Vitreous Materials*, Tite, M.S., Shortland, A.J. (eds). Oxford University School of Archaeology, Oxford.

Tite, M.S., Shortland, A.J., Bouquillon, A., Kaczmarczyk, A., Vandiver, P.B. (2008b). Faience production in the near East and the Indus Valley. In *Production Technology of Faience and Related Early Vitreous Materials*, Tite, M.S., Shortland, A.J. (eds). Oxford University School of Archaeology, Oxford.

Tite, M.S., Shortland, A.J., McCarthy, B., Paynter, S. (2008c). Production of glazed pottery and Brickwork in the Near East. In *Production Technology of Faience and Related Early Vitreous Materials*, Tite, M.S., Shortland, A.J. (eds). Oxford University School of Archaeology, Oxford.

Tite, M.S., Pradell, T., Shortland, A. (2008d). Discovery, production and use of tin-based opacifiers in glasses, enamels and glazes from the late iron age onwards: A reassessment. *Archaeometry*, 50, 67–84.

Tite, M.S., Watson, O., Pradell, T., Matin, M., Molina, G., Domoney, K., Bouquillon, A. (2015). Revisiting the beginnings of tin-opacified Islamic glazes. *J. Archaeol. Sci.*, 57(80), 80–91.

Van Keuren, S., Neff, H., Agostini, M.R. (2013). Glaze-paints, technological knowledge, and ceramic specialization in the fourteenth-century pueblo southwest. *Archaeol. Anthropol. Sci.*, 32(4), 675–690.

Vandiver, P., Bouquillon, A., Kerr, R., Scott, R. (1997). The technology of early overglaze enamel from the Chinese imperial and popular kilns. *Techne*, 6, 25–34.

VanValkenburgh, P., Kelloway, S.J., Dussubieux, L., Quilter, J., Glascock, M.D. (2015). The production and circulation of indigenous lead-glazed ceramics in northern Peru during Spanish colonial times. *J. Archaeol. Sci.*, 61, 172–185.

VanValkenburgh, P., Kelloway, S.J., Privat, K.L., Sillar, B., Quilter, J. (2017). Rethinking cultural hybridity and technology transfer: SEM microstructural analysis of lead glazed ceramics from early colonial Peru. *J. Archaeol. Sci.*, 82, 17–30.

Waksman, S.Y., Bouquillon, A., Cantin, N., Katona, I. (2007). The first Byzantine "Glazed White Wares". In *The Early Medieval Technological Context, in Archaeometric and Archaeological Approaches to Ceramics (EMAC'05)*, Waksman, S.Y. (ed.). BAR International Series 1691, Oxford.

Waksman, S.Y., Bouquillon, A., Cantin, N., Katona, I. (2008). Approche archéométrique des premières "Byzantine glazed white ware" et de productions glaçurées romaines et romaines tardives. In *Rei Cretariae Romanae Fautorum Acta 40*, Biegert, S. (ed.), 531–536.

Walton, M.S. (2004). A materials chemistry investigation of archaeological lead glazes. PhD Thesis, University of Oxford, Oxford.

Walton, M.S. and Tite, M.S. (2010). Production technology of Roman lead-glazed pottery and its continuance into late antiquity. *Archaeometry*, 52(5), 733–759.

Wang, Y., Zhou, Y., Yang, Z., Cui, J. (2019). A technological combination of lead-glaze and calcium-glaze recently found in China: Scientific comparative analysis of glazed ceramics from Shangyu, Zhejiang Province. *PLoS One*, 14(7), 1–11.

Wood, N. (1999). *Chinese Glazes, their Origins, Chemistry and Recreation*. A&C Black, London.

Živković, J., Power, T., Georgakopoulou, M., Carvajal López, J.C. (2019). Defining new technological traditions of Late Islamic Arabia: A view on Bahlā Ware from al-Ain (UAE) and the lead-barium glaze production. *Archaeol. Anthropol. Sci.*, 11(9), 4697–4709.

Exhibition catalogs

Crépin-Leblond, T., Gerbier, A., Madinier-Duée, P. (2016). *Masséot Abaquesne – L'éclat de la faïence à la Renaissance*. Réunion des Musées Nationaux, Paris.

Musées de Marseille (1995). *Le Vert et le Brun – De Kairouan à Avignon, céramiques du Xe au XVe siècle*. Musées de Marseille, Réunion des Musées Nationaux.

Pannequin, B. and Lahaussois, C. (1999). *L'art de la terre vernissée du Moyen-Age à l'an 2000*. Musée National de la céramique à Sèvres, Réunion des Musées nationaux.

Rosen, J. and Crépin-Leblond, T. (eds) (2000). *Images du pouvoir – Pavements de faïences en France du XIIIe au XVIIe siècle*. Musées de Brou et Réunion des Musées Nationaux, Paris.

11

Paste–Glaze Interaction

Laurent CORMIER and Marie GODET
IMPMC, CNRS, Sorbonne University, MNHN, IRD, Paris, France

A vitreous coating on ceramic has several functions: a practical one, by waterproofing and protecting the earthenware object, but also an esthetic and symbolic one. The extent and the nature of the interface connecting the two materials – fired clay of the body and glassy material of the decoration – play an essential role in the final composition of the glaze and many physicochemical properties of the finished object. Relying on the pioneering and very detailed works of Molera et al. (2001), Walton (2004), Ben Amara (2002) and the recent works of Roisine (2018) and Godet et al. (2019), this chapter proposes a review of the state of knowledge for lead-rich glazes[1].

11.1. Context

The use of lead in glazes offers several technological and esthetic advantages, such as low melting temperatures facilitating their application on the ceramic object (lead acts as an efficient flux), strong brightness and a significant range of colors depending on the added metallic oxides.

---

1. Acknowledgements: this project was financed by the Région Île-de-France (DIM Heritage and Ancient Materials) and Sorbonne Université. Our thanks go to N. Menguy and J.-M. Guigner (IMPMC) for their help in TEM data acquisition. We also thank the FIB-SEM platform of IMPMC (support Région Île-de-France SESAME 2006 No. I-07-593/R, INSU-CNRS, INP-CNRS, Sorbonne Université). This chapter benefited from fruitful exchanges with Anne Bouquillon, Daniel Caurant, Patrice Lehuédé and Gauthier Roisine.

Lead in Glassy Materials in Cultural Heritage,
coordinated by Anne BOUQUILLON and Patrice LEHUÉDÉ.

Several methods of application are known, from the simplest to the most complex ones (see Chapter 10): (1) direct application of a lead component, most often litharge (PbO) and also sometimes galena (PbS), white lead ($2PbCO_3 \cdot Pb(OH)_2$) or minium (Pb_3O_4) ; (2) application of a mixture of lead oxide PbO and silica (sand, pebbles, siliceous clay); (3) PbO-SiO_2 mixture previously melted to form a frit that, after a fine milling, will be applied; (4) addition of a mixture containing variable amounts of alkali.

Depositing these mixtures is sometimes facilitated by putting them in suspension in water and by adding clay or organic binder to the glaze mixture (Arabic gum or starch, for example). The formulation thus obtained is applied by immersing the object into the mixture, by pouring the mixture on the object or by paintbrush. The object to be glazed may be raw (not fired) paste or a paste that was subjected to a first firing (referred to as biscuit). In practice, the glaze mixture will be more easily applied to a fired body which would be less affected by the absorption of water in the suspension and by drying (Benedetto et al. 2004). After applying the mixture, the paste–glaze system is fired most often between 700°C and 1,000°C for lead glazes. The presence of fluxes in large amounts (lead sometimes associated with alkali), and having low melting points, facilitates the formation of a liquid phase of low viscosity that penetrates the pores in the surface of the paste and reacts with the minerals constituting the ceramic substrate (Costa et al. 2014), which may lead to their partial dissolution in the glassy phase or to their transformation. The result is a transfer of matter between the glaze and the substrate. When cooling, the thin liquid layer forms the glaze, which is essentially a glass that may include crystalline inclusions.

The reactions between the paste and the molten glaze mixture create a contact zone that may be qualified as “strategic”. Indeed, they strongly influence the physical, chemical and mechanical properties of the glaze, such as crack resistance, mechanical strength, chemical durability, thermal shock resistance, the maturation behavior of the glaze and sometimes even the esthetic appearance (Olupot et al. 2010). Their study reveals valuable information on the technological processes used throughout history.

This chapter presents a review of the parameters influencing the glaze–ceramic interfaces by describing the chemical processes involved, the characterizations of the microstructure and crystalline phases. This knowledge makes it possible to address essential questions concerning the elaboration processes and to better understand their influence on the final properties of the ceramic.

The results are illustrated by a study of the interface between replica glazes of Palissy type and a biscuit composed of a kaolinitic clay of Provins commercialized under the reference RP299 (79.09% SiO_2-19.18% Al_2O_3-0.02% K_2O-0.48% MgO-0.62% CaO-0.06% TiO_2-0.55% Fe_2O_3) mixed with 10% Fontainebleau sand. The frits used to obtain the lead glazes are transparent, colored with iron, with and without aluminum: 31.8% PbO-58.7% SiO_2-6.9% Al_2O_3-2.6% Fe_2O_3 (composition A) and 34.2% PbO-63.0% SiO_2-2.8% Fe_2O_3 (composition B). All the compositions are in mol% unless otherwise mentioned. The elaboration of these replicas has been previously described by Godet et al. (2019). These results are compared to those obtained for archeological shards of glazed ceramics from Palissy's workshop in Paris (Tuileries).

11.2. Paste–glaze interface

The paste–glaze interface is the site of complex dissolution–diffusion processes unfolding in three stages (Molera et al. 2001): (1) dissolution of the ceramic body in contact with the lead molten glaze; (2) diffusion of the chemical species from the clay body to the glaze (alkali, alkali-earth, aluminum, silicon) and from the glaze to the clay substrate (lead, potential chromophores); (3) formation and growth of crystalline phases. The glaze may furthermore penetrate the ceramic bodies having open porosity.

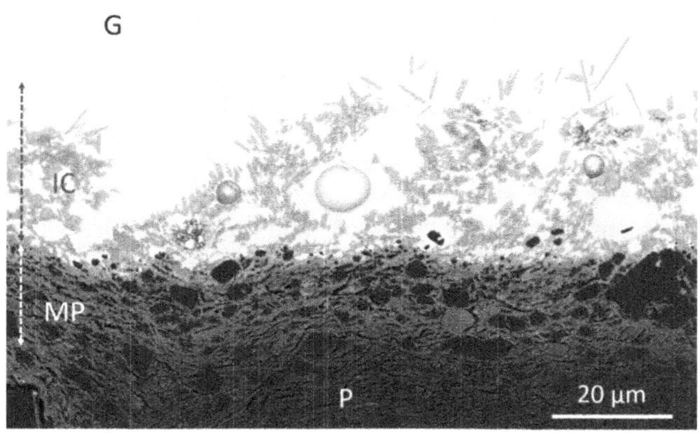

Figure 11.1. *Paste–glaze interface of a Palissy-type replica (composition A on kaolinitic biscuit) obtained after firing at 1,000°C for 2 h and cooled at 50°C·h^{-1} (SEM image, backscattered electrons). G = glaze, IC = interfacial crystals, MP = modified paste, P = paste. For a color version of this figure, see www.iste.co.uk/bouquillon/lead.zip*

The interface corresponds to a band of specific structure and composition, differing both from the ceramic substrate and from the glaze. Its total extension must therefore be assimilated to the layer of devitrification crystals and the zone of the paste modified by lead penetration (Figure 11.1) (Godet et al. 2019). It should however be noted that certain authors may only take into account the devitrification layer (Coentro et al. 2017).

The body-glaze frontier is often irregular and diffuse, and its exact delimitation is delicate to estimate. It is possible that crystals get detached from the interface and "float" in the thickness of the glaze. Bubbles may also be present at the interface. The thickness and the morphology of this layer depend on many parameters that will be developed in the next section.

11.3. Factors affecting the paste–glaze interface

Several parameters influence the physicochemical characteristics and the development of the interface (Parmelee and Buckles 1942; Molera et al. 1999; Molera et al. 2001; Kara and Stevens 2002; Benson 2003; Olupot et al. 2010): first of all, the composition of the glaze and of the clay substrate, then the mode of application (deposition on already fired or raw clay, raw or prefritted glaze); firing temperatures, firing time at high temperature, cooling rate also have a role in the diffusion of chemical elements between the paste and the glaze, and in the nature and quantity of crystals developing at the interface.

11.3.1. *Composition of the body (Ca-rich clays or siliceous ones) and its state (raw or fired)*

In the heritage materials, the types of pastes are globally ranged into three large categories: calcareous pastes, non-calcareous and highly aluminous pastes, siliceous pastes. Each of these groups is then subdivided depending on the other components it contains. The interfaces are generally larger on calcareous pastes, due to the good reactivity of calcium with the molten glaze; the presence of potassium for aluminous clays also facilitates the development of the interface with respect to a pure kaolinitic clay[2] (Molera et al. 2001). The interaction of the glaze with the ceramic body results from physicochemical processes; researchers have initially proposed

2. Kaolinite is a clay without cation in the interfolliar space.

that it was more significant in the raw pastes (Thomas et al. 1945). Indeed, the components of the raw paste can react with the constituents of the glaze mixture, while for a fired paste they would have already reacted with other components of the body and they would be less available to react with the glaze. This crucial and intensely debated point will be discussed in detail in section 11.7.1.

11.3.2. *Composition of the glaze mixture*

The objective of the reproduction tests is often the study of the effect of the glaze mixture composition (lead only, mixture of lead + silica or alkali-lead mixture opacified or not) and, to a lesser extent, the effect of the state of the glaze mixture (fritted or not, granulometry) on the development of the interface. Higher lead contents of the glaze mixture facilitate the development of interfaces (Molera et al. 2001). Conversely, the presence of aluminum induces a less developed interface (Figure 11.2). It is clearly visible on our replicas: the interface with the glaze containing Al_2O_3 (composition A) measures around 40 µm while that with the glaze without alumina (composition B) exceeds 60 µm.

Moreover, it is known that the addition of sodium, in the form of carbonate, in a lead glaze mixture enhances the dissolution/diffusion process between the glaze and the earthenware and, consequently, the thickness of the interface zone (Ben Amara 2002).

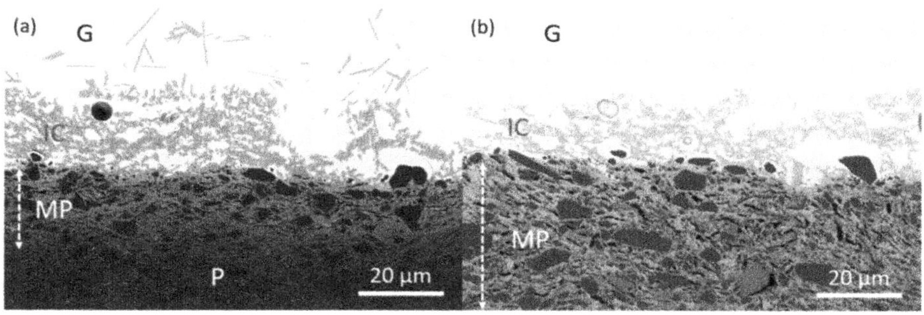

Figure 11.2. *Effect of the glaze mixture composition on interfaces with a kaolinitic biscuit obtained after firing at 1,000°C for 2 h and cooled at 50°C·h⁻¹ (SEM images, backscattered electrons). For a color version of this figure, see www.iste.co.uk/bouquillon/lead.zip*

COMMENT ON FIGURE 11.2.– *(a) Glaze mixture with aluminum (composition A). (b) Glaze mixture without aluminum (composition B). G = glaze, IC = Interfacial crystals, MP = modified paste, P = paste. Adapted from Godet* et al. *(2019).*

11.3.3. *Viscosity and solubility of the body and glaze components*

Viscosity and reactivity of the mixtures are at the origin of two mechanisms operating during firing (Walton 2004). The application of only PbO on a ceramic object gives a very fluid liquid that, according to certain authors, reacts preferentially with the quartz grains available in the paste. Low viscosity facilitates the infiltration of liquid into the clay body and the dissolution of remaining minerals. In a PbO + SiO_2 mixture, the two oxides react first, forming a liquid of relatively high viscosity. This liquid shallowly penetrates the ceramic matrix; it also dissolves the present mineral phases, but to a lesser extent. Typical values of viscosity of 10^4–10^5 Pa.s between 900°C and 1,000°C are reported in the literature (Tite et al. 1998; Roisine 2018).

Another element, aluminum, increases strongly the viscosity. Glazes with very high lead content generally have Al_2O_3 contents ranging between 2 and 7 weight% (Roisine 2018). In the presence of alumina, the glaze tends less to crystallize given the limited mobility of chemical species that would have more difficulties in getting structurally ordered to form a crystal. As illustrated in Figure 11.2, the penetration of lead into the paste is significantly reduced in the presence of Al (Godet et al. 2019), thus confirming the observations of Molera (Molera et al. 2001) and Bouquillon for the more aluminous glazes of post-Palissy ceramics (Bouquillon et al. 2018). Aluminum therefore reduces the interface reactions and acts as a glaze stabilizer.

11.3.4. *Thermal history of the ceramic body-glaze system*

Temperature and melting duration affect the diffusion of chemical elements; cooling rates influence crystal growth. As shown on our reproductions (Figures 11.3–11.5), the thickness of the interface increases with higher firing temperature, longer maturation time and slower cooling (larger crystals with more facets, Figure 11.5) (Godet et al. 2019). These observations are systematically found among all authors.

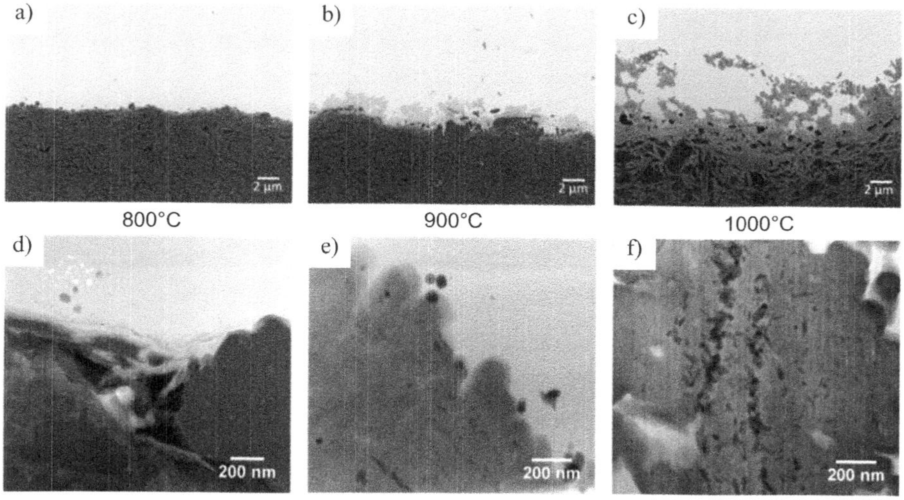

Figure 11.3. *Effect of temperature on the paste–glaze interface of Palissy-type replicas (composition A on kaolinitic biscuit) fired without plateau and quenched. Observations at micro-scale (a–c, SEM images, backscattered electrons) and nano-scale (d–f, STEM-HAADF images (chemical contrast images))*

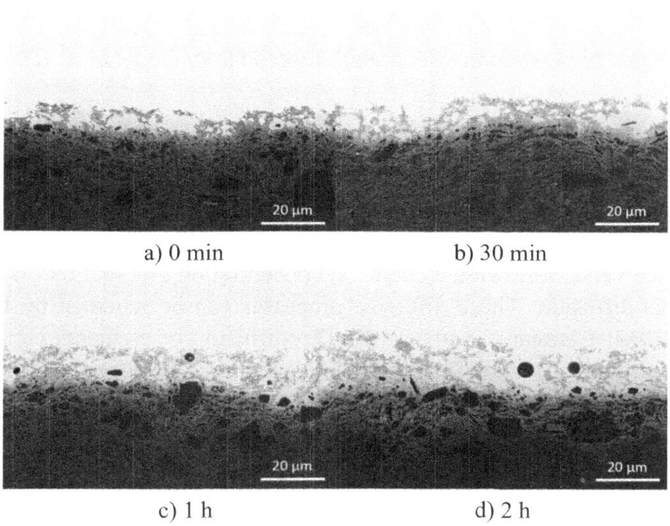

Figure 11.4. *Effect of the firing time on the paste–glaze interface of Palissy-type replicas (composition A on kaolinitic biscuit) fired at 1,000°C and quenched (SEM images, backscattered electrons)*

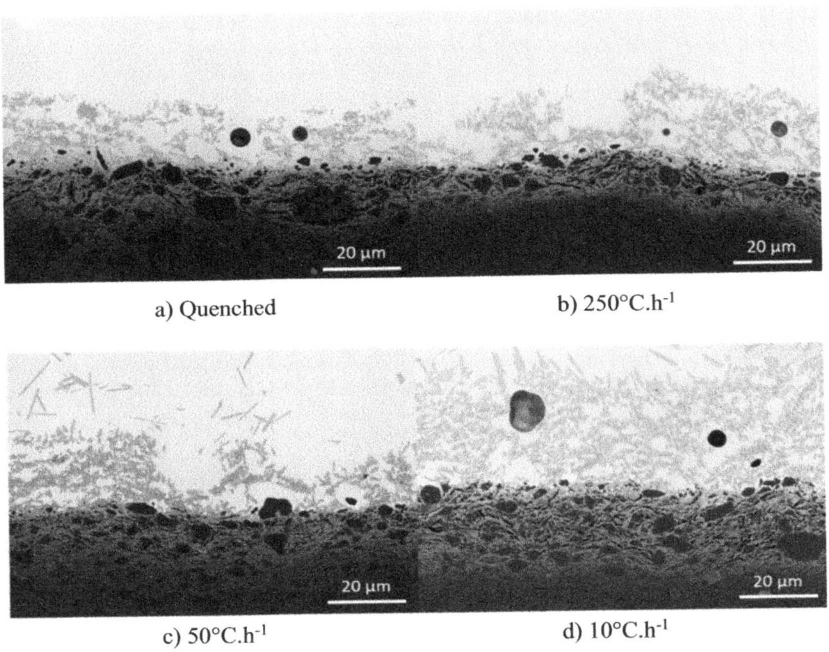

Figure 11.5. *Effect of cooling rate on the paste–glaze interface of Palissy-type replicas (composition A on kaolinitic biscuit) fired at 1,000°C for 2 h (SEM images, backscattered electrons). Adapted from Godet et al. (2019)*

11.4. Cross-diffusion of chemical elements

When temperature increases, the glaze becomes fluid and reacts with the underlying clay body. Then chemical species migrate from the clay substrate to the glaze and vice versa, following a chemical concentration gradient that is the driving force for this diffusion. These diffusive processes can be modeled by Fick's laws (Claireaux 2014; Claireaux et al. 2016). The diffusion of elements continues until the accumulation of species diffusing in the glaze drives a flat compositional profile, following a long firing or a slow cooling (Molera et al. 2001). The appearance of crystals renders this perspective on diffusion more complex.

Diffusion profiles can be used to determine the spatial distribution of elements (Molera et al. 2001; Ben Amara 2002; Walton 2004; Benedetto et al. 2004): from the ceramic body to the glaze, the diffusing elements are Si, Al, Fe, K, Ti, Mg and Ca for calcareous clays; conversely, Pb diffuses from the glaze to the ceramic body that does not contain it.

This chemical diffusion changes the composition of the glaze with respect to the initially applied mixture and modifies the distribution of elements near the interface, driving the formation of new phases and controlling the final morphology of the interface (Kara and Stevens 2002).

The diffusion of Al from a kaolinitic paste toward a glaze without aluminum is highlighted in Figure 11.6 for a glaze of composition B and a 41.4PbO-58.6SiO_2 glaze. Several glazes present final compositions close to that of eutectic in the PbO-SiO_2-Al_2O_3 ternary system (Tite et al. 1998; Chen et al. 2001): 61.2PbO-31.7SiO_2-7.1Al_2O_3. Molera obtained a comparable result for re-creations from various mixtures of oxides, with a final composition 42–45 weight% PbO, 30–35 weight% SiO_2, 10 weight% Al_2O_3 and 10–15 weight% other cations (K, Fe and Ca) (Molera et al. 2001).

Profiles of chemical diffusions are obtained on archeological samples such as the glaze of a Palissy ceramic shard presented in Figure 11.7 (transparent honey glaze on red paste). The diffusion profile of Al decreases with the distance from the interface and reaches a maximum in its proximity (~10 μm), which coincides with a minimum in the concentration of Si. Pb and Fe contents are quasi-constant in the thickness of the glaze, with a decrease in Pb at the paste–glaze interface, indicating its partial penetration in the paste.

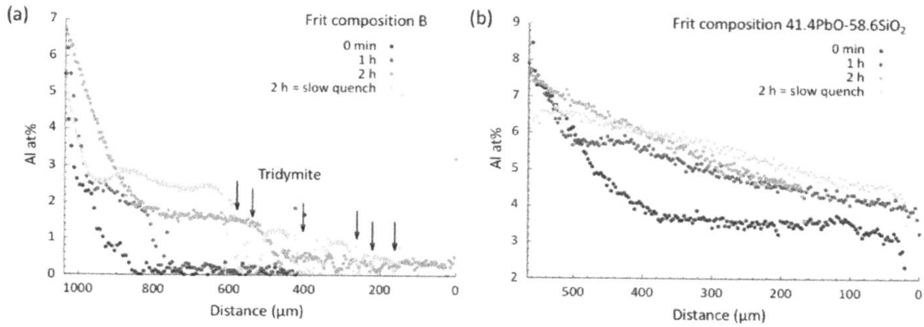

Figure 11.6. *Profiles of Al concentration measured by energy-dispersive X-ray (EDX) spectroscopy from the surface of the glaze (on the right) to the first interface crystals (on the left) for an annealing at 1,000°C with various firing times. For a color version of this figure, see www.iste.co.uk/bouquillon/lead.zip*

COMMENT ON FIGURE 11.6.– *(a) Glaze of composition B with crystallization of tridymite (located by arrows). (b) Binary glaze 41.4PbO-58.6SiO_2. For a firing time of 0 min, Al has already diffused during the temperature rise to 1,000°C. Adapted from Roisine (2018).*

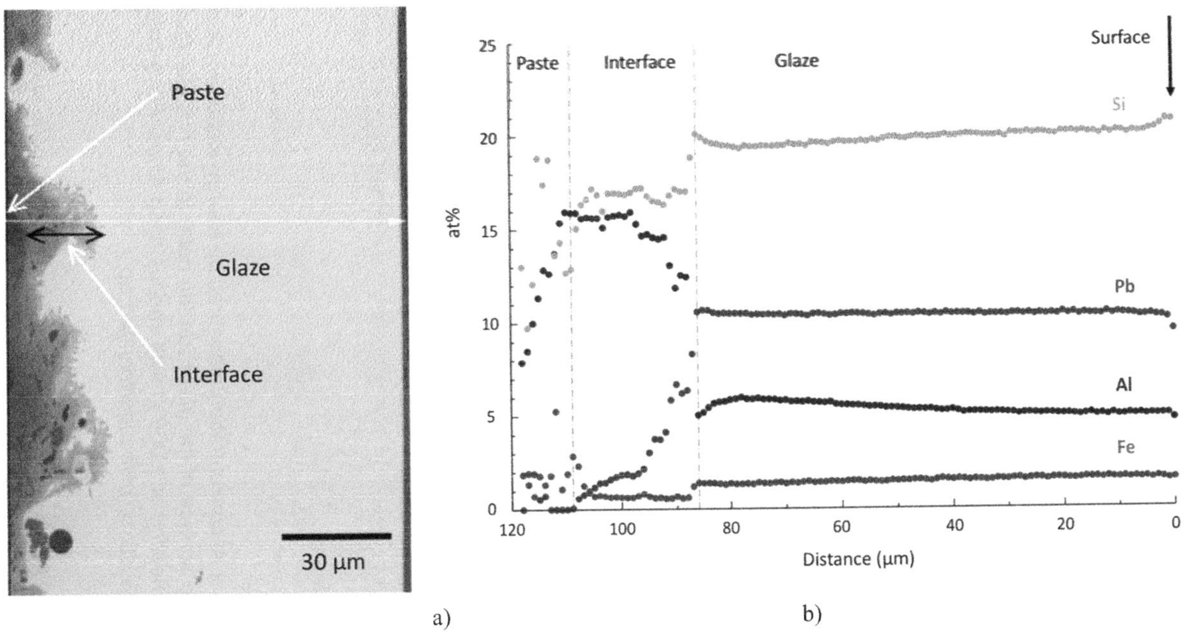

Figure 11.7. *Glaze of a Palissy ceramic shard (inventory no. EP341). (a) Interface micrograph (SEM image, backscattered electrons). (b) Concentration profiles measured by EDX with an origin corresponding to the surface of the glaze. Adapted from Roisine (2018). For a color version of this figure, see www.iste.co.uk/bouquillon/lead.zip*

The chemical profiles of glazes strongly depend on the thermal process employed and on the type of clay body. The diffusion of chemical elements from the clay body to the glaze is more significant for long firings or slow cooling (Molera et al. 2001). These conditions correspond to those observed in traditional wood furnaces, explaining that the diffusional equilibrium is reached and a quasi-uniform chemical composition is often observed in the glazes of archeological samples (Benedetto et al. 2004).

The local modifications of glaze composition, due to the diffusion during the firing of Al, K, Ti, Mg and Ca (in the case of Ca-rich pastes) and Fe, facilitate the crystallization of certain phases at the interface (see section 11.6).

11.5. Morphology of the interface

We studied the evolution of the morphology of the interface between a Palissy-type replica glaze (composition A) and a kaolinitic paste biscuit as a function of the firing temperature (Figure 11.3), the firing time (Figure 11.4) and the cooling rate (Figure 11.5). An increased digestion of the clay body is highlighted for longer firing times (Figure 11.4) or lower cooling rates (Figure 11.5), which can be explained by the diffusion, which is all the more significant as the time is longer. The interface appears very irregular and frequently contains newly formed crystals (mainly Pb feldspars described in section 11.6).

These first data obtained on re-creations were complemented by the study of an archeological shard of Bernard Palissy (inventory no. EP2319) by a transmission electron microscope (TEM). This shows the heterogeneity of the interface zone at micro and nanometric scales (Figure 11.8). Interfacial crystals (IC in Figure 11.8(a)) were studied by STEM-HAADF (scanning transmission electron microscopy-high-angle annular dark-field) imaging on a thin section obtained by FIB (focused ion beam) cutting. They have a complex morphology: a wide majority includes rod-shaped nanoparticles and some of them are surrounded by a lead-richer border (Figure 11.8(b)). The layer of crystals seems connected to the modified paste by an intermediate area where dark "parallel lines" can be observed within a clearer matrix. TEM observations reveal that these lines are composed of the same rod-shaped nanoparticles as those present inside the crystals (Figure 11.8(c)). The chemical and structural analysis (detailed in section 11.6) shows that the crystals are lead feldspars ($PbAl_2Si_2O_8$) and mullite ($3Al_2O_3$-$2SiO_2$) nanoparticles. Remarkably, the replicas realized show a morphology similar to the one observed on the Palissy sample (Godet et al. 2019).

The interface present in the Hispano-Moresque ceramic tiles, on a calcareous body, was studied by Coentro et al. (2014, 2017), revealing large thickness variations (30–100 µm). For the thickest interfaces, a large number of small mineral inclusions spread through the glaze until reaching the surface. According to the samples studied by Coentro et al. the interface may present a Ca or K enrichment whose origin was not specified. Similar to the samples studied by Bouquillon et al. (2018), glazes of different colors and therefore of different chemical compositions present smaller or larger interfaces with more or fewer crystals: green, amber or black glazes seem more reactive than blue or white glazes. However, the influence on the reactivity of the interface of chromophore agents or the initial minerals containing them has not been studied yet.

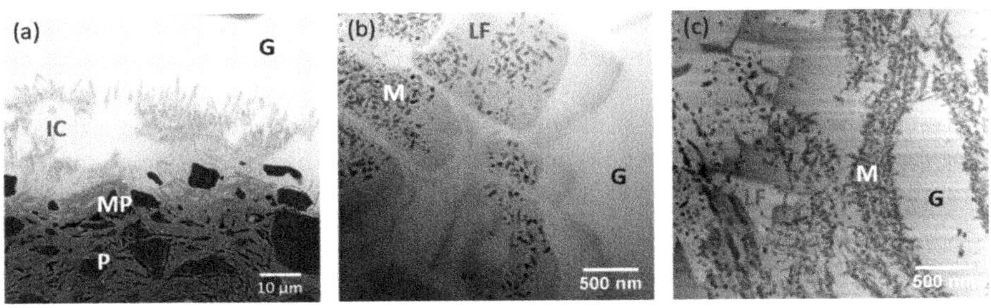

Figure 11.8. *Morphology of the paste–glaze interface in an archeological sample of Bernard Palissy (EP2319), (a) at micrometer scale (SEM image, backscattered electrons), (b and c) STEM-HAADF images of interfacial crystals, IC on (a). G = glaze, IC = interfacial crystals, MP = modified paste, P = paste, LF = lead feldspars, M = mullite*

11.6. Identification of crystalline phases at the interface

Crystalline phases can be observed both in the thickness of the glaze and in the proximity of the body–glaze interface. Three types of crystals can be identified: (1) relic minerals corresponding to unmolten raw materials used to obtain the glaze, (2) newly formed phases crystallizing in the thickness of the glaze and (3) other newly formed phases resulting from the physicochemical interactions between the paste and the applied glaze mixture.

As already shown, the substrate plays a major role in the development of the interface by bringing sufficient local concentrations of certain elements (Al, K, Ca), allowing crystal precipitation (case no. 3). Newly formed phases may provide important information as to how the glaze was elaborated. It is however not always easy to determine if they are actually associated with the paste–glaze interaction, as some of them, formed at the interface, migrate afterward in the thickness of the glaze.

The exact identification of these phases, chemical composition and crystalline structure, is often difficult especially when the crystals have a nanometric size. Their characterization then requires high spatial resolution techniques. In addition, compositions may vary from one crystal to another, even within the same crystal, due to large solid solutions (see section 11.6.1).

11.6.1. *Lead feldspar*

The most commonly detected newly formed minerals are feldspars, corresponding to compositions covering a large solid solution between the potassium and lead endmembers ($KAlSi_3O_8$-$PbAl_2Si_2O_8$). Lead atoms can replace potassium in the structure of feldspars, such as sanidine $KAlSi_3O_8$ (Molera et al. 1993), and in that of phyllosilicates due to the compatibility of cationic radii (r_{Pb2+} = 1.20 Å, r_{K+} = 1.33 Å) (Fortina et al. 2005). Similarly, sodium (r_{Na+} = 1.02 Å) and calcium (r_{Ca2+} = 1.00 Å) have close ionic radii and may therefore replace potassium.

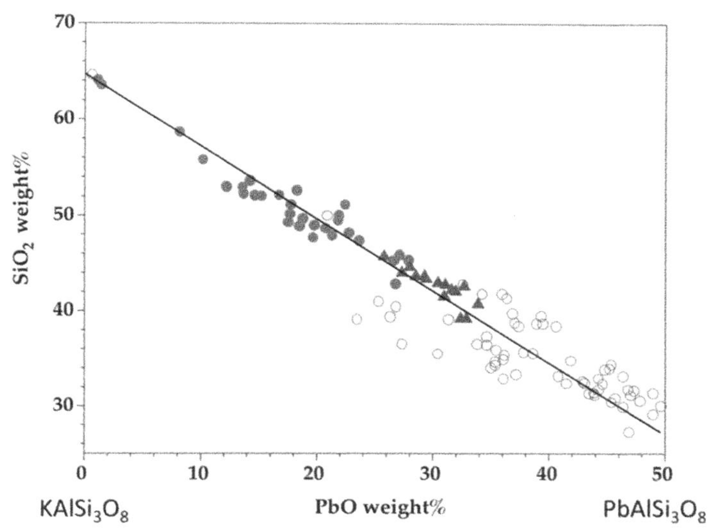

Figure 11.9. *SiO_2 content versus PbO content in the feldspar crystals developed at the interface between a PbO-SiO_2 glaze mixture and either an illitic clay (solid dots), or on a kaolinitic clay (open dots). For a color version of this figure, see www.iste.co.uk/bouquillon/lead.zip*

COMMENT ON FIGURE 11.9.– *For the interface with illitic clay, the crystals near the interface (blue solid dots) are more K-enriched than those farther from the interface (triangles). According to Molera* et al. *(2001).*

The formation of lead-potassium feldspars (Figure 11.9), with low lead content, is typical for glazes deposited on a potassium-rich clay substrate (Molera et al. 1993; Pradell et al. 2010). Lead sanidine crystals were frequently identified in lead glazes (Molera et al. 1993; Pradell et al. 2010; Coentro et al. 2017); they seem to occur for Na-poor clays (Pradell et al. 2010) and, occasionally, by epitaxial growth on potash feldspar crystals (Molera et al. 1993). Sanidine is a high-temperature feldspar of nominal composition $KAlSi_3O_8$, with a maximal degree of Si/Al disorder (Si and Al are randomly distributed on tetrahedral sites; Brown and Parsons 1989).

Figure 11.10 displays the analysis by electron diffraction of a lead feldspar crystal whose structure is $PbAl_2Si_2O_8$ containing traces of K, Ca and Ti in a Palissy sample (inventory no. EP2319).

Lead enters the sites of potassium and the $K^+ \leftrightarrow Pb^{2+}$ substitution (or Ca^{2+} or Na^+) is compensated by the $Si^{4+} \leftrightarrow Al^{3+}$ substitution. These substitutions explain why complex chemical compositions were mentioned (EDX measurements). A $(Na_{0.05}K_{0.13}Pb_{0.69}Ca_{0.05})Al_{1.62}Si_{2.08}O_8$ composition was found on honey-glazed ceramics by Bernard Palissy (2018). Post-Palissy glazes yielded a $[(Na,K)_{0.51}(Ca,Pb)_{0.62}Al_{1.18}Si_{2.68}O_8]$ composition (Bouquillon et al. 2018). Moreover, lead sanidine of average formula $(K_{0.85}Pb_{0.12}Ca_{0.03})Al_{1.08}Si_{2.92}O_8$, was identified in Hispano-Moresque glazed ceramics (Molera et al. 1993).

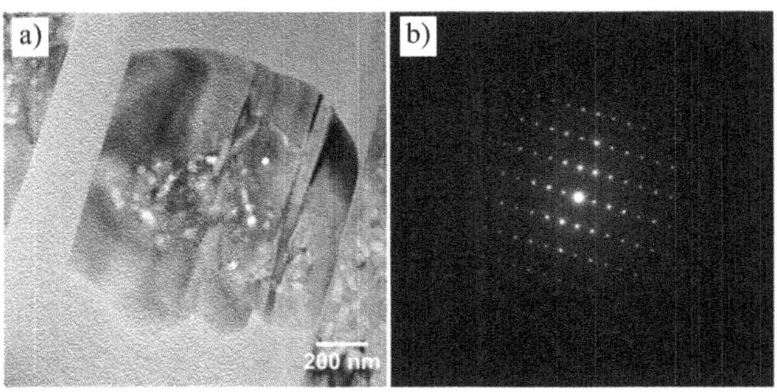

c)

d_{hkl}	$PbAl_2Si_2O_8$
6.46	001
3.41	-112
2.73	-311
2.09	-401
1.77	-441
1.70	-224
1.55	223
1.37	-553

Figure 11.10. *Analysis of a lead feldspar crystal (K,Ca)PbAl2Si2O8 by electron diffraction observed at the interface of an archeological shard by Bernard Palissy (EP2319)*

COMMENT ON FIGURE 11.10.– *(a) Bright field TEM image (the brighter points inside correspond to mullite crystals, see Figure 11.8), (b) selected area electron diffraction (SAED) diagram of the crystal, (c) interplanar spacings* d_{hkl} *extracted from (b) and corresponding diffraction planes.*

This wide variability of composition, from one sample to another, from one crystal to another or even inside the same grain, depends on the glaze composition in the area where crystallization occurs. Therefore, for an illitic substrate, the crystals near the surface of the glaze are richer in Pb, while those near the interface are richer in K, as illite contains potassium in the interfoliar space (Figure 11.9) (Molera et al. 2001). Illite may serve as nucleation seeds for Pb-K feldspars, which would explain why more feldspar crystals are present for an illitic body compared to a kaolinitic body.

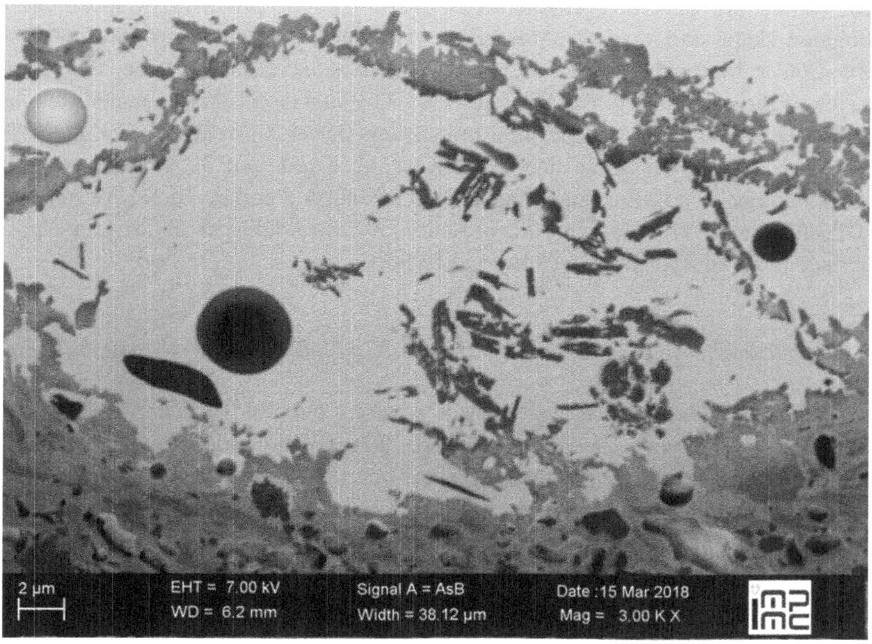

Figure 11.11. *Zoned feldspars observed at the paste–glaze interface of a Palissy-type replica (composition A on kaolinitic biscuit) fired at 1,000°C without plateau and quenched. Light gray crystals are lead feldspars and dark gray crystals are anorthite (lime feldspar). It is important to note the presence of a zonation on certain crystals (SEM image, backscattered electrons)*

The crystallization of Pb feldspars, $PbAl_2Si_2O_8$, occurs at low temperatures. In a mixture composed of PbO and precursors (kaolinite, mullite) and fritted at high temperature, they were detected from 700°C (Lu 2013). This phase is jointly formed with other phases of intermediate composition up to 950°C, due to insufficient diffusion of chemical elements from the precursors. From 950°C, a temperature commonly accessible in the furnaces of ceramists, only the feldspar phase $PbAl_2Si_2O_8$ is obtained (Benna et al. 1996; Chen et al. 2001; Lu 2013). Its presence in an interface may therefore be associated with firing temperatures ranging between 700°C and 1,000°C. Moreover, the cooling rate is an important parameter in the development of these crystals, suggesting their growth during temperature decrease (Molera et al. 2001; Roisine 2018).

Pb feldspars at the ceramic body/glaze interface are present whatever the type of paste. They are generally characterized by a tabular, prismatic, trapezoidal or elongated shape and variable dimensions of several dozen microns (Fortina et al. 2005, 2007). Moreover, these feldspar crystals may leave the interface, float to the surface of the glaze and grow (Walton 2004). Crystals then have elongated (tabular, needle-like and skeleton) shapes and dimensions below 2 µm (Fortina et al. 2007). Finally, zonations on some feldspars were observed and EDX measurements indicate a core of potash feldspar and edges of anorthic composition (Fortina et al. 2007). Figure 11.11 illustrates the presence of zoned feldspar at the paste–glaze interface in a Palissy-type replica fired at 1,000°C.

11.6.2. *Wollastonite ($CaSiO_3$), anorthite ($CaAl_2Si_2O_8$), Ca-Mg pyroxenes*

Wollastonite phase is less systematically present than feldspars and is most of all observed for calcareous clays (Pradell et al. 2010; Walton and Tite 2010; Coentro et al. 2017). Its crystallization occurs in areas rich in Si and Ca and poor in Al, often in the proximity of large grains rich in CaO (Pradell et al. 2010; Coentro et al. 2014). In the presence of Al, plagioclases (here feldspars rich in Ca such as anorthite) may form (Fortina et al. 2007; Pradell et al. 2010; Coentro et al. 2014). Large anorthite grains along the interface were interpreted as resulting from a long firing at high temperatures and low cooling rate (Fortina et al. 2007). Trapezoidal or prismatic morphologies and dimensions of the order of dozen microns have been reported (Fortina et al. 2005; Maggetti and d'Albis 2017). Furthermore, micrometric anorthite crystals were observed on interfaces of Palissy-type replica between a non-calcareous kaolinitic biscuit and an iron-colored glaze fired at 1,000°C (Figure 11.12).

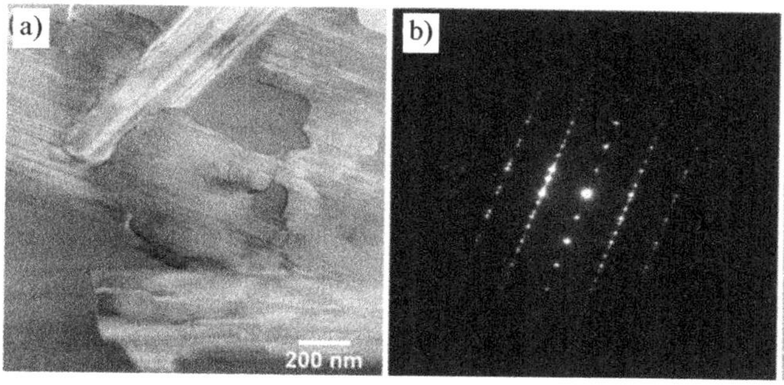

c)

d_{hkl}	$CaAl_2Si_2O_8$
6.37	020
4.04	-202
3.76	112
3.43	-222
3.03	1-32
2.84	132
2.22	-224
2.12	-1-54

Figure 11.12. *Analysis of an anorthite crystal (Pb)$CaAl_2Si_2O_8$ observed by TEM at the paste–glaze interface of a Palissy-type replica (composition A on kaolinitic biscuit) fired at 1,000°C without plateau and quenched. (a) Bright field-TEM image, (b) selected area electron diffraction (SAED) diagram of the crystal in the center, (c) interplanar spacings d_{hkl} extracted from (b) and corresponding diffraction planes*

Similarly, Ca-rich pyroxenes appear for calcareous substrates (Molera et al. 2001; Pradell et al. 2010). They may form from the interface, being in this case enriched in Al (Pradell et al. 2010). They would possibly float in the glaze without directly participating in the formation of the interface (Molera et al. 2001). For ceramic bodies rich in Ca and Mg, Ca-Mg silicates such as diopside ($CaMgSi_2O_6$) are observed (Pradell et al. 2010; Coentro et al. 2017), with variable forms (prismatic to elongated) and dimensions of the order of several dozen microns (Fortina et al. 2005). Temperatures close to 1,000°C were suggested for their formation.

11.6.3. *Phases rich in Na*

Various phases rich in Na were revealed by μ-XRD (Pradell et al. 2010): nepheline (Na,K)$AlSiO_4$, carnegieite $NaAlSiO_4$ and albite $NaAlSi_3O_8$. These phases require significant Al and Na contents and form at the interface (or in the ceramic body for fritted silica pastes) during firing of glazes with low contents of Al_2O_3. Nepheline forms around 975°C (indicating firing temperatures close to 1,000°C) but nucleating agents (such as Sn) may significantly reduce crystallization temperatures (Pradell et al. 2010).

11.6.4. *Mullite ($3Al_2O_3$-$2SiO_2$)*

This phase, rarely observed, was identified on clay substrates very rich in Ca and Mg (Pradell et al. 2010) and also in ceramic shards of Palissy (lead-rich glaze on kaolinitic substrate) or re-creations (frit on non-calcareous kaolinitic substrate) (Godet et al. 2019). In this latter study, we revealed the presence of nanometric mullite grains within the lead feldspar crystals or in the modified paste area (Figures 11.8, 11.10 and 11.13). Mullite results of transformation of metakaolinite during the first firing of the paste, before depositing the glaze mixture. The digestion of mullite by the molten glaze may be slow and the lead feldspar crystals may grow around these mullite relics. An alternative hypothesis is that an excess of silica and alumina, resulting from the digestion of the paste, precipitates in the form of mullite within lead feldspar crystals. The presence of mullite indicates the firing of the paste at temperatures near 950–1,000°C.

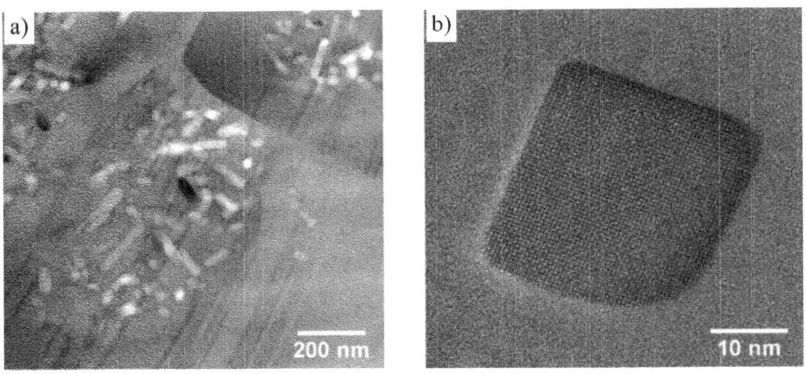

Figure 11.13. *Mullite nanocrystals observed within a lead feldspar $PbAl_2Si_2O_8$ microcrystal in an archeological shard of Palissy (EP2319). (a) Bright-field TEM image and (b) high-resolution TEM (HRTEM) image of one mullite monocrystal*

11.6.5. *Phases rich in colorant or opacifier*

Many crystals containing coloring elements (Fe, Mn) or opacifiers (Sn) were mentioned in the literature as being present in the close proximity of interfaces. The crystallization and evolution of the composition occurring at the interface may facilitate the formation of these phases that may also appear independently in the volume of the glaze. Future studies are expected to clarify the influence of the interface on the formation of these phases.

Lead glazes on potteries (13–19th centuries) show the formation of hexagonal hematite crystals (α-Fe_2O_3) of several dozen microns precipitating preferentially at the body-glaze interface, resulting from an iron oversaturation from the pottery (Di Febo et al. 2017a, 2017b). Due to the very well formed facies, they can be differentiated from the initial hematite grains from the paste. Their crystallization occurs between 915°C and 950°C (Roisine 2018; Godet et al. 2019). They are sometimes associated with melanotekite crystals ($Pb_2Fe^{3+}{}_2Si_2O_9$) that nucleate around hematite grains and have skeletal morphologies that are normally related to phases formed under rapid cooling conditions. In the presence of Mn, brought to color the glaze, a complete solid solution, melanotekite-kentrolite ($Pb_2Mn^{3+}{}_2Si_2O_9$), is possible (Di Febo et al. 2017b). The presence of melanotekite makes it possible to estimate a firing temperature between 900°C and 925°C. The role of hematite as an efficient nucleating agent is also manifest in the crystallization of cristobalite and the clearly visible hematite-cristobalite intergrowths (Beauvoit 2017; Di Febo et al. 2017b). Melanotekite and cristobalite can be considered as technological markers associated with certain workshops (Di Febo et al. 2017b).

Manganese is added to the glaze for its action on the color, depending on its oxidation state. Bustamite ($CaMnSi_2O_6$) is a manganese pyroxene, observed near the interface in the glazes of Hispano-Moresque tiles (Coentro et al. 2014). This phase precipitates from the liquid for temperatures near 950°C (Pradell et al. 2013), following a reaction between MnO brought by the glaze and the elements (Ca) of the ceramic paste. Hausmannite ($Mn^{2+}Mn^{3+}O_4$), braunite ($Mn^{2+}Mn^{3+}{}_6SiO_{12}$) and jacobsite ($Mn^{2+}Fe^{3+}O_4$) crystals may also be present, but their formation cannot be related to the interface with certainty, as they may occur from a reaction between manganese and the molten glaze (Pradell et al. 2013; Coentro et al. 2014; Di Febo et al. 2017b).

Tin is classically used as glaze opacifier, generally to obtain cassiterite SnO_2 (Tite et al. 1998). Its presence may induce specific crystallizations at the interface, such as malayaite ($CaSnSiO_5$), which was identified by μ-Raman (Coentro et al. 2014). The presence of transition ions (such as copper in red glazes) may also facilitate the appearance of this phase (Piña et al. 2005; Coentro et al. 2014).

11.7. Interface as an indicator of elaboration processes

Examination of the interface may be a particularly interesting source of information to differentiate the technical processes related to the initial glaze mixture, glaze application modes, either directly on the air dried ceramic body or after a first firing, or to the firing protocols (firing temperature and duration, cooling rate).

11.7.1. *Simple or double firing, raw or fired substrate*

The authors do not agree on this crucial point. Certain authors state that the interface will be more developed when the glaze mixture is deposited on a non-fired clay (Molera et al. 2001). Indeed, a higher reactivity of the raw clay minerals, during the application of the glaze mixture and during firing, would induce a facilitated diffusion of chemical elements from the raw paste to the glaze, leading to a significantly higher concentration of feldspar crystals at the body–glaze interface (Thomas et al. 1945). Conversely, clays fired a first time would be more stable and therefore less likely to react with the glaze (Molera et al. 1997, 2001; Tite et al. 1998). Hence, thin interfaces (thickness < 5–10 μm) would indicate a double firing process, while large interfaces (thickness > 30 μm) would prove a single firing process.

Other authors nuance or even contest this conclusion, highlighting a lack of chemical or microstructural resolution or multifactor causes that may influence the thickness of the interface (see section 11.3) (Ben Amara 2002; Walton 2004; Ben Amara and Schvoerer 2006). Separating the effects of various variables proves to be very difficult. Recent reproductions also confirm the difficulty of correlating the thickness of the interface with the raw or fired state of the substrate (Figure 11.14) (Roisine 2018). Either on biscuit or on raw paste, the microstructures and the size of the interface (~25 μm) are very similar.

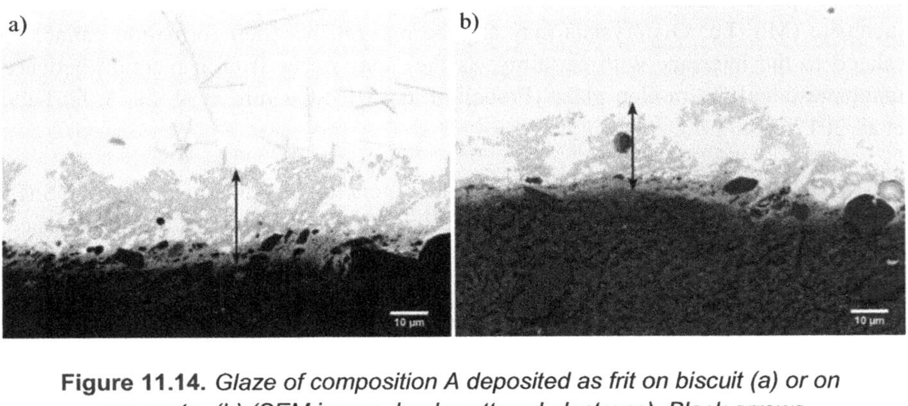

Figure 11.14. *Glaze of composition A deposited as frit on biscuit (a) or on raw paste (b) (SEM image, backscattered electrons). Black arrows indicate the thickness of the interface (~25 μm)*

As a conclusion, though the paste–glaze interaction often produces a more significant interface for a raw substrate in the reproductions, the most recent observations do not allow drawing an a posteriori conclusion on the raw or fired state of the clay substrate at the moment of lead glaze application.

11.7.2. *Application of PbO or PbO + SiO_2 mixtures*

The study of the interface also enables us to develop hypotheses concerning the initial composition used for the glaze mixture.

In order to distinguish between the application of only lead oxide and the application of a mixture of lead oxide and silica, a proposal was made to take into account the diffusion of chemical elements from clay to glaze. For this purpose, the lead oxide and copper oxide (or another colorant) contents are subtracted from the global composition of the glaze, which is then renormalized at 100% (Tite et al. 1998; Walton 2004; Walton and Tite 2010). The resulting concentrations for each element should (1) be similar to those of the clay body for a glaze made only with PbO and (2) present a constant composition profile throughout the thickness of the glaze. Conversely, for a PbO-SiO_2 mixing, the silica concentration of the glaze should be above that of the clay body, and the alumina (and other oxides) concentration should be below (attention should however be paid to the case when SiO_2 is brought to the glaze mixture by the addition of a clay similar to that of the object; in this case, the corrected composition would tend to be similar to that of the clay body). According to this hypothesis, steep concentration profiles should appear close to the paste–glaze interface.

Contrasted diffusion profiles depending on the type of glaze mixture applied have been observed in reproductions. They show a strong gradient of concentration close to the interface for PbO + SiO_2 compared to a flatter profile for PbO (Walton and Tite 2010). However, many re-creations that do not use PbO only do not show a gradient of concentration in the thickness of the glaze, as a sufficiently long period of firing and cooling produces a uniform distribution of elements. It is also rare to observe steep profiles of concentration close to the interface for archeological samples, though this was observed in some Roman glazes from Canosa (Benedetto et al. 2004). The diffusion due to paste–glaze interaction would therefore offer little information allowing a differentiation of the applied glaze mixture. However, the technique of compositions renormalized after subtraction of PbO and CuO contents shows in certain cases convincing correlations allowing a differentiation between these two methods according to the spatial or temporal origin of objects (Walton and Tite 2010; Giannossa et al. 2015).

11.7.3. *Firing protocol*

The development of the interface may also bring information on the firing protocol. Thus, a thick interface, a large extent of the crystallization of anorthite at the body–glaze interface and the growth of cristobalite have been interpreted as resulting from the firing at very high temperatures for a long firing and cooling time (Fortina et al. 2005). Conversely, a thin interface, with newly formed mineral phases of small dimensions, indicates a short firing/cooling time. These criteria are obviously highly qualitative and many other parameters may influence the results and limit the reliability of the conclusions. As part of the study conducted on Bernard Palissy's glazed ceramics, the study of the interface of several archeological shards compared to those of re-creations under controlled thermal conditions allowed us to propose a firing temperature between 900°C and 950°C with a plateau of several hours and a slow cooling in the furnace (Roisine 2018).

11.7.4. *Al_2O_3 origin: glaze mixture or paste?*

Aluminum either originates from the addition of clay to the glaze mixture in order to improve its suspension, or from the minerals of the ceramic body. Recreations of glazes that do not initially contain alumina (by depositing a frit or a mixing of oxides) show after firing a Al_2O_3 content comparable to those of mixtures containing it (Molera et al. 2001; Beauvoit 2017; Roisine 2018). This reflects the significant diffusion of Al moving from the paste to the glaze and the difficulty of bringing information on the original recipe of the glaze mixture by only observing the profile of Al concentration.

However, the examination of the interfaces of iron-colored glazes of Palissy show thicknesses of 10–30 μm, notably with a thickness of modified paste (Pb presence) limited to 10 μm (Godet et al. 2019). The low diffusion of Pb in the paste may correspond to the role played by Al (see section 11.3) limiting the diffusive phenomena and would be explained by the voluntary addition by Palissy of an aluminous phase to the initial glaze mixture. Once molten, the latter would be more viscous, limiting the interpenetration of glazes.

A relevant parameter for revealing the diffusion of certain chemical elements from the paste to the glaze and providing information on the recipes used would be to follow the oxide ratios, for example, Al_2O_3/TiO_2 (Beauvoit 2017). If the alumina of the glaze originates exclusively from the paste, the Al_2O_3/TiO_2 ratio must be the same in the glaze as in the paste.

11.8. Influence of interface on the physical properties

The reactions occurring at the glaze/body interface may have a significant impact on several properties of the object.

11.8.1. *Thermal expansion*

The main factor influencing the stress relations between glaze and ceramic body is the difference in their thermal expansion coefficient (TEC) (Kavanova et al. 2017). The two materials having a different expansion behavior, permanent stresses in the body and glaze appear during cooling (Figure 11.15). A compression of the glaze by the ceramic is generally sought-for by adjusting the TEC of the glaze so that it is slightly below that of the shard (see Chapter 1, section 1.1.3). When cooling, glaze retracts less than ceramics, which drives the compression of the glaze by the ceramic substrate (Peterson et al. 2007). If the TEC of the glaze is higher than that of the shard, a tensile stress appears in the glaze. Since glazes resist less to tensile stress than to compression stress, the glaze may crack (known as crazing) during the final phase of cooling (Schurecht and Pole 1930).

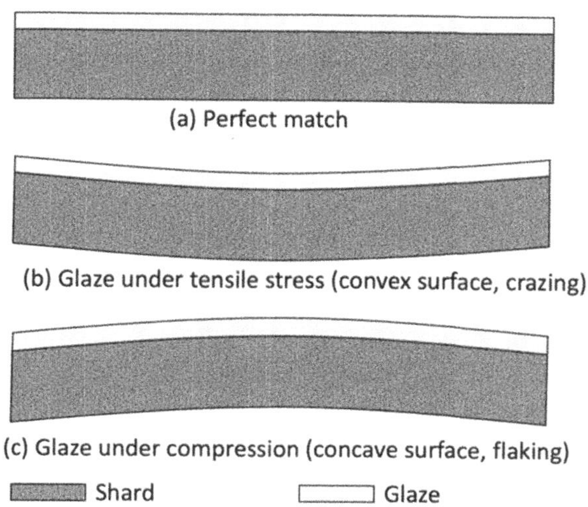

Figure 11.15. *Schematic representation of stresses and deformations resulting from differences in thermal expansion coefficients between the paste and the glaze*

Stresses in a direction parallel to the interface, resulting from different TECs, do not depend on the development of the interface. On the other hand, a more diffuse interface leads to a better resistance to delamination: the effect of TEC differences is then less dramatic.

11.8.2. *Adherence defects*

A proper bond between the glaze and its substrate is essential for the long-term conservation of the object. Adherence is governed by two parameters:

– Interaction with the paste during the deposit of the glaze mixture. When the glaze is deposited, the water contained in the glaze mixture solution may be absorbed by the pores of the substrate by capillarity. The biscuits present a porosity allowing water to infiltrate, facilitating the holding of the deposit before firing. On the other hand, raw bodies have water-saturated pores that render the application of the glaze more difficult.

– Paste–glaze cohesion after firing. The more significant the paste–glaze interaction, the better the adherence of the glaze to its substrate. A calcareous clay inducing thicker interfaces leads to improved adherence (Ben Amara 2002).

Glaze defects have various causes (Kavanova et al. 2017): improper adjustment of TEC or technological processes and changes in the ceramic body (hydration, crystallization of salts and biological growth such as mold).

11.8.2.1. *Flaking*

When there is no interface or when the existing one is under-developed, flaking (unsticking of glaze fragments) may occur. This phenomenon may originate from an excessive compression of the glaze. A thin interface often results from improper application of the glaze mixture and is specific to biscuits, whose reactivity is lower than that of a raw paste, as already mentioned. The glaze is less sensitive to compression because of the development of the interface that firmly connects it with the shard.

11.8.2.2. *Cracking*

Glaze cracking has several causes.

When the glaze cools much more rapidly than the ceramic, given its adhesion to the paste, which prevents it from retracting, tensile stresses transiently appear, which may drive the formation of cracks. However, after complete cooling of the object, cracks reseal and are therefore slightly visible.

Due to improper adjustment of TEC, as previously indicated (section 11.8.1), a network of cracks may appear when cooling (crazing). This phenomenon is generally prevented, but in certain cases it may, on the contrary, be sought for, for its decorative effect (Plešingerová and Kovalčíková 2003).

Cracks may also occur several days or even several years after cooling, particularly for porous bodies. Indeed, rehydration or crystallization of soluble salts in the pores of the ceramic induces a swelling of the paste exerting a tensile stress on the glaze, leading to undesired crazing. It is a very frequent pathology for wall ceramic tiles placed in wet environments, but the influence of the interface on this phenomenon has yet to be further studied. This type of defect is affected by pre-existing stresses in the glaze.

For thick interfaces, the crack may propagate in the paste, while its tendency may be to turn in parallel to the surface for underdeveloped interfaces, leading to a possible flaking of the glaze (Mimoso et al. 2012; Costa et al. 2014).

11.8.3. *Mechanical resistance*

The mechanical resistance of ceramic objects is higher when they are glazed (Haberko and Haberko 1975). For example, a study on glazes deposited on fine China porcelain shows that the mechanical resistance of the object is improved when the duration and melting temperature parameters are significant (Ichiko 1994). The observations reveal that the origin of fractures is localized at the body–glaze interface (mainly near pores) or slightly below this region; no origin in the glaze was found (except for significant stresses when bubbles are present) (Benson 2003).

Several explanations are proposed to justify the improved mechanical properties (Haberko and Haberko 1975):

– a high-quality glaze has few defects at the surface and in its thickness, which limits the appearance of fractures;

– compression stresses on the glaze (when its TEC is below that of the ceramic body) oppose the propagation of fractures and strengthen the system (Duke et al. 1968);

– a diffuse interface (chemical diffusion, intergrown crystals) limits the concentrations of stresses in the vicinity of interface defects. For example, more crystal growth enhances the resistance of the material (Benson 2003);

– the nature of certain crystalline phases (at the interface or within the glaze) may influence the properties of the material (as for glass-ceramics).

11.8.4. *Color*

The color of glazes is generally obtained by adding chromophores (mainly transition elements) to the initial glaze mixture. Sometimes, a significant diffusion from the paste to the glaze may yield a coloration, particularly for iron (green shade for Fe^{2+} and yellow-brown for Fe^{3+} in a lead glaze) (Molera et al. 1997). For transparent glazes, the perceived color is the combination of intrinsic contributions of the glaze, interface, and paste.

The significant development of the interface area influences the color of the glazed ceramic (Ben Amara 2002). The presence of hematite crystals in the paste or at the interface modifies the perception of the initial color of the glaze (Molera et al. 1997). The crystals at the interface (Pb feldspars) will have a dispersive effect on the light and appear as a white layer under optical microscope. For transparent glazes, a significant intermediate layer makes it sometimes possible to dissimulate the underlying clay body playing in practice a role equivalent to a white engobe (Benedetto et al. 2004).

11.9. Conclusion

The interface between paste and glaze is the site of complex physicochemical processes during the firing of a glazed ceramic. The diffusion of chemical elements from the paste to the glaze and vice versa leads to the penetration of lead into the substrate and to the formation of a layer of several dozen microns of thickness of devitrification crystals (lead feldspars) with heterogeneous morphologies at micro and nanoscale. The extension of the interface and the nature of crystallizations depend on many physicochemical and thermal factors. Certain aspects would require additional studies for a proper evaluation of their impact: the influence of chromophores and opacifiers on the crystallization at the interface may promote certain phases usable as thermal markers; the diffusion of Al or other minor elements from the paste would allow a better understanding of the initial compositions of the glaze mixtures. This is why the paste–glaze interface is a very interesting marker in the exploration of the technological processes used by the ceramic artists.

11.10. References

Beauvoit, E. (2017). Céramiques glaçurées de la Renaissance : les glaçures "miel" de Bernard Palissy. Master's Thesis, Université Bordeaux Montaigne, Bordeaux.

Ben Amara, A. (2002). Céramique glaçurées de l'espace méditerranéen (IXe–XVIIIe siècles AP J.C.). PhD Thesis, Université de Bordeaux 3, Bordeaux.

Ben Amara, A. and Schvoerer, M. (2006). Interactions between lead glazes and bodies: Research on the mode of application of the glazing mixture. In *Inst. Fernando El Católico 34th Int. Symp. Archaeom.* Institución Fernando el Católico, Zaragoza.

Benedetto, G.E.D., Acquafredda, P., Masieri, M., Quarta, G., Sabbatini, L., Zambonin, P.G., Tite, M., Walton, M. (2004). Investigation of Roman lead glaze from Canosa: Results of chemical analyses. *Archaeometry*, 46, 615–624.

Benna, P., Tribaudino, M., Bruno, E. (1996). The structure of ordered and disordered lead feldspar ($PbAl_2Si_2O_8$). *Am. Mineral.*, 81, 1337–1343.

Benson, J.L. (2003). Effect of glaze variables on the mechanical strength of whitewares. Master's Thesis, Alfred University, New York.

Bouquillon, A., Castaing, J., Barbe, F., Crepin-Leblond, T., Tilliard, L., Paine, S.R., Christman, B., Heuer, A.H. (2018). French decorative ceramics mass-produced during and after the 17th century: Chemical analyses of the glazes. *Archaeometry*, 60, 946–965.

Brown, W.L. and Parsons, I. (1989). Alkali feldspars: Ordering rates, phase transformations and behaviour diagrams for igneous rocks. *Mineral. Mag.*, 53, 25–42.

Chen, S., Zhao, B., Hayes, P.C., Jak, E. (2001). Experimental study of phase equilibria in the PbO-Al_2O_3-SiO_2 system. *Metall. Mater. Trans. B*, 32, 997–1005.

Claireaux, C. (2014). Mobilité atomique dans les silicates vitreux et fondus. PhD Thesis, Université Pierre et Marie Curie, Paris.

Claireaux, C., Chopinet, M.-H., Burov, E., Gouillart, E., Roskosz, M., Toplis, M.J. (2016). Atomic mobility in calcium and sodium aluminosilicate melts at 1200°C. *Geochim. Cosmochim. Acta*, 192, 235–247.

Coentro, S., Trindade, R.A.A., Mirão, J., Candeias, A., Alves, L.C., Silva, R.M.C., Muralha, V.S.F. (2014). Hispano-Moresque ceramic tiles from the Monastery of Santa Clara-a-Velha (Coimbra, Portugal). *J. Archaeol. Sci.*, 41, 21–28.

Coentro, S., Alves, L.C., Relvas, C., Ferreira, T., Mirão, J., Molera, J., Pradell, T., Trindade, R.A.A., Da Silva, R.C., Muralha, V.S.F. (2017). The glaze technology of Hispano-Moresque ceramic tiles: A comparison between Portuguese and Spanish collections. *Archaeometry*, 59, 667–684.

Costa, M., Cachim, P., Coroado, J., Rocha, F., Velosa, A. (2014). Ancient wall tiles – The importance of the glaze/ceramic interface in glaze detachment. *Mater. Sci.*, 20, 108–113.

Di Febo, R., Molera, J., Pradell, T., Vallcorba, O., Capelli, C. (2017a). Technological implications of neo-formed hematite crystals in ceramic lead glazes. *STAR Sci. Technol. Archaeol. Res.*, 3, 366–375.

Di Febo, R., Molera, J., Pradell, T., Vallcorba, O., Melgarejo, J.C., Capelli, C. (2017b). Thin-section petrography and SR-μXRD for the identification of micro-crystallites in the brown decorations of ceramic lead glazes. *Eur. J. Mineral.*, 29, 861–870.

Duke, D.A., Megles, J.E., Macdowell, J.F., Bopp, H.F. (1968). Strengthening glass-ceramics by application of compressive glazes. *J. Am. Ceram. Soc.*, 51, 98–101.

Fortina, C., Santagostino Barbone, A., Turbanti Memmi, I. (2005). Sienes "archaic" majolica: A technological study of ceramic bodies and coatings. *Archaeometry*, 47, 535–555.

Fortina, C., Memmi Turbanti, I., Grassi, F. (2007). Glazed ceramic manufacturing in southern Tuscany (Italy): Evidence of technological continuity throughout the medieval period (10th–14th centuries). *Archaeometry*, 50, 30–47.

Giannossa, L.C., Fico, D., Pennetta, A., Mangone, A., Laviano, R., De Benedetto, G.E. (2015). Integrated investigations for the characterisation of Roman lead-glazed pottery from Pompeii and Herculaneum (Italy). *Chem. Pap.*, 69, 1033–1043.

Godet, M., Roisine, G., Beauvoit, E., Caurant, D., Majérus, O., Menguy, N., Dargaud, O., Bouquillon, A., Cormier, L. (2019). Multi-scale investigation of body-glaze interface in ancient ceramics. *Heritage*, 2, 2480–2494.

Haberko, M. and Haberko, K. (1975). Effect of glaze on strength of high-tension porcelain. *Ceramurg. Int.*, 1, 28–32.

Ichiko, T. (1994). A consideration about the glazing to the bone China. *J. Ceram. Soc. Jpn.*, 102, 471–475.

Kara, A. and Stevens, R. (2002). Interactions between an ABS type leadless glaze and a biscuit fired bone china body during glost firing. Part I: Preparation of experimental phases. *J. Eur. Ceram. Soc.*, 22, 1095–1102.

Kavanova, M., Klouzkova, A., Klouzek, J. (2017). Characterization of the interaction between glazes and ceramic bodies. *Ceram. Silik.*, 61, 267–275.

Lu, X. (2013). Application of X-ray diffraction to identify the phases formed during lead stabilization and resource recovery. PhD Thesis, University of Hong Kong, Hong Kong.

Maggetti, M. and d'Albis, A. (2017). Phase and compositional analysis of a Sèvres soft paste porcelain plate from 1781, with a review of early porcelain techniques. *Eur. J. Mineral.*, 29, 347–367.

Mimoso, J.M., Pereira, S., de Lurdes, M., dos Santos Silva, A. (2012). A tour of façade azulejos decay in Lison. In *Proc. Int. Congress Azulejar.* Universidade de Aveiro, Aveiro.

Molera, J., Pradell, T., Martinez-Manent, S., Vendrell-Saz, M. (1993). The growth of sanidine crystals in the lead of glazes of Hispano-Moresque pottery. *Appl. Clay Sci.*, 7, 483–491.

Molera, J., Vendrell-Saz, M., García-Vallés, M., Pradell, T. (1997). Technology and colour development of Hispano-Moresque lead-glazed pottery. *Archaeometry*, 39, 23–39.

Molera, J., Pradell, T., Salvadó, N., Vendrell-Saz, M. (1999). Evidence of tin oxide recrystallization in opacified lead glazes. *J. Am. Ceram. Soc.*, 82, 2871–2875.

Molera, J., Pradell, T., Salvadó, N., Vendrell-Saz, M. (2001). Interactions between clay bodies and lead glazes. *J. Am. Ceram. Soc.*, 84, 1120–1128.

Olupot, P.W., Jonsson, S., Byaruhanga, J.K. (2010). Study of glazes and their effects on properties of triaxial electrical porcelains from Ugandan minerals. *J. Mater. Eng. Perform.*, 19, 1133–1142.

Parmelee, C.W. and Buckles, P.E. (1942). Study of glaze and body interface. *J. Am. Ceram. Soc.*, 25, 11–15.

Peterson, M., Bernardin, A.M., Kuhnen, N.C., Riella, H.G. (2007). Evaluation of the Steger method in the determination of ceramic-glaze joining. *Mater. Sci. Eng. A*, 466, 183–186.

Piña, C., Arriola, H., Nava, N. (2005). Study of malayaite and malayaite cobalt pigment. *Hyperfine Interact.*, 161, 93–97.

Plešingerová, B. and Kovalčíková, M. (2003). Influence of the thermal expansion mismatch between body and glaze on the crack density of glazed ceramics. *Ceram.-Silikáty*, 47, 100–107.

Pradell, T., Molera, J., Salvadó, N., Labrador, A. (2010). Synchrotron radiation micro-XRD in the study of glaze technology. *Appl. Phys. A*, 99, 407–417.

Pradell, T., Molina, G., Molera, J., Pla, J., Labrador, A. (2013). The use of micro-XRD for the study of glaze color decorations. *Appl. Phys. A*, 111, 121–127.

Roisine, G. (2018). Céramiques glaçurées de Bernard Palissy : à la recherche des secrets d'un maître de la Renaissance. PhD Thesis, PSL Université, Paris.

Schurecht, H.G. and Pole, G.R. (1930). Method of measuring strains between glazes and ceramic bodies. *J. Am. Ceram. Soc.*, 13, 369–369.

Thomas, E., Tuttle, M.A., Miller, E. (1945). Study of glaze penetration and its effect in glaze fit. *J. Am. Ceram. Soc.*, 28, 52–62.

Tite, M.S., Freestone, I., Mason, R., Molera, J., Vendrell-Saz, M., Wood, N. (1998). Lead glazes in antiquity: Methods of production and reason for use. *Archaeometry*, 40, 241–260.

Walton, M.S. (2004). A materials chemistry investigation of archaeological lead glazes. PhD Thesis, University of Oxford, Oxford.

Walton, M.S. and Tite, M.S. (2010). Production technology of Roman lead-glazed pottery and its continuance into late antiquity. *Archaeometry*, 52, 733–759.

12

Weathering of Ancient Lead Glazes

Anne BOUQUILLON
C2RMF, IRCP, Paris, France

There is abundant research concerning ancient alkaline or alkali-lime glasses weathered in atmospheric environment, immersed or buried. The studies on the corrosion processes of these glasses have developed since they echoed subjects related to the ageing of glassy matrices that were of interest for the long-term storage of radioactive wastes. These ancient glasses are then considered as "natural equivalents", allowing the assessment of the validity of long-term behavior models for glassy waste, which were established on the basis of accelerated ageing without considering the "time" parameter. These research studies are also widely used by glass manufacturers concerned about the durability of their productions or the positive impact of cleaning agents on the preservation of daily-life objects.

Observations differ for glasses or glazes containing lead. Investigations focused on the weathering of lead glassy materials started only in the 1990s, in a context of exacerbated awareness of the effects of lead on health. Studies dedicated to the leaching of crystal glasses during use or cleaning are thus supported by crystal glass manufacturers, and to a lesser extent by those interested in the processes of inerting residues of incineration fumes and household waste (RPIFHW); lead-rich glasses or glazes, as the metallurgical slags of blast furnaces whose vitrified part is rich in heavy metals (Pb, C, Ni, etc.) noxious for the environment (Seignez et al. 2007), serve as natural equivalents. Knowledge accumulates, research projects are getting organized, PhD theses are defended (Chaulet 2000; Ben Kacem 2019). These

Lead in Glassy Materials in Cultural Heritage,
coordinated by Anne BOUQUILLON and Patrice LEHUÉDÉ.

research projects will be widely developed in Part 5 of this volume (Chapters 13 and 14).

It is beyond the scope of this chapter to exhaustively review all the types of weathering encountered in ancient glassy materials containing lead. We chose to present three types of representative weathering, from the first stages to major degradations, illustrated by examples encountered in our studies of lead glazes having been subjected to prolonged burial; these complex, multiple, more or less aggressive environments actually offer the possibility to observe the widest variability of weathering structures and best reveal the weaknesses or, on the contrary, the durability of glazes. After having presented the main characteristics, a specific outline directly related to these research efforts will be provided in a text box. A last section will rapidly present the attempts to protect by sol-gel methods slightly weathered lead glasses.

12.1. Slightly weathered glazes

Certain glazes, despite prolonged burial, do not seem to present obvious signs of weathering, except for a frosted surface aspect related to an often mechanical wear of the surface. This can be observed on highly lead-rich transparent glazes of the medieval tiles discovered during the excavations at Château de Vincennes (Bouquillon and Mille 2009, p. 197). These tiles were buried for over 800 years in a dumping ground. Yellow glazes (Figure 12.1(a)) are characterized by high contents of lead (50–70% PbO), silica (SiO_2) and alumina (5–8% Al_2O_3). The total (alkaline + alkaline-earth) represents less than 2%; iron (1–2% Fe_2O_3) is the coloring agent.

The clay body/glaze interface is most often very well developed with many crystallizations of lead feldspars (Figure 12.1). Very locally, at the surface, more or less adhering to the glaze, fine secondary crusts can be detected, which are rich in phosphorus, chlorine, and calcium and present variable lead contents. These secondary precipitations, by the presence of lead, suggest a low leaching of the lead contained in the glaze and a reaction with the elements of the surrounding ground (here a lime mortar) and interstitial solutions. These characteristics of a limited weathering in a quite aggressive burial environment are quite often and mainly found for compositions of glazes similar to those described above: glazes of Byzantine objects (Waksman et al. 2008) or everyday ceramic objects from the Renaissance (pots, coquemars, etc.) coated with a high-lead transparent glaze of green or brown color, as the transparent glazes of certain objects created by Bernard Palissy. Chemical composition has an essential role here. Even though these alumina-rich and slightly alkaline compositions really seem more resistant to weathering, which is confirmed by the Ballidon experiment on a high lead optical glass that did not seem weathered after 16 years of burial (Davison 2003, p. 290), hasty generalization should be avoided. The same object coated with the same glaze may present weathering levels that are not necessarily homogenous (Bouquillon et al. 2008): the variations of composition of the surrounding environment, the proximity of other objects, and even the shape of the object are all parameters that render the study more complex. On the other hand, certain glazes known however for their brittleness seem well preserved despite an unfavorable burial environment, as we observed for certain alkali-lead glazes from Nishapur (Ranté et al. 2005).

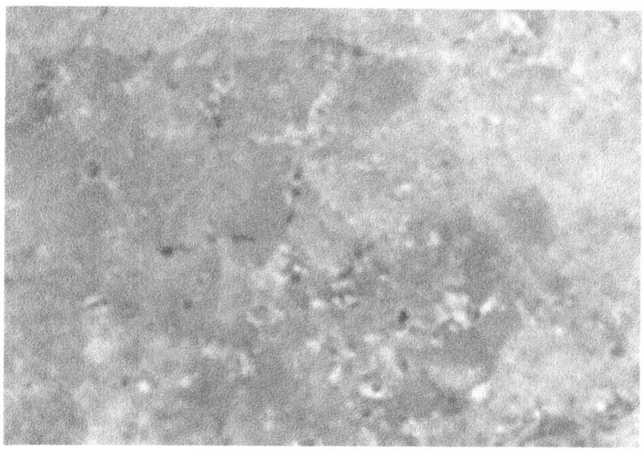

a)

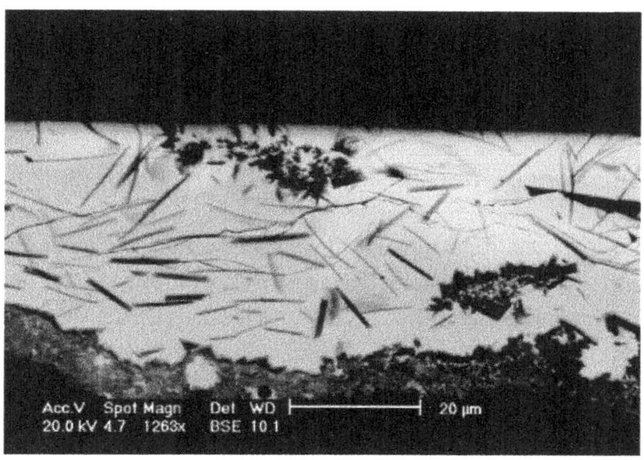

b)

Figure 12.1. *Example of a slightly weathered glaze: (a) macrophotography of the surface; (b) stratigraphic cross-section. Needle-like black crystals are lead feldspars. C2RMF/G.Devés*

In the heritage field, and more specifically in museum collections, preservation of art works is an essential priority. Conservation optimization requires the ability to identify and characterize various weathering states, pay attention to their evolution, but above all to understand when, why and how materials get weathered. In a museum, regular check of the sanitary state of the works of art is recommended. While for glasses and glazes, a stable environment, whose temperature and hygrometry are controlled, is often sufficient for proper conservation, this climate control is sometimes difficult to implement long-term without taking into account possible environmental disturbances (changes of showcase, extreme climate events, floods, etc.) that can cause or worsen damaging weathering.

This is why we worked on a set of portable non-destructive techniques allowing regular in situ follow-up of the surface characteristics of glasses, particularly of lead glasses (Bouquillon et al. 2007; Charbonneau et al. 2007). Weathering is first signaled by an increase in surface roughness that can be measured by a reflectance spectroscopy system (Charbonneau et al. 2007). Then the hydration of glass surfaces is observed; it is detected in several ways: (1) by near-infrared spectrometry, a technique allowing the quantification of hydration by evaluating the surface of O-H bands; the thickness of the hydrated layer is determined by measuring the distance between interference bands related to variations of reflection of light incident on air/pristine glass and pristine glass/weathered glass interfaces, (2) indirectly, by portable X-ray fluorescence spectrometry, by following the evolution of the ratio between M and L lines of lead; during leaching, the M/L ratio decreases, as the lead present on the surface, marked by M lines, is leached, while the lead deeper in the glass core (L lines) remains unchanged at first. These protocols have been validated by traditional techniques (RBS, SEM, Raman spectrometry).

Box 12.1. *Identify the first weathering stages of a lead-rich glassy material*

12.2. Weak iridescences and pinholes

Most of the time, ancient glazes have an iridescent appearance or a quite heterogeneous and irregular surface. Electron micrographs of the surface show the superposition of several different “layers” (Figure 12.2). When examining the characteristics of microstructures in stratigraphic cross-sections, two cases are observed: in the first case, a quite regular and homogeneous layer appears darker on the electron micrographs (backscattered electrons mode); it also extends along the micro cracks or around the pinholes (Figure 12.3(a)) and develops even around the heterogeneities in the glaze, leaving bubbles or unmolten quartz grains, for example. A very high magnification observation sometimes reveals the presence of very fine lamellae, which will be developed in the next section. On top of this weathered layer there are sometimes more or less thick heterogeneous crusts, containing lead-rich crystallized components (most often phosphates or carbonates).

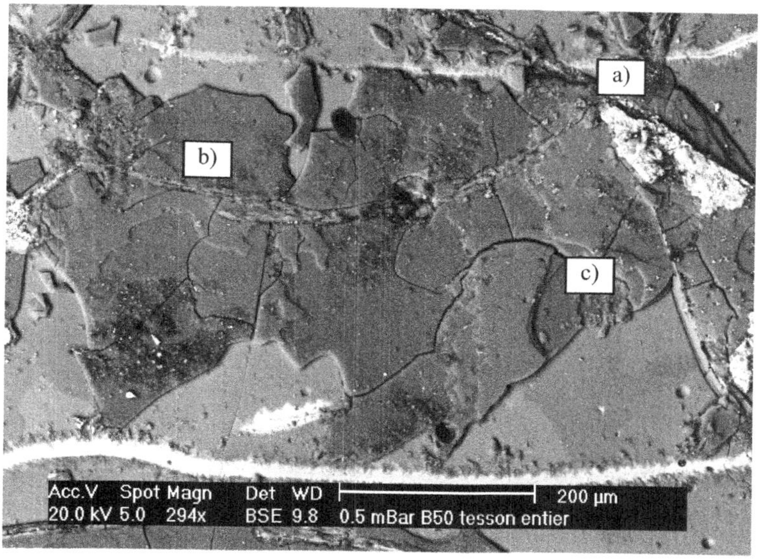

Figure 12.2. *Detail of the surface of a Renaissance ceramic shard observed by SEM. (a) Encrustation, (b) lead-depleted hydrated layer and (c) pristine glass. C2RMF/S.Vialle*

This is a quite simple stratigraphy, compatible with what is found in the experimental weathering schemes of modern glasses: pristine glass, hydrated silica layer that is highly depleted in glass modifiers (lead, alkali, coloring oxides) and secondary phases precipitated at extreme surface.

In the second case, which is rarer, the weathered area seems to develop in a multifaceted way inside the glaze, below an apparently intact surface, while small islands of pristine glaze are preserved (Figure 12.3(b)). This can be observed, for example, in the glazes of Nasrid architectural elements of Alhambra in Granada studied (Coquinot et al. 2012).

In both cases, the chemical characteristics of the weathered layer remain the same: the contents of lead, alkali and copper (for green or blue glaze) decrease compared to a "pristine" glaze, and, on the other hand, silica and alumina persist. In a tin oxide opacified glaze, cassiterite microcrystals are not affected by weathering and are easily visible in the hydrated layer (Figure 12.3); when these areas are studied by Raman microspectrometry, significant bands can be noted around 3,000 cm^{-1}, signaling strong hydration (Bouquillon et al. 2008). As mentioned previously, precipitations at the surface in what is referred to as crust are enriched in phosphorus, calcium and lead.

a)

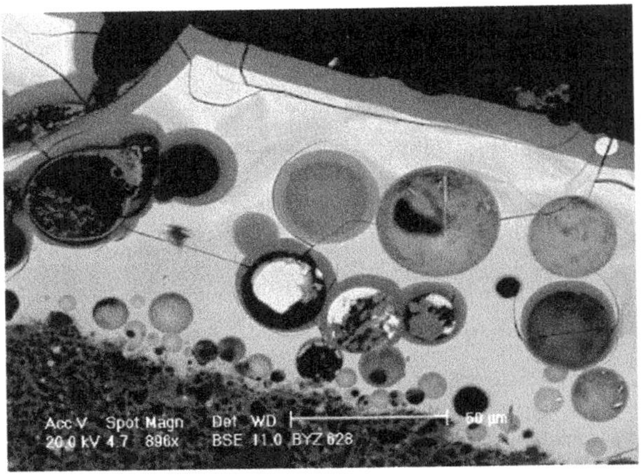

Byz 628 – Pristine glaze PbO: 74%; SiO_2: 17.5 .; alk + CaO: 1.5%; Al_2O_3: 6%; Fe_2O_3: 2.5%. There are many bubbles.

b)

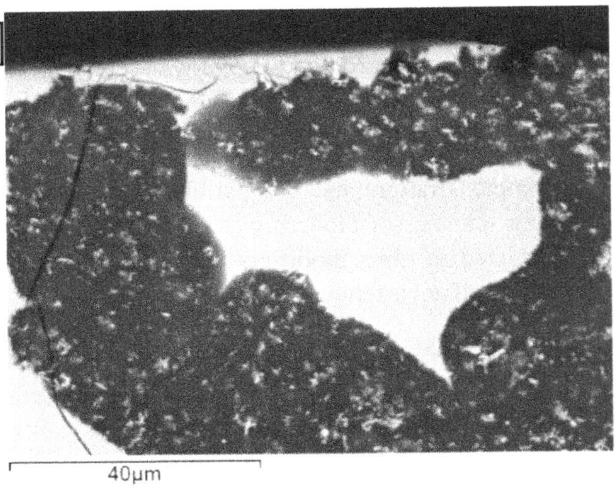

Inv **787-1921-2** – Pristine glaze PbO: 50.5%; SiO_2: 39%; alk + CaO: 5%; SnO_2: 3.5%; Al_2O_3: 0.5%.

Figure 12.3. *Stratigraphic cross-section of two glazes: (a) Byzantine object; (b) Nasrid vase. C2RMF/A. Bouquillon*

COMMENT ON FIGURE 12.3.– *(a) Weathering is signaled by a fine dark border at the surface, along cracks and around bubbles. Very clear areas can be noted, corresponding to lead concentrations (carbonates) in certain bubbles. (b) The weathered areas have more complex shapes and are highlighted by the presence of cassiterite microcrystals. SEM images, BSE mode. C2RMF/A. Bouquillon.*

The weathering processes described here correspond to what is known for all the glasses and are not specific to lead glassy matrices. They will be explained in detail in Chapter 13. It can simply be said that it is a process involving several steps, starting with the hydration of surface layers of glass and being followed by exchanges between the surrounding environment and the glass components. Finally, there is a surface deposit of secondary minerals resulting from the interactions between leached elements and those present in the environment (phosphates, carbonates, sulfates, etc.). This explains here the presence of lead phosphates or lead carbonates in surface encrustations.

12.3. Advanced weathering in the context of burial

Prior to the final stage of flaking of the glaze and its definitive disappearance, weathering is visually marked by total opacification, strong iridescence, loss of colors, very rough surface and an often significant development of secondary crusts containing most of the time gypsum, lead phosphates and/or carbonates. The microstructures of the altered areas of the glaze *s.s.* are characterized by a series of micronic lamellae that develop either in parallel with the surface, or along fractures perpendicular to the surface, around the bubbles or still along the interface between clay body and glaze, that is, in brittle zones allowing the passage of fluids from the burial ground. The existence of inclusions in the glassy phase (cassiterite microcrystals, opacifiers, or unmolten quartz crystals) seems to facilitate the appearance of these laminated microstructures (Figure 12.4).

Energy dispersive X-ray spectroscopy analysis shows that lamellae seem sometimes enriched in lead or, on the contrary, enriched in silica, alkali having practically disappeared throughout the laminated zone. Raman spectrometry analysis of the weathered layer indicates an intensification of hydroxyl bands toward 3,000–3,500 cm^{-1} sign of strong hydration.

Associated with lamination, other less frequent alteration patterns are observed: vermicular figures (see Figure 12.4B(c)) and lead nodes (probably carbonates) of variable diameters reaching sometimes approximately 10 μm (Figure 12.4B(c)). This example of brown tin glaze colored with manganese, coming from a medieval tile discovered in the same context as the one illustrated in Figure 12.1, shows a significant weathering level and its complexity. In the vermiculated zones, phosphorus and sulfur increase jointly with manganese (coloring element of the glaze), suggesting the presence of newly formed pyrrhomorphite.

It is important to note that the chemical compositions of these strongly weathered glazes are variable, but most of them have significant contents of alkaline elements and often heterogeneous microstructures, particularly due to opacifying microcrystals. These compositions seem therefore less resistant to weathering.

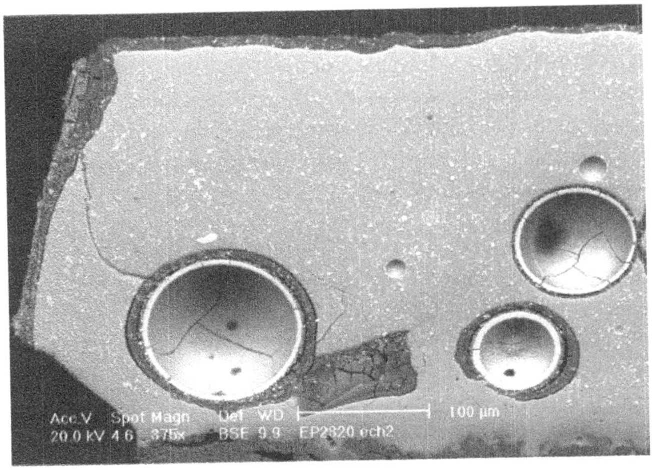

a) *Palissy* – Pristine glaze PbO: 42.7%; SiO_2: 33%; alk + CaO: 7.5%; Al_2O_3: 0.9%; SnO_2: 15%.

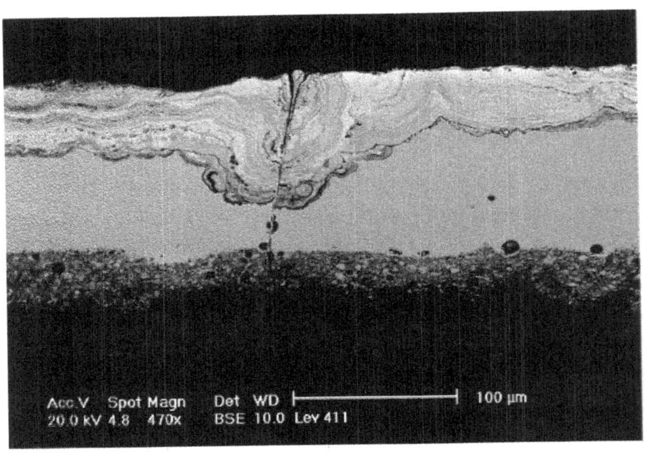

b) *Saint-Syméon – LEV 411* – Pristine glaze PbO: 68%; SiO_2: 28.5%; alk+ CaO: 1.5%; Al_2O_3: 0.5%; CuO: 1.5%.

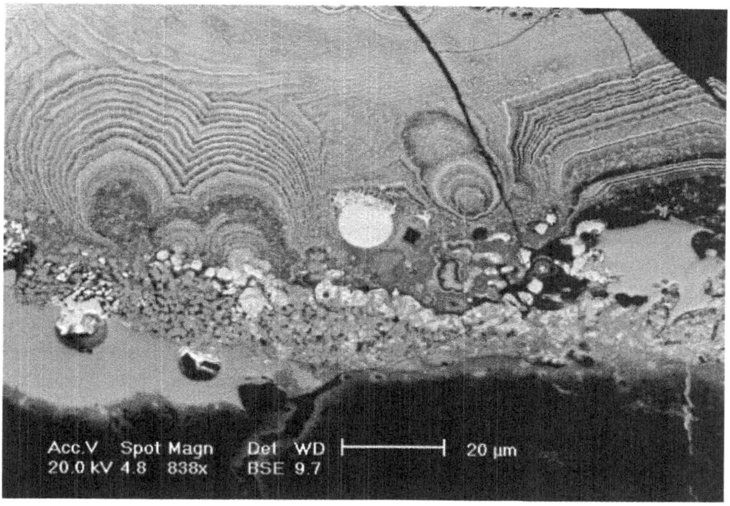

c) Vincennes – Pristine glaze PbO: 62%; SiO_2: 19.5%; alk + CaO: 0.5%; Al_2O_3: 3.5%; MnO: 1%; SnO_2: 10%.

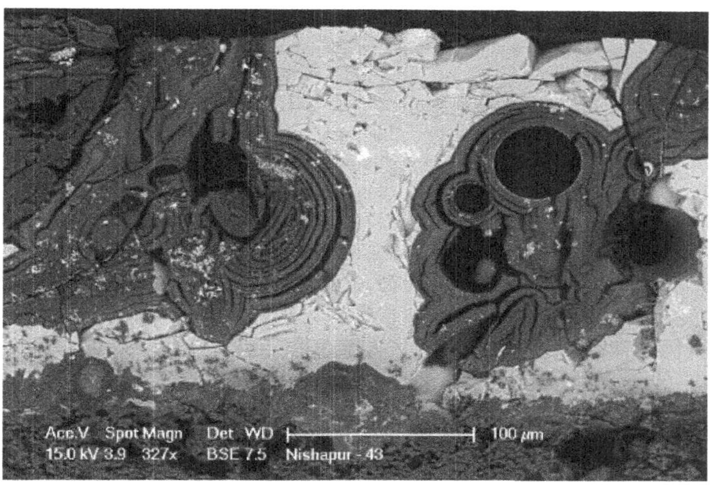

d) Nishapur-43 – Pristine glaze PbO: 36%; SiO_2: 46%; alk + CaO: 8%; Al_2O_3: 1.1%; CuO: 2%; SnO_2: 4.3%.

Figure 12.4. *Examples of laminated structures on opacified alkali-lead glazes (a) or on transparent lead glazes (b) C2RMF/A. Bouquillon and C. Doublet. Example of laminated structures on opacified alkali-lead glazes (c and d). C2RMF/A. Bouquillon and C. Doublet*

Laminated layer structures are very common in weathered ancient glasses and glazes, irrespective of their initial composition, as proven by the frequent mentions or illustrations in archaeometry articles; these structures appear independently of the conservation environment, for glassy materials that were buried, which is the case here, as well as for those that were immersed or still for those that remained in atmospheric context (particularly stained glasses). They are not specific for lead-containing glassy matrices, being present in all glasses, irrespective of their composition. Despite this wide presence, and the efforts made by many researchers to understand them, the processes governing their appearance are still subjected to discussion. Since the end of the 19th century, Brewster (1863) indicated that this zone must be a porous material. One century later, Brill formulated the hypothesis that it was a manifestation of cyclic weathering of glass depending on seasons and the variations of humidity in the glass conservation environment. This hypothesis was rejected after the first attempts at re-creating these structures in laboratory under strictly controlled conditions (Römich 2003). To our knowledge, the only experiments that successfully re-created these leaves concerned Si-Na-B glasses (Dohmen et al. 2013; Geisler et al. 2015) or Si-Ca-K glasses with $K_2O > 25\%$ (Römish et al. 2003) weathered over a long period of time (over 3 months) and/or by accelerating the reactions by increasing the temperature of the experiment. Our own experiments of laboratory weathering of glasses with various lead contents, in various acidic, basic or neutral media, at temperatures ranging between 20 and 90°C over 3-month durations have never led to the formation of such weathering patterns (Bonnet et al. 2007).

Thanks to the improvement of analytical techniques and particularly of imaging techniques, significant progress was made in the fine characterization of weathering layers. All the authors now agree on admitting that lamellae are widely constituted of silica nanoparticles of more or less significant size, that their porosity plays an essential role, that the interface between the pristine glass and the weathered layer is a strategic zone. The phenomena of diffusion between this interface zone and the surrounding environment are complex, and everyone agrees that pH is a key parameter. But there are diverging opinions on the mechanism: according to some, the weathered layer should be considered as a colloidal glass (Schalm and Anaf 2016, p. 10) and the laminae illustrate alternations of nanoparticles of more or less significant size constituting more or less compact layers of variable porosity; the presence of elements other than silica (Ca, Al, P) would correspond to the migration of charged solutions through the silica layer and to differential precipitations according to the encountered porosities and redox conditions (Geisler et al. 2015; Schalm and Anaf 2016). Others rather consider a complex series of episodes of hydrolyses, diffusion and precipitations, in the weathered phase, mechanism disturbed by cracks in this layer facilitating the circulation of new weathering solutions and significant local precipitations (Lombardo et al. 2013); others still think of self-organization within the weathered film (Dal Bianco et al. 2004).

Box 12.2. *Enigma of lamination*

12.4. Protection of lead-glazed weathered objects by sol-gel methods

Certain decoration practices allow a natural preservation of glazes, as it can be noted in certain lustred ceramics (see Box 12.3), but most of the time, people in charge of heritage collections should be offered easy-to-implement solutions, which do not modify the visual appearance, friendly toward the original materials and time resistant. The prior cleaning of the object is already a very delicate phase (Fekrsanati et al. 2000; Abd-Allah 2013), particularly for lead glasses (Davison 2003, p. 194) highly reactive with the most commonly used products such as EDTA, for example. Consequently, preserving glasses and glazes in a perfectly stable environment may sometimes be sufficient. However, in an attempt to find other solutions, research is being conducted to try and stabilize the weathered layer by sol-gel methods as it is done for alkali or alkali-lime glasses. But the presence of lead in the glassy matrix poses a problem.

Lustre: ancient decoration technique that blocks weathering

Glass weathering can be blocked by artificially modifying the composition of its surface, for example, by high-temperature dealkalization treatment of the first micrometers. This is a common operation in the glass industry, and more recently also in the lead crystal glass industry (Rybarikova 2001) (see Chapter 13). But decorations relying on similar mechanisms, lustres, invented more than one millennium ago on glazes with (or without) lead had the same preservation effects. Lustre corresponds to a layer of extreme surface of sub-micronic thickness containing metallic copper and/or silver nanoparticles of size ranging between 2 and 50 nm, sometimes exceeding 100 nm (Perez-Arantegui 2001; Chabanne et al. 2012; Sciau 2012); they are distributed in one or several layers of the order of 100 nm to 1 or several micrometers. The area thus treated gains a specific metallic colored aspect in specular reflection. The lustring mixture is composed of silver and/or copper salts, clays, and vinegar laid by paintbrush on a glaze. It is first of all fired under oxidizing or neutral atmosphere then undergoes a short plateau under reducing atmosphere within highly constrained ranges of temperatures, between 450 and 600°C. These are processes of interdiffusion of copper and/or silver with alkali, potassium and/or sodium followed by a reduction of Ag^{+} and/or Cu^{+} ions, then by crystal seeding and growth that govern lustre development (Pradell et al. 2008). It can be noted on certain shards that glazes are strongly weathered, but in the lustred zones, the preservation of the decoration and the underlying glaze is far better. RBS (Rutherford backscattering spectrometry) experiments on weathered lustred ceramic shards clearly indicate it: in the non-lustred zones of the shard, the retreat of the lead front, indicating a surface depletion by leaching, is significant, while in the lustred zones or in the immediate proximity of the decoration, this depletion is significantly reduced, even quasi-nonexistent. The following explanation was proposed: the exchange of sodium and potassium with silver and copper followed by firing is sufficient to create a barrier to hydration, thus improving the resistance of the underlying glaze (Pichon et al. 2010).

Such an observation was also made on alkali glasses colored with silver yellow in which the coloration is obtained by an exchange between sodium and silver and the precipitation of metallic silver nanoparticles (Beltran et al. 2020).

Box 12.3. *Lustres as protection against weathering*

Indeed, the routine application protocol requires an acid solution of TEOS (tetraethyl orthosilicate) in ethyl alcohol (C_2H_5OH) and deionized water. But the presence of hydrochloric acid in the solution attacks the lead glass (Bonnet et al. 2003; Bertoncello et al. 2004; Chapter 13) and iridescences, and even changes of colorations appear very rapidly at the surface of treated glasses. To avoid the use of acid, the protocol had to be adapted by using the catalysis properties of metallic ions, here Pb^{2+} ions, susceptible to accelerate the polymerization (*gelation*) in sol-gel reactions (Patra and Gangli 1992). In the case of lead glasses and glazes, lead comes directly from the underlying glass; it migrates into the TEOS + alcohol + water solution at the moment of application[1] (Bertoncello et al. 2006).

The results are very encouraging, not only for glasses containing lead, but also for other types of glasses that are highly sensitive, for example with high content of potash or soda–lime. For the latter, lead is then introduced in the form of lead nitrate ($Pb(NO_3)_2$) in the precursor, according to a molar ratio $Pb(NO_3)_2/TEOS = 3.6 \times 10^{-6}$. The film deposited by dipping in single layer (thickness = approximately 60 nm) (Bertoncello 2006, p. 318) or in multiple layers is homogeneous, colorless, and regularly distributed throughout the treated surface. Moreover, according to the studies initiated in 2008, it seems that the new formulation presents a quite satisfactory resistance to accelerated ageing tests (Dal Bianco et al. 2008), comparable to that of other traditional sol-gel layers containing HCl.

This type of coating has certainly several advantages: the treatment is conducted at ambient temperature; it is invisible to the naked eye and does not modify the colors. Restoration and conservation experts are however very reluctant to routinely use it on museum objects; the process is irreversible and therefore does not follow the rules of the current deontology. Therefore, the question of the development of such techniques in the domain of museum heritage remains open.

1. It is important to note that the role of lead in the polymerization process is not yet fully understood and several hypotheses coexist.

12.5. Conclusion

As far as weathering studies are concerned, the presence of lead in the glassy matrix has several advantages: it is easier to detect and measure than a significant number of other elements, it is generally present only in the glaze and not in the substrate. Moreover, these glazes are widespread and therefore offer "natural equivalents" corresponding to very wide ranges of ages, environments and compositions.

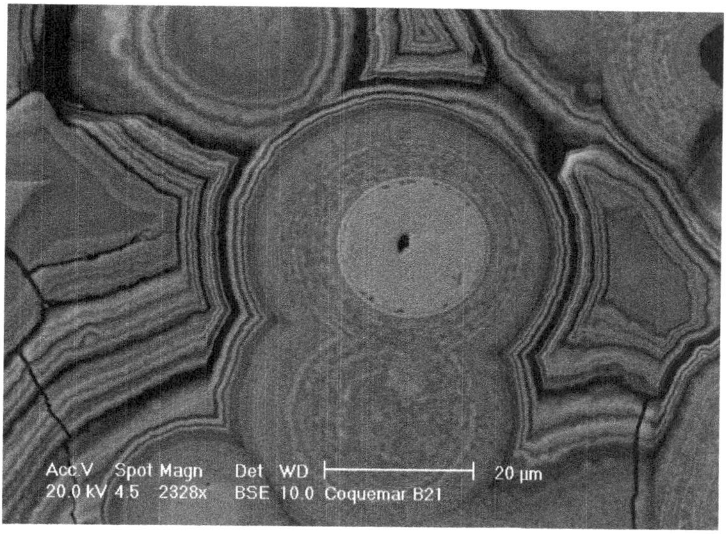

Figure 12.5. *Detail of a complex weathered zone. It is important to note the presence of more or less lead-enriched laminae (appearing clear on the image), a highly lead-rich node, in whose proximity a granular phase develops. C2RMF/S. Vialle*

Research projects conducted on the weathering of ancient glazes (lead-rich or not) that were subjected to a burial phase are extremely complex, as the characteristics of the weathered zones are generally highly diversified (Figure 12.5). Nevertheless, as far as lead matrices are concerned, we can rely on new and high-performance techniques, feasibility studies and most of all on the progress of knowledge on weathering in aqueous medium of these glasses (see Chapter 13). It is a path to be explored in the coming years in order to better understand the development of these laminated zones and their evolution in the conservation context of museums.

12.6. References

Abd-Allah, R. (2013). Chemical cleaning of soiled deposits and encrustations on archaeological glass: A diagnostic and practical study. *J. Cult. Herit.*, 14(2), 97–108.

Beltrán, M., Schibille, N., Brock, F., Gratuze, B., Vallcorba, O., Pradell, T. (2020). Modernist enamels: Composition, microstructure and stability. *J. Eur. Ceram. Soc.*, 40, 1753–1766.

Bertoncello, R., Milanese, L., Bouquillon, A., Dran, J.C., Mille, B., Salomon, J. (2004). Leaching of lead silicate glasses in acid environment: Compositional and structural changes. *Appl. Phys. A*, 79(2), 193–198.

Bertoncello, R., Milanese, L., Dran, J.C., Bouquillon, A., Sada, C. (2006). Sol-gel deposition of silica films on silicate glasses: Influence of the presence of lead in the glass or in precursor solutions. *J. Non-Crystal. Solids*, 352(4), 315–321.

Bonnet, C., Bouquillon, A., Turrell, S., Deram, V., Mille, B., Salomon, J., Thomassin, J.H., Fiaud, C. (2003). Alteration of lead-silicate glasses due to leaching in heated acid solutions. *J. Non Crystal. Solids*, 323, 214–220.

Bouquillon, A. and Mille, B. (2009). Les carreaux de pavements de quelques sites médiévaux d'Ile de France – Analyse des matériaux. In *Terres cuites architecturales médiévales et modernes*, Chapelot, J., Chapelot, O., Rieth, B. (eds). CRAHM, Caen.

Bouquillon, A., Turrell, S., Robinet, L., Charbonneau, C. (2007). Des glaçures Renaissance au cristal moderne – Comportement du plomb dans les verres altérés. *L'actualité chimique*, 40–46, 312–313.

Bouquillon A., Robinet L., Elias M., Charbonneau C., Katona I., Pagès-Camagna S., Calligaro T., Huet N., Sarzeau N., Pallot-Frossard I. et al. (2008). Méthodes portables non destructives d'analyse de l'altération des verres au plomb. *Techne Hors série – Programme National de Recherche sur la Connaissance et la Conservation des matériaux du Patrimoine Culturel*, 104–113.

Chabanne, D., Aucouturier, M., Bouquillon, A., Darque-Ceretti, E., Makariou, S., Dectot, X., FaÿHallé, A., Miroudot, D. (2012). Ceramics with metallic lustre decoration. A detailed knowledge of Islamic productions from 9th century until Renaissance. *Matériaux & Techniques*, 100(1), 47–68.

Charbonneau, C., Elias, M., Frigerio, J.M. (2007). Non-destructive study of lead glasses' alteration by reflectance spectroscopy. *Optics Communications*, 270(2), 280–289.

Chaulet, D., Bouquillon, A., Thomassin, J.H., Le Coustumer, P., Deves, G. (1999). Altération des verres au plomb : analogues archéologiques et expérimentation. Influence des hétérogénéités chimiques et texturales. In *Waste Stabilization and Environment*, Méhu, J., Grelier-Volatier, L., Burleigh, T. (eds). Elsevier, Amsterdam.

Chopinet, M.-H. and Lehuédé, P. (2010). Les problèmes d'altération rencontrés sur des verres industriels. *Verre*, 16(2), 20–27.

Coquinot, Y., Bouquillon, A., Doublet, C., Déléry, C. (2012). Réflexions sur les techniques de fabrication et les lieux de productions d'une sélection de céramiques à décor de lustre métallique d'époque nasride conservées par le Museo de la Alhambra, le Victoria and Albert Museum et le Louvre. In *1st International Conference of the European Network of Museums of Islamic Art*. REMAI, Alhambra.

Dal Bianco, B., Bertoncello, R., Bouquillon, A., Dran, J.C., Milanese, L., Roehrs, S., Sadad, C., Salomon, J., Voltolina, S. (2008). Investigation on sol-gel silica coatings for the protection of ancient glass: Interaction with glass surface and protection efficiency. *J. Non Cryst. Solids*, 354(26), 2983–2992.

Davison, S. and Newton, R.G. (2003). *Conservation and Restoration of Glass*. Butterworth-Heinemann, Oxford.

Dohmen, L., Lenting, C., Fonseca, P.O., Nagel, T., Heuser, A., Geisler, T. (2013). Pattern formation in silicate glass corrosion zones. *Int. J. Appl. Glass Sci.*, 4(4), 357–370.

Fekrsanati, F., Hildenhagen, J., Dickmann, K., Troll, C., Drewello, U., Olaineck, C. (2000). UV-laser radiation: Basic research of the potential for cleaning stained glass. *J. Cult. Herit.*, 1(2), 155–160.

Geisler, T., Nagel, T., Kilburn, M.R., Janssen, A., Icenhower, J.P., Fonseca, R.O.C., Grange, M., Nemchin, A.A. (2015). The mechanism of borosilicate glass corrosion revisited. *Geochim. Cosmochim. Acta*, 158, 112–129.

Lombardo, T., Gentaz, L., Verney-Carron, A., Chabas, A., Loisel, C., Neff, D., Lero, E. (2013). Characterisation of complex alteration layers in medieval glasses. *Corrosion Science*, 72, 10–19.

Mirguet, C., Fredrickx, P., Sciau, P., Colomban, P. (2008). Origin of the self organisation of Cu°/Ag° nanoparticles in ancient lustre potter – A TEM study. *Phase transit.*, 81(2/3), 253–266.

Patra, A. and Ganguli, D. (1992). Effects of Pb^{2+}-doping on alkoxide-derived silica sols and gels. *J. Non Cryst. Solids*, 139, 165–171.

Perez-Arantegui, J., Molera, J., Larrea, J., Pradell, A., Vendrell-Saz, T., Borgia, I., Brunetti, B., Gariati, F., Fermo, P., Mellini, M. et al. (2001). Lustre pottery from the thirteenth to the sixteenth century: A nanostructured thin metallic film. *J. Am. Ceram. Soc.*, 84(2), 442–446.

Pichon, L., Aucouturier, M., Beck, L., Lebon, M., Bouquillon, A., Lehuédé, P., Polverinos, A., Castaing, J. (2010). Imagerie quantitative en analyses par faisceaux d'ions appliquées aux objets du patrimoine. *Actes du colloque "Matériaux"*, Nantes, 1–12.

Pradell, T., Molera, J., Smih, A.D., Climent-Font, A., Tite, M.S. (2008). Technology of Islamic lustre. *J. Cult. Herit.*, 9, 123–128.

Rante, R., Collinet, A., Bouquillon, A., Coquinot, Y., Doublet, C. (2012). Pottery study and analyses. In *Nishapur Revisited: Stratigraphy and Ceramics of the Qohandez*, Collinet, A., Rante, R. (eds). Oxbow, Oxford.

Römich, H. (2003). Studies of ancient glass and their application to nuclear-waste management. *MRS Bulletin*, 28(7), 500–504.

Rybarikova, L., Dvorska, L., Hradecka, H., Jiricek, P. (2001). Surface treatment of lead glasses for reducing the leaching of lead. *Ceramics*, 45(1), 31–34.

Schalm, O. and Anaf, W. (2016). Laminated altered layers in historical glass: Density variations of silica nanoparticle random packings as explanation for the observed lamellae. *J. Non Cryst. Solids*, 442, 1–16.

Sciau, P. (2012). Nanoparticles in ancient materials: The metallic lustre decorations of medieval ceramics. In *The Delivery of Nanoparticles*, Hashim, A.A. (ed.). InTech., London.

Seignez, N., Gauthier, A., Bulteel, D., Buatier, M., Recourt, P., Damidot, D., Potdevin, J.L. (2007). Effect of Pb-rich and Fe-rich entities during alteration of a partially vitrified metallurgical waste. *J. Hazard. Mat.*, 149(2), 418–431.

Verità, M. (2006). Modern and ancient glass: Nature, composition and deterioration mechanism. In *The Materials of Cultural Heritage in their Environment*, Lefevre, R. (ed.). Edipuglia, Bari.

Waksman, S.Y., Bouquillon, A., Cantin, N., Katona, I. (2008). Approche archéométrique des premières "Byzantine Glazed White Ware" et de productions glaçurées romaines et romaines tardives. *Rei Cretariae Romanae Fautorum Acta*, 40, 531–536.

PART 5

Weathering of Lead Glasses and Standards

13

Lead Leaching in Industrial Crystal Glasses: Role of Chemical Composition, Structure and Surface Treatments

Frédéric ANGELI[1], Léa BRUNSWIC[1], Thibault CHARPENTIER[2] and Stéphane GIN[1]

[1] *CEA, DES, ISEC, DE2D, University of Montpellier, Marcoule, France*

[2] *NIMBE, CNRS, CEA, Paris-Saclay University, Gif-sur-Yvette, France*

Starting with the 17th century, lead oxide was added to alkali silicate to obtain a material known as crystal glass, conferring a specific brightness and sonority to the glass. Furthermore, it expands the range of temperatures within which the glass melt is malleable, which offers the possibility to make objects of particularly complex and original geometry, especially for the luxury industry. Crystal glass designation has been regulated since 1969 (see details in Chapter 14) to ensure the quality of the material, relying on criteria of density, refractive index and lead content. In particular, the mass content of lead oxide (PbO) must be higher than or equal to 24% so that the glass may be considered crystal glass. Currently, it corresponds to the reference content for industrially used crystal glass.

Lead exposure is a source of concern for the environment and for human health (Graziano et al. 1996; Patrick 2006). However, when inserted into the glassy network in a crystal glass, lead cannot be attributed to the degree of dangerousness

Lead in Glassy Materials in Cultural Heritage,
coordinated by Anne BOUQUILLON and Patrice LEHUÉDÉ. © ISTE Ltd 2023.

that it would have if it were fully absorbed. It is therefore important to study its behavior since glass may be exposed to conditions in which it may be extracted from the glassy matrix, especially during lengthy contact with a liquid. This is particularly a matter of concern for materials in contact with food products (Geueke et al. 2018) and their recycling (Baeyens et al. 2010). For certain applications, surface treatments are used in industry to prevent lead release.

According to the European regulation REACH (Registration, Evaluation, Authorization and Restriction of CHemicals) (see Chapter 14), in-force since 2006, manufacturers are required to file an application for marketing authorization for each usage in question. If risk is not controlled, the authorization may be denied. Authorizations, including usage restrictions, may be modified, suspended or revoked. It is the manufacturer's responsibility to prove that risks associated with its usage are under control.

Moreover, the practical application of the requirements for food contact differs depending on the materials (plastics, metals, ceramics). The field of ceramics is currently regulated by Directive 84-500, which includes lead and cadmium migration limits (see Chapter 14). These limits rely on the dosage of leached elements according to a standardized test (ISO 7086) that is currently considered a reference (24 hours in a 4% acetic acid solution at a temperature of 22°C) but could evolve to closer to the actual conditions of use (repeated contacts). At a national scale, the recommendations of DGCCRF concern, according to this test protocol, glasses that are in contact with commodities, food products and beverages for lead, cadmium, arsenic, cobalt and aluminum.

It is therefore essential to understand the behavior of lead in the manufactured objects in the conditions of use, especially if they contain liquids for consumption. Since the 1990s, many studies have been conducted to study the leaching of lead present in glass or ceramics in reactive environments, especially when in contact with beverages (such as alcoholic drinks or spirits) or acidic aqueous solutions that represent the most unfavorable case in relation to lead release. Acetic acid was chosen by the International Organization for standardization as a reference leaching solution in the standardized tests, and it was widely used, as well as nitric acid, in scientific tests (Graziano and Blum 1991; Ahmed and Youssof 1997a, 1997b, 1997c; Schultz-Münzenberg et al. 1998; Guadagnino et al. 2000, 2002; Bonnet et al. 2003; Bertoncello et al. 2004; Mizuno et al. 2005, 2008; Rahimi et al. 2009; Sadrnezhaad et al. 2009; Rahimi and Sadrnezhaad 2012). In commercial crystal glasses, it is generally observed that the release of alkali is faster than that of lead, in both cases being proportional to the square root of time (Ahmed and Youssof 1997b, 1997c; Schultz-Münzenberg et al. 1998; Bertoncello et al. 2004). Their release kinetics increase with temperature and acetic acid concentration (Ahmed and Youssof 1997b). In the case of successive contacts, the release of elements decreases when reintroduced in solution (Ahmed and Youssof 1997a), highlighting the barrier role of a leached layer enriched in network forming elements.

The structural arrangements resulting from the incorporation of lead oxide in a network of alkaline silicates may significantly modify the macroscopic properties of glass, especially its chemical durability (Mizuno et al. 2005). Unlike the majority of metallic cations, lead has the specificity of being able to form a glass within a very wide range of concentrations (Fajans and Kreidl 1948) that may reach over 80 mol% of PbO for a rapidly quenched glass (Feller et al. 2010; Alderman et al. 2013). While its structural role seems to depend on its concentration in the glass and on the chemical composition of the latter, the precise contents at which these modifications occur remain difficult to evaluate and are still being discussed. A detailed presentation of the structural evolution of the PbO environment can be found in Chapter 3.

Recent results are presented in this chapter on the behavior during leaching in aqueous medium of a typical composition of a commercial lead crystal glass. In order to better describe the mechanisms for lead release over time, a full analysis was conducted by comparing the structural properties and chemical durability of glasses. To better understand the relation between the lead content in glass and its release in solution, various glasses containing 10–50 mol% PbO were studied. Their chemical composition is presented in Table 13.1. The reference Pb10 glass corresponds to the chemical composition of a commercial lead crystal glass containing approximately 10 mol% lead oxide (28% mass). The other glasses were manufactured by replacing lead with silicon and removing the 0.2% antimony present in the industrial glass introduced to facilitate the elimination of gas inclusions in the glass and whose influence on the studied mechanisms is considered negligible.

Many leaching data were obtained under various experimental conditions on commercial glass and are summarized in this chapter. Additional experiments allowing reaching various kinetic regimes for short- and long-term alterations, with or without change of solution, from several hours to 1 year of leaching, are described. Experiments are conducted in 4% v/v acetic acid, commonly used as a reference solution (Hight 1996; Ahmed and Youssof 1997b; Guadagnino et al. 2000). The whole series of glasses was leached over short durations to show the direct influence of the structure and chemical composition on the mechanisms of lead crystal alteration.

	Pb10	Pb20	Pb30	Pb40	Pb50
SiO_2	77.1	70.8	61.8	51.4	39.4
PbO	10.6	19.7	27.6	38.8	49.5
K_2O	11.3	9.0	10.1	9.2	10.5
Na_2O	0.8	0.5	0.5	0.6	0.6
Sb_2O_3	0.2	–	–	–	–

Table 13.1. *Glass composition analyses (oxide molar percentage)*

13.1. Influence of lead content on crystal glass structure

Glassy network polymerization can be accessed by probing the silicon that constitutes the framework of the glassy network. High-resolution nuclear magnetic resonance (NMR) spectroscopy (with magic angle sample spinning) (Eckert 1992, 2018; Edén 2012) of ^{29}Si can be used to obtain this information, as shown in Figure 13.1. On industrial glass (Pb10), two contributions are observed at –94 ppm and –106 ppm, which correspond to Q^3 silicon entities (silicon tetrahedra with three bridging bonds with other silicon and a Si-O- non-bridging bond) and Q^4 entities (four bridging bonds), in agreement with the data obtained on binary glasses of lead silicate (Fayon et al. 1998; Shrikhande et al. 2001). The displacement toward strong chemical shifts (to the left in the figure) with the increase in lead concentration proves the strong depolymerization of the network until reaching a majority of Q^2 species and approaching Q^1 species for the Pb50 glass.

The proportions of various Q^n species extracted from these spectra are presented in Table 13.2. For the Pb10 glass, the proportion of Q^3 units is 59% and that of Q^4 is 41%. The calculated number of non-bridging oxygens (NBOs) generated by these proportions is 25.7%. The direct calculation of NBOs based on the glass composition, considering that two NBOs are formed per entity of Na_2O, K_2O and PbO, yields 25.6% (formula [13.1]), very close to the experimental value:

$$\mathrm{NBO} = \frac{2(\mathrm{Na_2O}+\mathrm{K_2O}+\mathrm{PbO})}{\sum \mathrm{O}} \qquad [13.1]$$

Table 13.2 proposes this comparison for all glasses. There is very satisfactory agreement for the less lead-rich glasses, that is, Pb10, Pb20 and Pb30 glasses. For high lead content glasses (Pb40 and Pb50), the presence of Pb-O-Pb oxygens (denoted $PbO_{cluster}$) should be considered to account for the experimental number of NBOs considering formula [13.2]:

$$\mathrm{NBO} = \frac{2(\mathrm{Na_2O}+\mathrm{K_2O}+\mathrm{PbO}-\mathrm{PbO_{cluster}})}{\sum \mathrm{O}} \qquad [13.2]$$

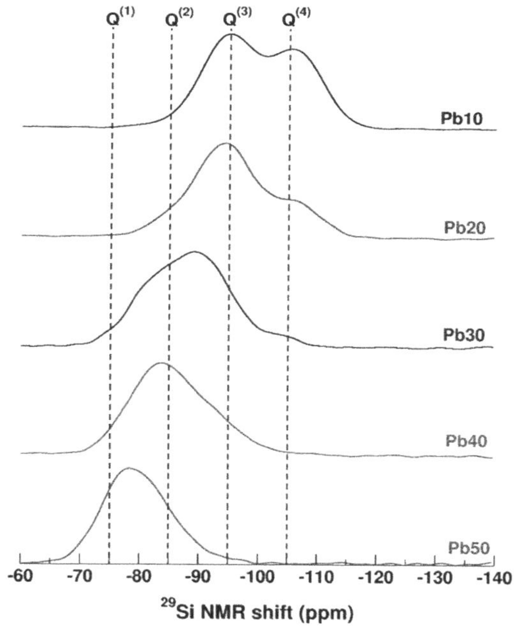

Figure 13.1. *^{29}Si MAS NMR spectra of the Pbx glass series. The dotted lines give the average positions of SiO_4 Q(n) entities (n = 1, 2, 3, 4)*

Glass	NBO		$Q^{(1)}$	$Q^{(2)}$	$Q^{(3)}$	$Q^{(4)}$	% Pb-O-Pb
	Calculation	Exp.					
Pb10	25.6	25.7	0	0	59	41	0
Pb20	34.2	35.7	0	9	69	22	0
Pb30	47.2	47.6	0	30	65	5	0
Pb40	59.3	59.3	0	65	35	0	3.7
Pb50	76.2	76.2	47	52	1	0	7.5

Table 13.2. *Comparison of the number of non-bridging oxygens (NBOs) calculated from the chemical composition and experimentally obtained for Pb10, Pb20 and Pb30 glasses and with adjustment to experimental data for Pb40 and Pb50 glasses to deduce the quantity of oxygen atoms in Pb-O-Pb form*

A part of the glasses of the (Pb10, Pb30 and Pb50) series was synthesized with isotopic enrichment in oxygen-17 to allow the study of the local environment of oxygen by ^{17}O NMR. Figure 13.2 compares the ^{17}O MAS spectra of these glasses and the MQMAS spectra of Pb10 and Pb30 glasses. The *multiple quantum MAS* (MQMAS) technique (Smith and van Eck 1999; Ashbrook 2009; Edén 2012, 2015; Ashbrook et al. 2018) is a two-dimensional experiment that allows coupling the MAS NMR spectrum to a second dimension, offering a better resolution. This dimension is denoted as the *isotropic dimension*, as the contribution to the line widening observed in MAS NMR is eliminated. Therefore, the width in this dimension reflects only the distribution of NMR parameters due to the inherent glass disorder. MQMAS experiment is specific to quadrupolar nuclei, with a nuclear spin $I > ½$, such as oxygen-17 ($I = 5/2$). For Pb10 glass, the NMR spectrum proves the distinct presence of three resonances characterizing three environments. A wide band in the interval from – 20 to 50 ppm corresponds to bridging oxygens Si-O-Si. Two narrower peaks centered at 70 and 110 ppm correspond to NBOs that can be attributed to Si-O^-(K^+) and Si-O^-(Pb^{2+},K^+) (Angeli et al. 2016b), respectively, to NBOs with potassium (only) and with potassium and lead for charge compensation. On the corresponding MQMAS spectrum (Figure 13.2(b)), these three contributions are clearly separated. Taking into account the low sodium content in the glass (below percent), its contribution, expected to be approximately 40 ppm (Angeli et al. 2011), cannot be observed.

Si-O(Pb^{2+},K^+) environments thus reflect the presence of a mix of lead and potassium (that interact directly with the same NBO), as was previously observed for other types of cations having quite close ionic radii (Lee and Stebbins 2003; Angeli et al. 2013). For the Pb10 glass, the quantification of each resonance coupled with knowledge of the chemical composition makes it possible to deduce that the NBO(Pb^{2+},K^+) environment (first coordination sphere) contains 38% potassium and 62% lead. In the binary glasses of lead silicates, a contribution related to Si-O^-(Pb^{2+}) bonds is noted at 151 ppm (Lee and Kim 2015); the Si-O^-(Pb^{2+},K^+) site at 119 ppm is positioned between the Si-O^-(K^+) site at 75 ppm and the Si-O^-(Pb^{2+}) site at 151 ppm. The higher the increase in the proportion of Pb with respect to that of K (from Pb10 to Pb50), the greater the position of the Si-O-(Pb,K) site is displaced toward the strongest chemical shifts to tend to the position of Si-O^-(Pb^{2+}) sites. Moreover, when the Pb content of the glass increases, there is no isolated K near the NBO; all of them are in the mixing sites with Pb.

For the lead-richest glass (Pb50), an additional contribution clearly appears at approximately 260 ppm. Lee and Kim (2015) studied the structural evolution of binary glasses of lead silicate by ^{17}O MQMAS NMR for high PbO contents close to orthosilicate compositions ranging from 60 to 71 mol%. They clearly show the presence of Pb-O-Pb sites in this range of chemical shifts.

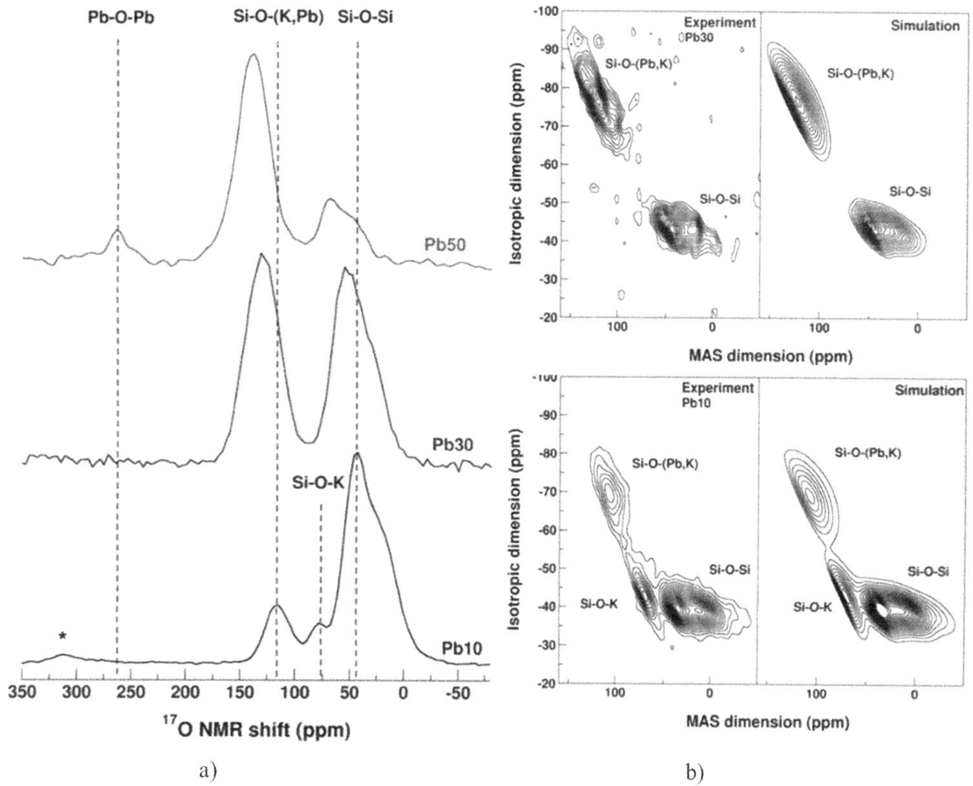

Figure 13.2. *(a) ^{17}O MAS NMR spectra of Pb10, Pb30 and Pb50 glasses and (b) ^{17}O MQMAS NMR spectra of Pb10 and Pb30 glasses (*spinning sidebands)*

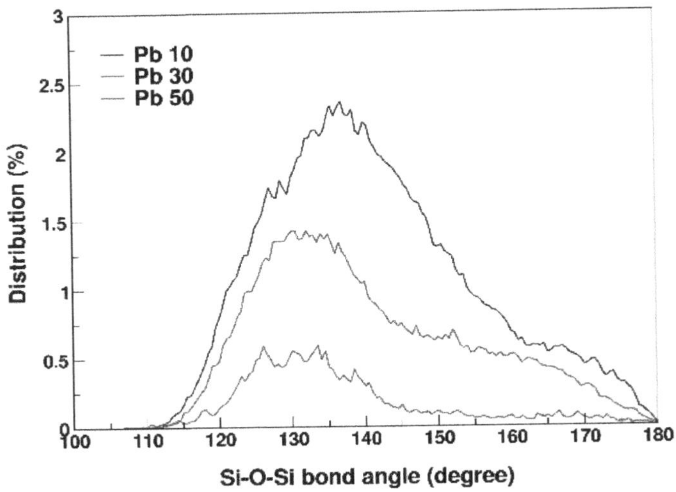

a)

Glass	Pb-NBO bond length (Å)	Si-O-Si angle (°)
Pb10	2.58	144.1
Pb30	2.61	142.5
Pb50	2.64	140.5

b)

Pb50	Pb-O-Pb	$Q^{(0)}$	$Q^{(1)}$	$Q^{(2)}$	$Q^{(3)}$	$Q^{(4)}$
Simulation	9%	4	43	56	6	0
Experiment	8%	0	47	52	1	0

c)

Figure 13.3. *Distribution of Si-O-Si bond angles for Pb10, Pb30 and Pb50 resulting from molecular dynamics simulations (a). The tables summarize the Pb-O lengths and the average Si-O-Si angles for various glasses (b) and make a comparison for Pb50 glass between the experimental and simulated proportions of Q^n species and of the proportion of Pb-O-Pb bonds (c). For a color version of this figure, see www.iste.co.uk/bouquillon/lead.zip*

Finally, it can be noted that the form and position of the Si-O-Si site on the ^{17}O MAS spectra evolves significantly with the lead content. The origin of these changes can be explained using atomistic simulation data of these glasses by classical molecular dynamics (MD) using potentials provided in the literature (Damodaran et al. 1990; Hemesath and René Corrales 2005; Hong et al. 2019). The simulation of the glass structure highlights both an increase in Pb-O$^-$ bond length and a decrease in Si-O-Si angle with increasing lead content (Figure 13.3). This evolution of the angular distribution may be at the origin of the variations in the position of $Q^{(n)}$ resonances on ^{29}Si MAS NMR spectra (Figure 13.2), as observed in glassy silica and alkaline silicates (Charpentier et al. 2009, 2013; Angeli et al. 2011).

The comparison of experimental and simulated proportions of Q^n species and the proportion of Pb-O-Pb bonds for Pb50 glass (Figure 13.3) shows good agreement with MD data. Pb50 glass is the only one (among the three studied compositions) on which the presence of atom grouping in the form of Pb-O-Pb at 8% was observed.

The extraction of only lead atoms interconnected by means of Pb-O-Pb bonds in a simulation box highlights the presence of connected Pb chains on the lead-richest composition (Pb50), in agreement with experimental data (Figure 13.4).

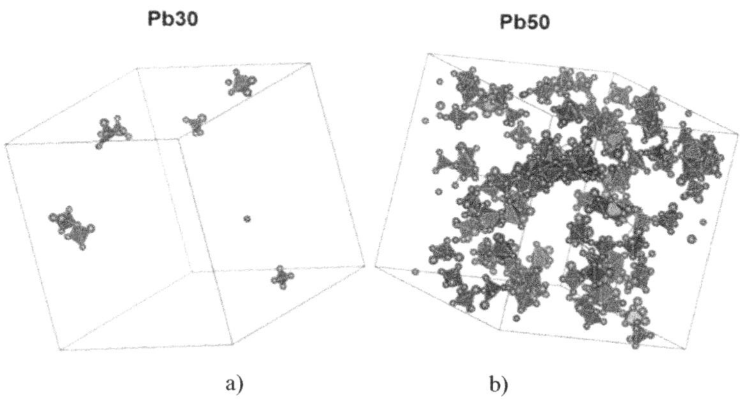

Figure 13.4. *3D visualization of lead atoms interconnected by Pb-O-Pb bonds obtained by molecular dynamics for Pb30 (a) and Pb50 (b) compositions. For a color version of this figure, see www.iste.co.uk/bouquillon/lead.zip*

13.2. Leaching mechanisms of lead glasses

This section focuses on the leaching behavior in solution of the series of lead glasses presented in Table 13.1. Similar to the first section, a more in-depth study was conducted for the commercial lead crystal glass Pb10. Additional experiments were conducted at various temperatures, on short durations, with and without change of solution, with or without saturation of the solution and over long periods of time.

The dosage of the elements released in solution aims at remounting the leaching kinetics. The elementary profiles of the altered layers were studied by time of flight secondary ion mass spectrometry (ToF-SIMS), and their structure was studied by solid-state NMR.

13.2.1. *Low reaction progress experiments*

Experiments that yield a maximal alteration rate, without feedback from the elements in solution resulting from glass leaching on the kinetics, are generally obtained under dynamic conditions with *high solution renewal* (Cailleteau et al. 2008).

Under these so-called initial rate conditions, measurements were conducted at 22, 40 and 70°C in 4 vol% acetic acid medium with the setup presented in Figure 13.5.

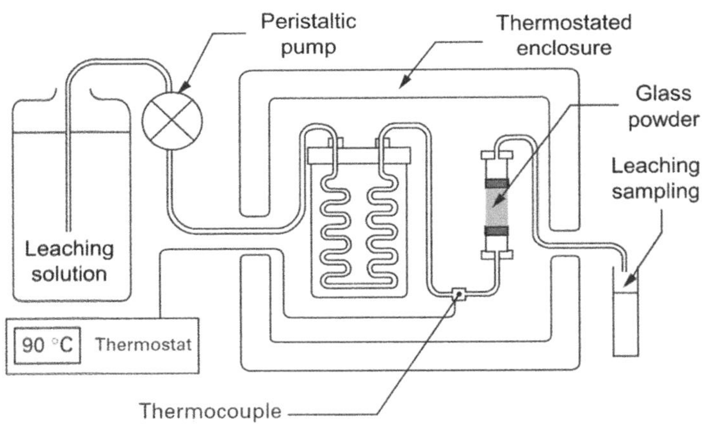

Figure 13.5. *Glass leaching protocol under dynamic conditions with high solution renewal*

The principle consists of passing over glass powder (size fraction 63–125 µm) a leaching solution at a high flow rate (Q). The dosage of the output elementary concentrations can be used to calculate the dissolution rate over several hours. The Q/S ratio (between the flow rate of the leaching solution and the glass surface area) is adapted as a function of the dissolution rate to have low silicon concentrations, below 100 µg/L, and thus avoid having an effect on the leaching kinetics. Flow rates range between 300 and 900 $mL \cdot j^{-1}$. The given rate values correspond to an average over three measurements. The evolution of the equivalent thicknesses of element *i*, *Eth(i)*, represents the depth over which element *i* is extracted from the weathering layer (in m) in time according to formula [13.3]:

$$Eth_i(t) = \frac{1}{\rho}\sum_{j=1}^{j=n}\frac{1}{x_i \times SSA \times m}\left[\frac{(C(i)_j + C(i)_{j+1})}{2} \times Q \times (t_{j+1} - t_j)\right] \qquad [13.3]$$

where *SSA* is the specific surface of the glass powder ($m^2.g^{-1}$), Q is the flow rate of the solution ($m^3 \cdot s^{-1}$), $C(i)_t$ is the i element concentration at *j*th sampling, ρ is the glass density, *m* is the glass mass, x_i is the mass fraction of element i in the glass and $t_{j+1} - t_j$ is the time between two samplings.

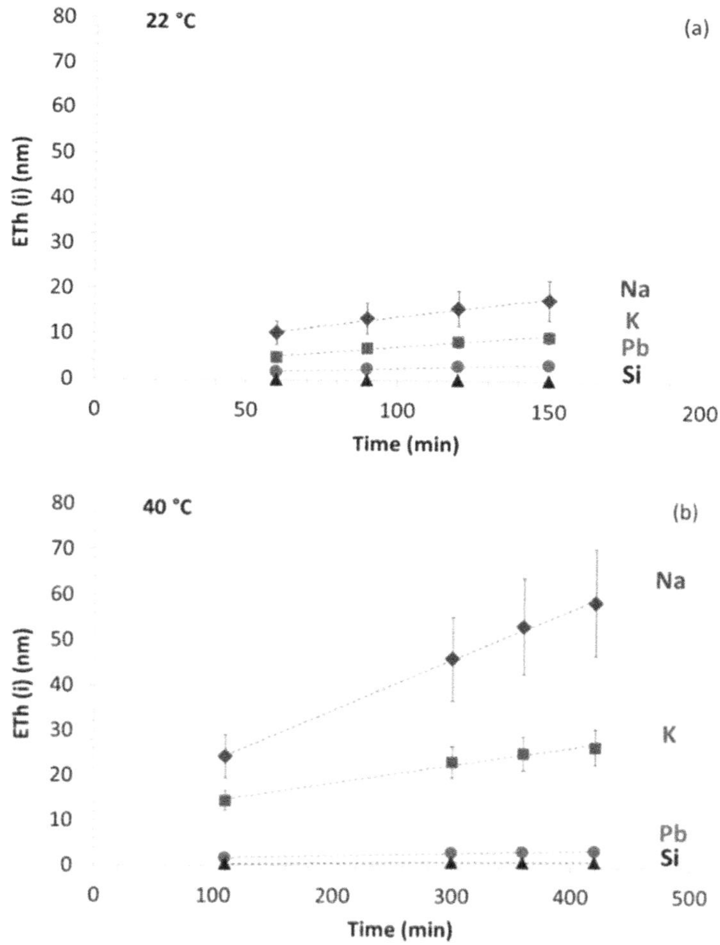

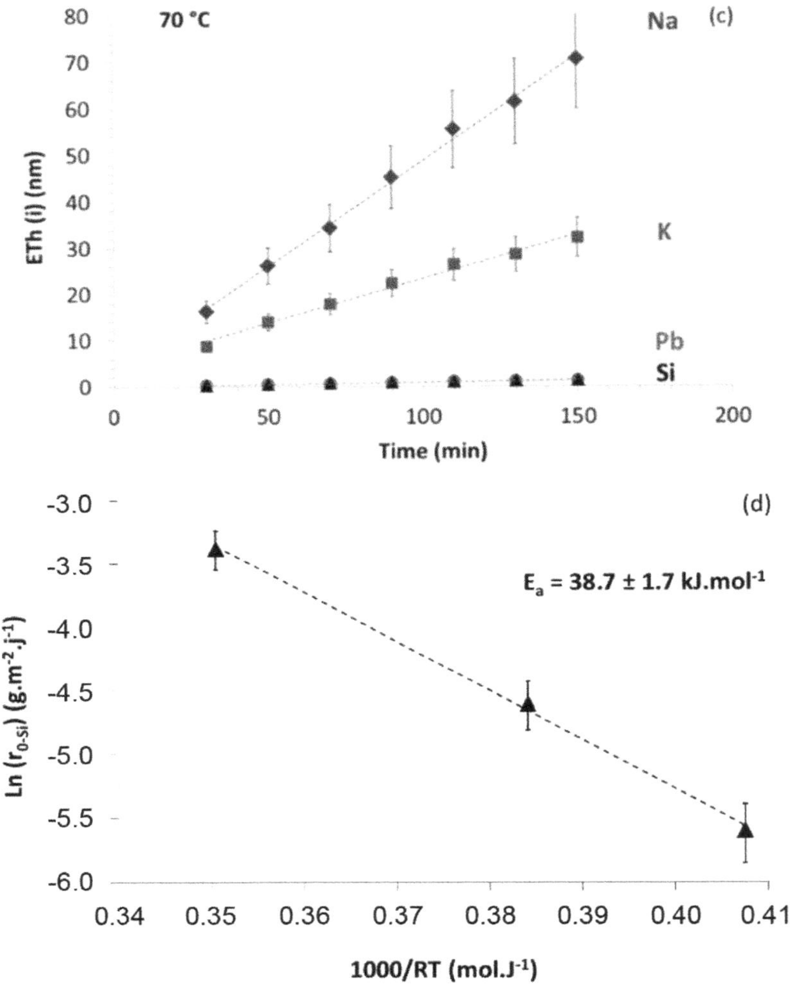

Figure 13.6. *Equivalent thickness of leached glass calculated from the release of Si, Pb, K and Na at 22°C (a) and 40°C (b) in acetic acid under the initial alteration rate regime for Pb10. Reproduced with permission of the American Chemical Society, Environ. Sci. Technol. 50 (Angeli et al. 2016). Equivalent thickness of leached glass calculated from the release of Si, Pb, K and Na at 70°C (c) in acetic acid under the initial alteration rate regime for Pb10. (d) Determination of the apparent activation energy of the leaching mechanisms under the initial rate regime based on Si release*

Figure 13.6 represents the equivalent alteration thicknesses at 22, 40 and 70°C calculated from the standardized mass losses of Si, Pb, K and Na. Sb release (not shown) is very close to that of Si. The dissolution is highly incongruent; Pb release is slightly higher than that of the Si network former, while K and Na are released far more rapidly.

The data can be used to determine from the silicon release an activation energy of the hydrolysis mechanisms under an initial rate regime of 38.7 ± 1.7 kJ·mol^{-1}. It is not possible to extract an activation energy for Pb, K and Na; the non-alignment of the three temperatures confirms in fact that the leaching of these elements is the result of both network hydrolysis and interdiffusion (H^+/(Pb^+,K^+, Na^+) exchange). At lower temperatures, the interdiffusion mechanism is favored, while at higher temperatures, hydrolysis predominates.

In order to focus on interdiffusion mechanisms, experiments were conducted in an initially saturated solution of amorphous silica (170 mg·L^{-1}) to block the hydrolysis of the glassy network (Gin et al. 2015). These experiments were conducted under static conditions in a Teflon® reaction vessel under the same acetic acid conditions as previously described. The equivalent thickness under these conditions is calculated as follows:

$$Eth_i(t) = \frac{1}{\rho}\left[Eth_i(t-1) + \frac{(C(i)_t - C(i)_{t-1}) \times V_t}{\rho \times SSA \times m \times x_i}\right] \qquad [13.4]$$

where *SSA* is the specific surface of the glass powder (m^2·g^{-1}), *Q* is the flow rate of the solution (m^3·s^{-1}), x_i is the mass fraction of silicon in the Pb10 glass, $C(i)_t$ is the *i* element concentration at moment *t*, ρ is the density of glass, *m* is the glass mass and x_i is the mass fraction of element i in the glass.

Figure 13.7 compares the equivalent thicknesses at various temperatures expressed as a function of the time square root. Due to the very low sodium concentration in the glass (0.2 mol%), its dosage in solution is possible only in the test at 70°C, which reveals that this element behaves similarly to potassium. A similar behavior of these two cations is expected at other temperatures.

The various cations seem well released in solution by a diffusive mechanism to the extent that a linear behavior as a function of the time square root is observed. Lead release is significantly below that of potassium. Lead behavior seems similar to that of multivalent elements, such as alkaline earths or rare earths, which leach more slowly than alkali elements (Angeli et al. 2013). Under these silicon-saturation conditions, it may be considered that the diffusion reactions are predominant with respect to hydrolysis reactions and therefore of dissolution of the glassy network. Fick's second law can then be applied to calculate the diffusion coefficients at the glass/solution interface from the concentrations in solution (Helebrant and Pekarkova 1996):

$$c_i = 2\frac{SSA \times m}{V} c_g \left(\frac{D_i t}{\pi}\right)^{\frac{1}{2}} \quad [13.5]$$

where c_i is the Pb or K concentration in solution, *SSA* is the specific surface area of the glass powder ($m^2.g^{-1}$), c_g is the Pb or K concentration in the glass, V is the volume of the solution, D_i is the diffusion coefficient of Pb or K and m is the mass of the glass.

Figure 13.7(d) presents the evolution of diffusion coefficients as a function of temperature for Pb and K. The resulting activation energies highlight a dependence on temperature that is far weaker for lead (37.2 ± 6.2 kJ·mol^{-1}) than for potassium (72.2 ± 7.6 kJ·mol^{-1}). At 22°C, the diffusion coefficient of lead is $(3.1 \pm 0.6) \times 10^{-22}$ m^2·s^{-1}. Depending on temperature, its release is between one and two orders of magnitude slower than that of alkali.

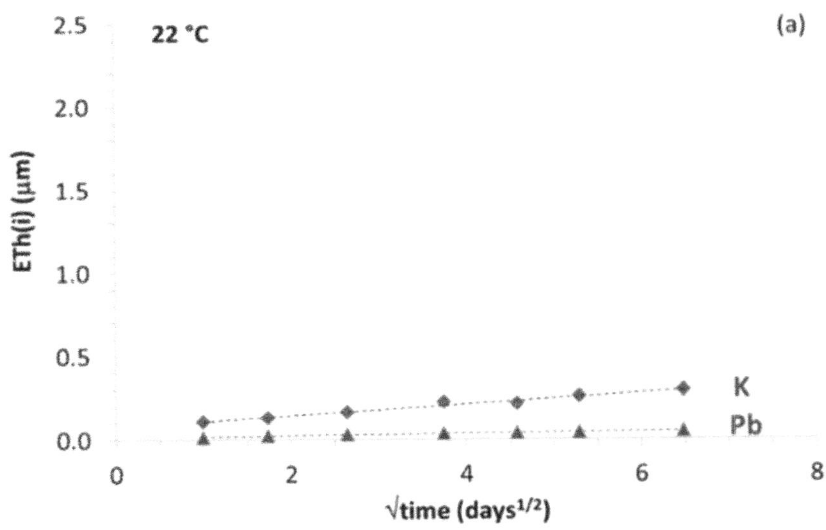

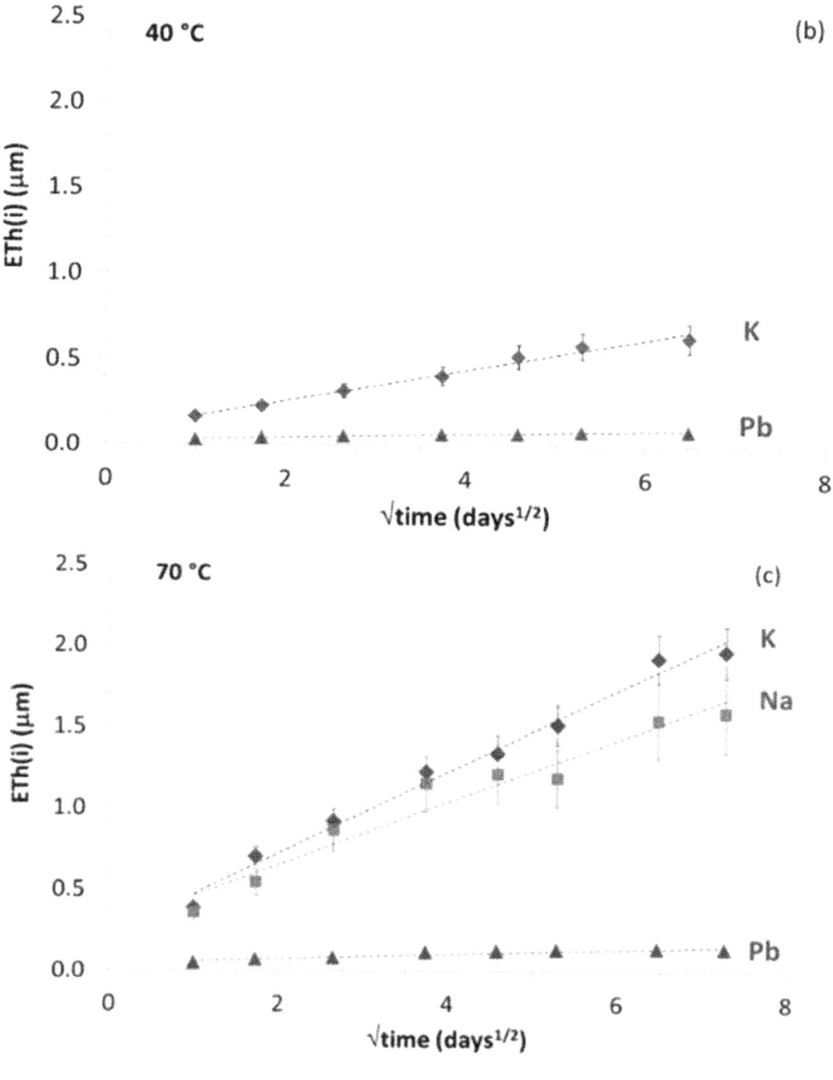
40 °C
(b)
ETh(i) (μm)
K
Pb
√time (days1/2)
70 °C
(c)
ETh(i) (μm)
K
Na
Pb
√time (days1/2)

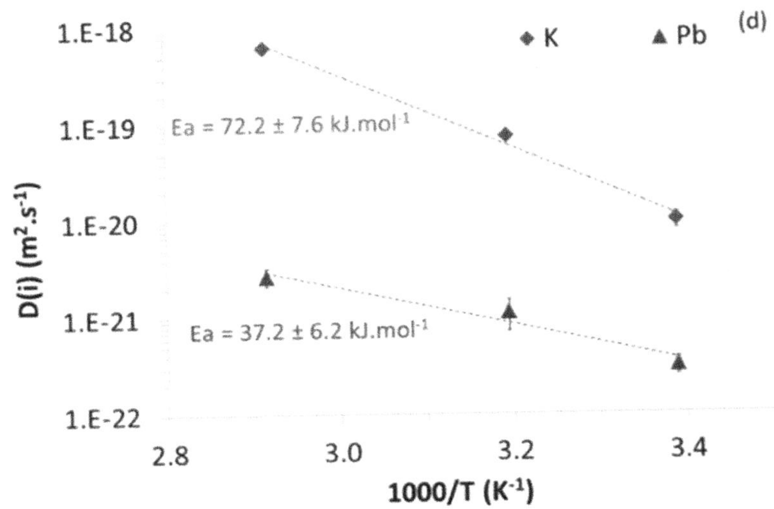

Figure 13.7. *Equivalent thicknesses of Pb10 glass leached at 22°C (a) and 40°C (b) in acetic acid in a solution initially saturated in amorphous silica (SA/V = 100 m^{-1}) as a function of the time square root. Reproduced with permission of the American Chemical Society, Environ. Sci. Technol. 50 (Angeli et al. 2016). Equivalent thicknesses of Pb10 glass leached at 70°C (c) in acetic acid in a solution initially saturated in amorphous silica (SA/V = 100 m^{-1}) as a function of the time square root. (d) Representation of the diffusion coefficients as a function of the inverse of temperature. The activation energy Ea is given by the slopes of the lines*

Figure 13.8 presents the elementary profiles of the alteration layers obtained by ToF-SIMS on glass leached at 70°C in acetic acid solution saturated in amorphous silica after 1, 7 and 53 days. The advancement of the alteration front in the pristine glass is clearly visible over time. At the interface between the pristine glass and the alteration layer, an advancement of the hydration front can be noted, which is reflected by a more significant penetration of hydrogen compared to the release of alkali. The profiles of sodium and potassium are very similar, in agreement with the concentrations measured in solution (Figure 13.7(c)). Lead is far less released than the alkali; it is depleted only in a fine external area of the alteration layer. The lead depletion depth after 53 days is 15 times smaller than the total depth of the alteration layer. Sb is fully retained throughout the depth of the layer.

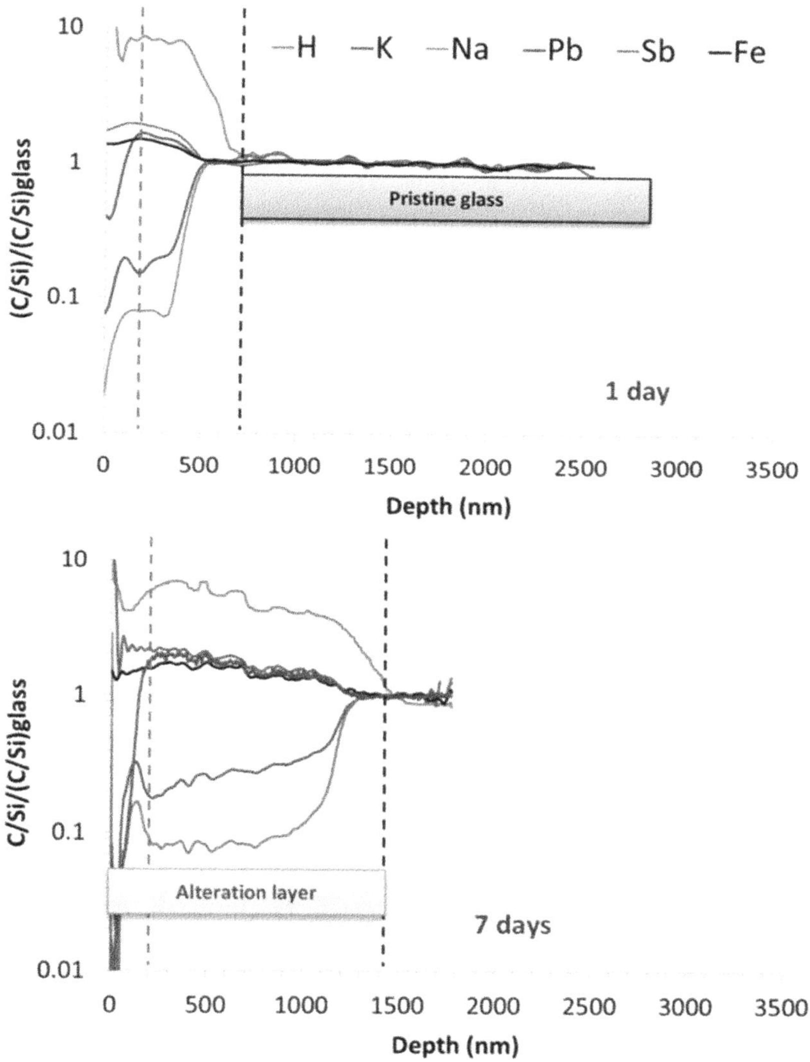
10
1
0.1
0.01
(C/Si)/(C/Si)glass
—H —K —Na —Pb —Sb —Fe
Pristine glass
1 day
0 500 1000 1500 2000 2500 3000 3500
Depth (nm)
10
1
0.1
0.01
C/Si/(C/Si)glass
Alteration layer
7 days
0 500 1000 1500 2000 2500 3000 3500
Depth (nm)

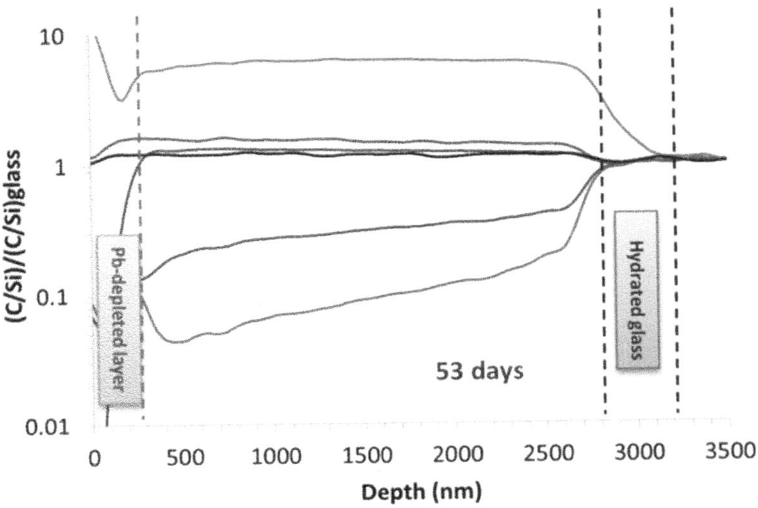

Figure 13.8. *Elementary profiles obtained by ToF-SIMS on Pb10 glass leached at 70°C for 1, 7 and 53 days in acetic acid solution saturated in amorphous silica. For a color version of this figure, see www.iste.co.uk/bouquillon/lead.zip*

COMMENT ON FIGURE 13.8.– *Data are standardized with respect to silicon, which is the least soluble element. All the elements are thus standardized to 1 in the pristine glass; lower contents indicate element depletion, while higher contents indicate enrichment. Reproduced with permission of the American Chemical Society, Environ. Sci. Technol., 50 (Angeli* et al. *2016).*

The set of glasses in Table 13.1 was leached at 40°C in acetic acid to extract the diffusion coefficients of various elements. The very high sensitivity to the chemical composition of glasses is clearly highlighted with variations of several orders of magnitude (Figure 13.9). These data are compared to those obtained on binary glasses of lead silicate, varying from 25 to 70 mol% of PbO, leached at 40°C in 0.1 N HNO_3 solution (Mizuno et al. 2005).

Although the composition of glasses may slightly differ (with or without alkali), as well as the pH of the solution (acetic acid with a pH close to 2 or nitric acid with a pH close to 1), the diffusion coefficients have the same order of magnitude as a function of Pb content in the glass. Roughly speaking, an increase in the diffusion coefficient of approximately one order of magnitude is observed when the quantity of PbO increases by approximately 5%. A composition of binary glasses at 25%

PbO seems to suggest a beginning of plateau for low Pb contents, but this behavior is not found in glasses containing alkali.

For the latter, only the industrial Pb10 glass is altered in a non-congruent manner with a more significant retention of Pb compared to the alkali. For the other glasses, Pb, Na and K are released at the same rate.

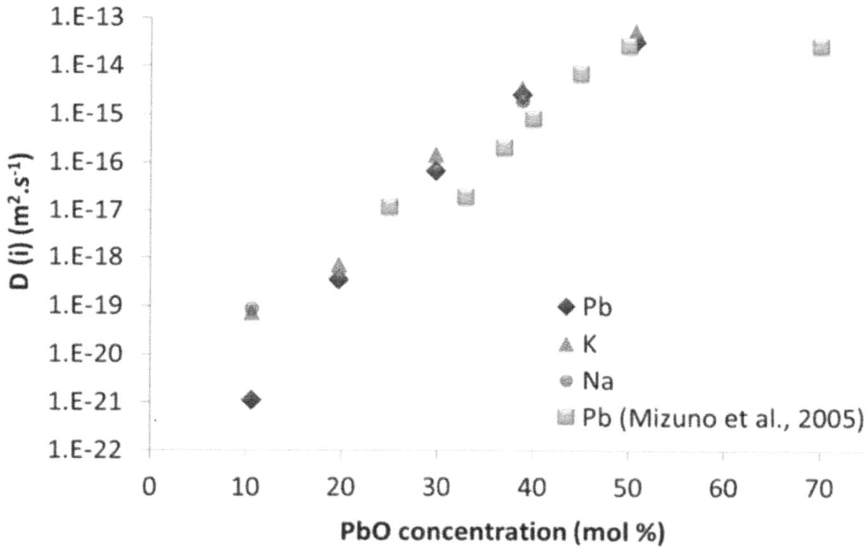

Figure 13.9. *Comparison between Pb diffusion coefficients obtained at 40°C on binary glasses of lead silicates leached in nitric acid (orange squares according to Mizuno et al. 2005) and Pb, K and Na diffusion coefficients at 40°C on glasses in Table 13.1 leached in acetic acid. For a color version of this figure, see www.iste.co.uk/bouquillon/lead.zip*

Lead behavior is closely related to the structural configuration it takes, both in the pristine glass and in the alteration layer. When the lead content in the glass is low, approximately 10 mol% of lead oxide (28% mass) is isolated in the glass, and it is more difficult to extract from the silicate network during leaching. It appears to preserve in the alteration layer an environment that is close to the one it had in the pristine glass, near silicon. Then, its release increases proportionally with its content, and starting with 20 mol% PbO, it is leached at the same rate as the alkali. Beyond 30%, lead clusters are formed, which may lead to percolation of the lead sub-network, which then rapidly passes into solution due to its high solubility.

13.2.2. *High reaction progress experiments*

In order to access higher reaction progress, which is expected over longer durations, experiments were conducted under static conditions at various *SA/V* ratios, varying from 10 to 10,000 m^{-1}. Experiments were conducted at 22°C in a Teflon reaction vessel with powders of 63–125 μm in size for the tests at 10 and 100 m^{-1} and with powders of 20–40 μm in size for tests at 1,000 and 10,000 m^{-1}. They were conducted in 4% v/v acetic acid for 1 year, without initial silicon saturation of the leaching solution.

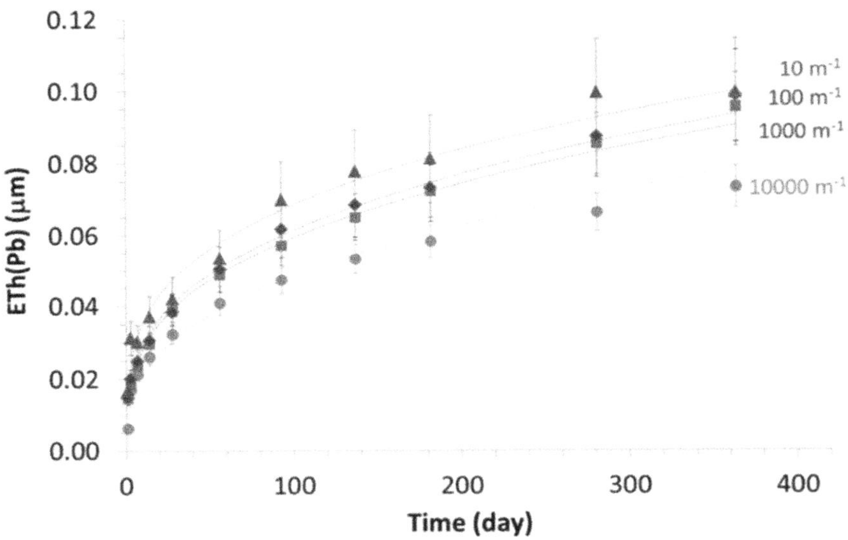

Figure 13.10. *Equivalent thickness of the leached glass calculated from lead release at 22°C in acetic acid for various SA/V ratios (10, 100, 1,000 and 10,000 m^{-1}). Reproduced with permission of the American Chemical Society, Environ. Sci. Technol. 50 (Angeli et al. 2016)*

Although there are four orders of magnitude differences between the surface ratios, Figure 13.10 shows relatively small differences for lead release. This indicates that the leaching process is still dominated by a diffusion mechanism, which depends on temperature and pH but does not depend on the *SA/V* ratio. The calculation of the retention factor of lead in the alteration layer compared to potassium ($Fr_{(Pb)} = 1 - (E_{th_{(Pb)}}/E_{th_{(K)}})$) yields quite similar values, with an average retention of 77% of lead in the alteration layer. It can be noted that Pb concentrations in solution remain far (at least three orders of magnitude) lower than the solubility of lead acetate (Apelblat and Manzurola 2001). This confirms that Pb release is not controlled by its limit of solubility but by diffusion mechanisms. Furthermore, it is verified that the evolution of Pb release is quite linear as a function of the time square root under various conditions.

The concentrations of all the elements of glass increased over time and with the *SA/V* ratio (Table 13.3). During the test conducted at the highest *SA/V* ratio (10,000 m^{-1}), the pH increases up to 3.3 at the end of the experiment for an initial value of 2.5. Under these conditions, the high alkali content in solution leads to a pH increase, which affects the leaching kinetics. Since the leaching kinetics decrease with increasing pH, the altered glass with the highest *SA/V* ratio is therefore slightly less leached than the others (Figure 13.10).

SA/V	10 m^{-1}						100 m^{-1}					
	Concentration (mg.L^{-1})						Concentration (mg.L^{-1})					
Days	Si	K	Na	Pb	Sb	pH	Si	K	Na	Pb	Sb	pH
1	0.262	0.463	0.104	0.139	0.00069	2.44	0.306	1.40	0.178	1.20	0.0042	2.42
3	0.058	0.717	0.080	0.258	0.00082	2.24	0.572	2.90	0.481	1.63	0.0057	2.16
7	0.067	0.554	0.065	0.250	0.00099	2.33	0.481	2.69	0.184	2.03	0.0069	2.30
14	<0.041	0.881	0.089	0.308	0.00116	2.40	0.643	6.76	0.383	2.50	0.0084	2.34
28	<0.041	0.854	0.093	0.350	0.00131	2.45	0.310	4.75	0.273	3.15	0.0096	2.40
56	<0.040	1.11	0.105	0.449	0.00158	2.38	0.520	6.36	0.399	4.14	0.0122	2.37
93	0.041	1.44	0.149	0.588	0.00173	2.39	0.312	8.07	0.456	5.08	0.0130	2.39
137	<0.041	1.79	0.131	0.655	0.00196	2.59	0.634	9.26	0.545	5.64	0.0146	2.46
182	0.073	2.25	0.264	0.684	0.00197	2.51	0.639	12.0	0.557	6.06	0.0180	2.47
281	0.075	2.60	0.174	0.848	0.00227	2.31	0.728	14.3	0.761	7.24	0.0200	2.35
365	0.091	3.11	0.145	0.851	0.00251	2.30	0.543	16.6	1.05	8.20	0.0249	2.29
SA/V	1000 m^{-1}						10000 m^{-1}					
1	0.853	13.7	0.480	11.9	0.020	2.44	1.89	102.5	4.88	50.1	0.014	2.59
3	0.858	19.3	0.541	15.2	0.031	2.20	2.85	194.5	8.59	138.8	0.084	2.50
7	1.07	27.6	1.62	19.4	0.040	2.34	2.96	259.5	12.0	174.3	0.108	2.70
14	1.55	36.7	3.28	24.3	0.047	2.38	4.20	338.4	16.2	214.9	0.126	2.82
28	0.701	43.7	1.69	31.7	0.055	2.46	2.55	450.8	21.8	265.9	0.143	2.97
56	1.53	63.3	2.88	40.0	0.064	2.47	6.20	634.7	33.0	336.9	0.172	3.06
93	0.982	77.8	3.52	47.1	0.084	2.50	5.57	801.0	37.7	390.0	0.191	3.17
137	2.21	87.5	3.74	53.6	0.090	2.59	10.8	896.1	43.7	441.5	0.210	3.27
182	1.90	99.8	4.90	60.0	0.091	2.63	10.6	984.6	54.5	484.9	0.231	3.33
281	2.45	124.4	8.38	70.5	0.105	2.53	14.1	1188	63.1	555.3	0.273	3.31
365	2.33	145.6	6.44	79.2	0.123	2.51	17.0	1366	73.4	615.4	0.333	3.31

Table 13.3. *The Si, K, Na, Pb and Sb concentrations (mg·L^{-1}) and pH evolution as a function of time for various SA/V ratios of Pb10 glass leaching in acetic acid at 22°C*

For the other *SA/V* ratios, the pH remained at 2.5, even after 1 year of leaching. The equivalent thicknesses *ETh (K, Na)* show that the release of these two elements is globally congruent, while Pb, Si and Sb are strongly retained in the alteration layer. Hence, for high reaction progress ($SA/V \geq 1{,}000$ m^{-1}), the Si and Sb retention rate is 0.99 and that of Pb is approximately 0.8.

The elementary profiles obtained by ToF-SIMS after 1 year of leaching at 22°C with *SA/V* ratios of 10 and 10,000 m^{-1} show that the various elements of glass have similar behavior for the two *SA/V* ratios (Figure 13.11) and are of the same type as those obtained at 70°C (Figure 13.8) in a medium initially saturated in silica. The alteration layer decomposes into three parts with nearly equal thicknesses for the two *SA/V* ratios, which confirms the fact that diffusion continues to control leaching at very high reaction progress.

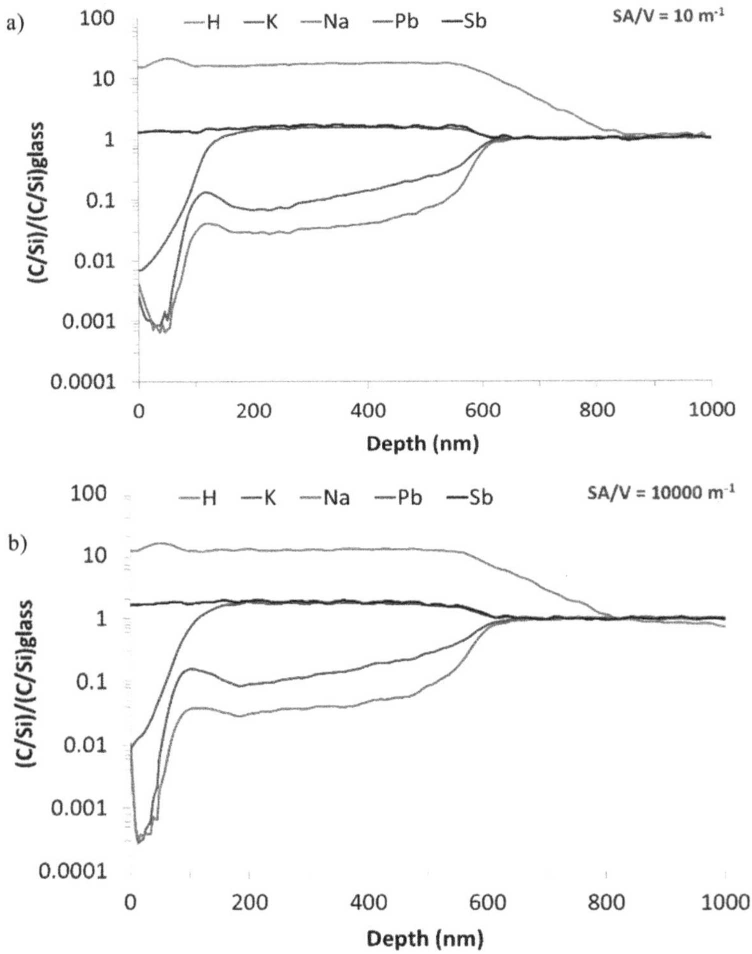

Figure 13.11. *Elementary profiles obtained by ToF-SIMS after 1 year of leaching for Pb10 glass at 22°C with SA/V ratios of 10 m^{-1} (a) and 10,000 m^{-1} (b) in acetic acid medium. For a color version of this figure, see www.iste.co.uk/bouquillon/lead.zip*

COMMENT ON FIGURE 13.11.– *Data are standardized with respect to silicon, which is the least soluble element. All the elements are thus standardized to 1 in the healthy glass; lower contents indicate element depletion, while higher contents indicate enrichment. Reproduced with permission of the American Chemical Society, Environ. Sci. Technol., 50 (Angeli et al. 2016).*

Whether the solutions are saturated or not in silicon, lead leaching is essentially controlled by diffusive mechanisms, hydrolysis mechanisms having a minor influence on lead release at 22°C.

13.2.3. *Structure of the alteration layer*

The structural evolution of the alteration layer is also a relevant parameter worth studying for a better comprehension of the long-term stability of the material. Pb10 glass was leached under static conditions at 70°C in acetic acid in a solution containing ^{17}O-enriched water to probe the alteration layer. The small size used (several microns) allows the rapid alteration of the majority of grains. A parallel experiment conducted in ^{16}O, used under the same conditions of *SA/V* ratio (but with a more significant volume of solution), allows the dosage of elements in solution and thus the calculation of the glass leaching rate. This makes it possible to stop the experiment when the entire glass is leached, relying on the entirety of sodium passed in solution. The composition of the alteration layer, deduced from analyses in solution, yields the following average molar composition: $88.7SiO_2 - 2.5K_2O - 7.9PbO - 1Sb_2O_3$.

Figure 13.12(a) compares the ^{29}Si MAS NMR spectra of the pristine glass and the altered glass for 35 days in ^{17}O. A very clear reorganization within the alteration layer is highlighted through the notable increase in the polymerization of the silicate network. Centered on Q^3-Q^4 species in the pristine glass, a majority of silicon atoms are found in the form of Q^4 units in the alteration layer.

The ^{17}O MAS NMR spectrum in Figure 13.12(b) shows that the majority of potassium was released in solution, while there was lead in the alteration layer. The line corresponding to the lead-potassium mixture site shifted toward strong chemical shifts and approached the Si-O-Pb site, showing an increase in the Pb/K ratio. This indicates that at the mixing sites, potassium extraction is preferred over lead extraction. The whole potassium in the vicinity of NBOs is released in solution (Si-O-K). Repolymerization drives a decrease in NBOs that pass from 26% in glass to 11% in the alteration layer.

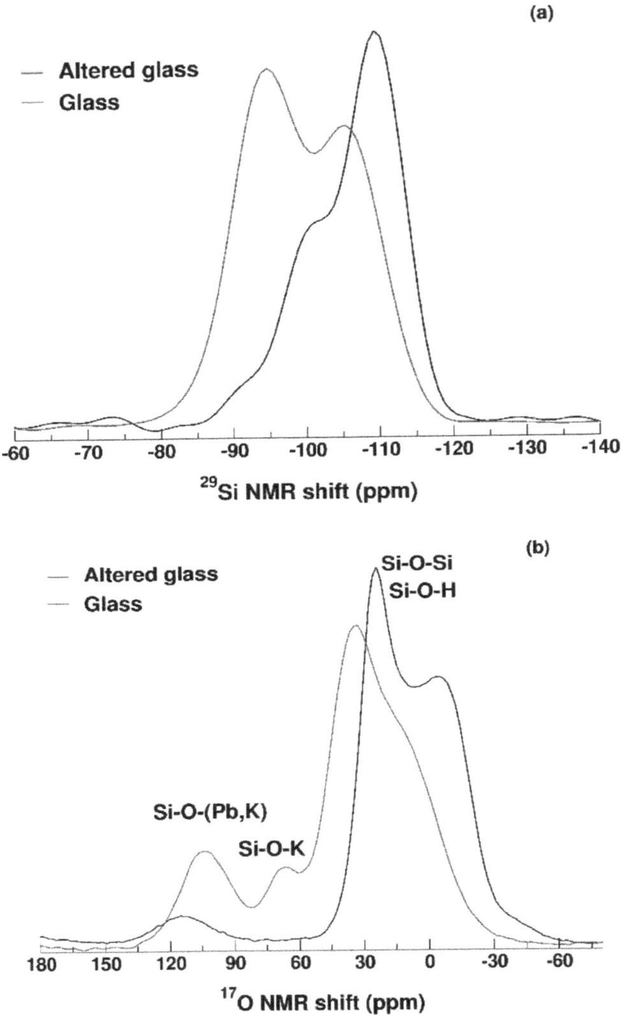

Figure 13.12. *(a) ^{29}Si MAS NMR spectra and (b) ^{17}O MAS NMR spectra of Pb10 glass in pristine glass and leached for 35 days at 70°C in an acetic acid solution enriched in oxygen-17. Reproduced with permission of the American Chemical Society, Environ. Sci. Technol., 50 (Angeli et al. 2016). For a color version of this figure, see www.iste.co.uk/bouquillon/lead.zip*

The set of experimental data was used to obtain the following formula for the estimation of the evolution of lead concentration in the presence of acetic acid as a function of time, temperature and the surface in contact with the liquid and the volume of the solution on the commercial crystal glass:

$$C_{Pb} = 3.4.10^{-2} \times \frac{SA}{V} \times \left(e^{-\frac{Ea}{RT}} \times t\right)^{\frac{1}{2}} \qquad [13.6]$$

where C_{Pb} ($mg \cdot L^{-1}$) is the lead concentration in solution, SA is the surface area of the sample (m^2), V is the volume of the solution (m^3), E_a is the activation energy (37,200 $J \cdot mol^{-1}$), R is the ideal gas constant (8.31 $J \cdot mol^{-1} \cdot K^{-1}$), T is the temperature (K) and t is the time (s).

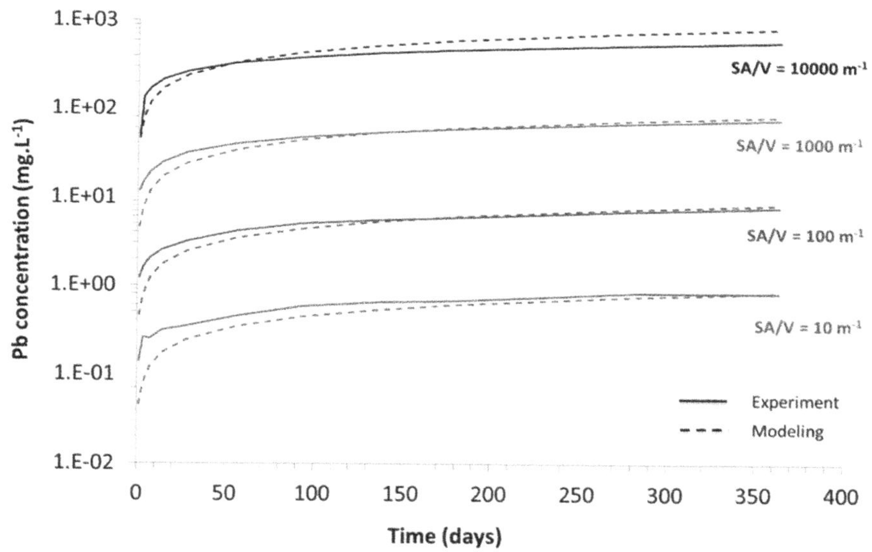

Figure 13.13. *Comparison between lead concentrations at 22°C in acetic acid over time for various SA/V ratios of experimental data (full line) and data calculated using equation [13.6] (dashed line). Reproduced with permission of the American Chemical Society, Environ. Sci. Technol., 50 (Angeli et al. 2016)*

Applying this equation to glasses leached for 1 year at 22°C shows good agreement with the experimental data (Figure 13.13). A more significant difference is observed in the test conducted with the highest *SA/V* ratio. As previously shown, these conditions generate an increase of approximately 1 pH unit due to the high alkali concentration in solution (Table 13.3). Given that equation [13.6] relies on experiments conducted at the pH of acetic acid, it yields higher calculated values than the experimental data in this case.

Equation [13.6] can be used to estimate the lead release for a lead crystal glass with Pb10-type composition according to the geometry of the container and the duration of storage of the solution. It is, however, important to note that the glasses that were the object of this study were not subjected to any surface treatment. In a certain number of objects, such as crystal decanter carafes, internal treatments of the surfaces allow a very significant reduction in lead release (Rybarikova et al. 2001). Some of these treatments are presented in section 13.3.

13.3. Industrial surface treatments limiting lead release by crystal glass

As soon as lead release by the crystal glass was evidenced in the 1990s and the mechanisms governing this phenomenon were understood, questions related to the passivity of these objects, which are highly valuable for our heritage, emerged. The stake is to preserve the specific properties of lead crystal glass: brightness, malleability over a relatively low and extended range of temperature, and sonority while ensuring its risk-free use for human health and the environment. Leaching studies revealed that a surface layer of several microns at most (Rigato et al. 1994; Bertoncello et al. 2004; Angeli et al. 2016) is at the origin of lead release, and the solutions considered consequently aim at modifying only the extreme surface of the crystal glass to leave most of the material intact. Several technical constraints must be considered in the search for an effective treatment of lead crystal glass. With a softening temperature of 500°C for a crystal glass with 24 mas% PbO, the working temperature of the treatment process must be lower to avoid modifying the original shape of the object. The treatments developed must also be compatible with the food safety of the container. Being aware of the challenge, the main French glassmakers and crystal glassmakers have empirically developed processes inspired by the glassmaking art, as well as innovative solutions allowing a reduction of the lead quantities released by their products. Several such processes are presented below.

13.3.1. *Acid polishing*

Inspired by the acid etching techniques used for decorating certain pieces, acid polishing was first used for an aesthetic purpose: to obtain the smoothest and brightest possible crystal glass pieces. This involves the successive use of two strong acids, the first one to attack the glass surface and the second one to stop the reaction and rinse the surface deposit generated by the action of the first acid. The first attack generally employs concentrated hydrofluoric acid and the second one sulfuric acid (Spierings 1993), the operations during the process being therefore extremely dangerous. Acid polishing modifies the structure of the surface glassy network and causes a loss of material, but this process is not particularly effective in relation to lead release. On the other hand, blocking lead migration by adding a surface layer was considered. Several processes for the deposition of this barrier layer were developed and patented by glass and crystal glassmakers.

13.3.2. *Incorporation of a surface glass during forming*

In 1989, Baccarat crystal glassworks filed a patent proposing the incorporation, during the first gathering, of a small amount of glass without lead at the core of the gob[1] of crystal glass before blowing (Dumas 1989). When forming the object, a thin film of lead-free glass, less reactive than the crystal glass, would then be stretched inside the object, protecting its surface. The patent gives no detail on the implementation of the process, and to our knowledge, it is no longer used today. The differences in physical properties between the two glassy materials, especially the thermal expansion coefficients, are strongly dependent on glass compositions and probably do not allow satisfactory formation of the objects.

13.3.3. *Sol-gel deposition of a glassy layer*

Another process that allows building a siliceous barrier layer at the surface is sol-gel deposition, which has the advantage that it does not require thermal treatment at high temperatures. In 1991, Mennig tested three compositions of protective layer: SiO_2, $SiO_2 + B_2O_3$ and $SiO_2 + B_2O_3 + Na_2O$ (Mennig 1991). The most effective deposited layer is that made from silica and boron oxides (SiO_2: 80 mas% and B_2O_3: 20 mas%), which are approximately 100–200 nm thick. This treatment can be used to reduce lead release by approximately one order of magnitude compared to a non-treated crystal glass. During deposition, an iridescence of the surface appears, which modifies the visual aspect of the items, but it disappears during annealing for gel densification at 475°C. The sol-gel method was once again used at the beginning of the 2000s in the heritage field to propose solutions for preserving the state of the surface of crystal glass objects (Bertoncello et al. 2003, 2006; Dal Bianco et al. 2008). While this field of application presents no specific constraint related to food toxicity, the treatment process should, on the other hand, remain reversible, according to the ethics of conservation (Bertoncello et al. 2004, 2006).

1. Gob: mass of glass melt to be formed by blowing.

13.3.4. *Dealkalinization treatments*

Dealkalinization consists of generating from the existing lead crystal glass a dense surface silica layer that acts as a barrier against lead diffusion. This result is obtained by leaching the alkali out of a small thickness of glass and by recondensing the remaining silica to obtain a silica layer that resists leaching and is not permeable to lead. Various dealkalinizing agents, such as aluminum chlorides or ammonium sulfates, are used in gaseous form to provide an optimal coating of the surface, which is necessary for the treatment of crystal glass articles of complex shapes that require a vaporization temperature below that of crystal glass softening. Considering this constraint, the process invented by Paul Cordié and patented in 1994 (Cordie 1994) developed the specific usage of ammonium iron sulfates or ammonium aluminum sulfates, which are deposited at the bottom of the article in the form of powder or in solution. The whole is heated at a temperature between 475 and 480°C for 15–180 minutes or until the surface is covered by a whitish coating. This coating is then removed by washing the surface with water or soapy water, which makes the treatment easier to apply at the industrial scale. This process, also known as SO_2-dealkalinization, modifies the properties of diffusion into the surface layer of the crystal glass and significantly slows lead leaching. The latter is detected at contents below several ppb after 24 h in acetic acid at ambient temperature (Cordie 1994). A desalkalinization process was also patented several years later by Françoise Rifqi for soda–lime–silica glasses used in computer components (Rifqi 1997). In this application, the dealkalinization process is preceded by chemical quenching or surface ion exchange to mechanically strengthen the glass surface.

13.3.5. *Acid and heat treatments*

The combination of acid treatment and heat treatment was more recently studied as a means to limit lead release in soils upon burial of the cullet resulting from cathode tubes (Yamashita et al. 2014; Akai et al. 2020), whose lead content is very close to that of crystal glass (23.1 mas% PbO). The idea is to use a part of the leaching mechanism to induce the conditions for forming an effective barrier layer. The glass is immersed in a solution of nitric, acetic or hydrochloric acid at a pH between 2 and 4, a silica-rich hydrated layer then forms by ion exchange between lead and the glass alkali, on the one hand, and the oxonium ions of the acidic solution, on the other hand, according to the previously described alteration process. An annealing stage between 350 and 500°C is then used to recondense the silanol groups in the layer, which gains barrier properties in relation to the acidic leaching and delays lead release. However, the effect of this treatment on alteration in the basic medium is very weak, the dissolution of the silica network is catalyzed under basic conditions, and the impact on lead release is practically zero. In 2020, Akai compared the effect of an acidic treatment and annealing with that of a treatment by SO_2-dealkalinization with ammonium sulfates and concluded that SO_2-dealkalinization further reduces potassium release but is less effective for lead than the acidic and annealing treatments.

13.3.6. *Metallic salt-based treatments*

Metallic salt-based treatments were initially developed by Saint-Gobain to prevent the atmospheric weathering of glazings, which led to filing a patent related particularly to a method for the application at ambient temperature via organometallic derivatives of zinc, thorium, barium or aluminum in an organic solvent (Rullier 1975). These works, related exclusively to soda-lime-silica glasses, reveal a specific interest for zinc-based components, which are very efficient. These components were used for a while to prevent the leaching caused by phosphate detergents in dishwashers, which occurs in basic conditions at a temperature of approximately 60°C. In the past few years, phosphate detergents have no longer been used, and glass protection agents, such as polyalkylimines, have been added to washing formulas. On the other hand, zinc and/or bismuth metallic salts are still considered curative treatments in the heritage domain. The adsorption of metallic ions at the glass surface allows the formation of an impermeable layer, a barrier against weathering. The precise mechanism was studied by Alloteau (Alloteau 2017; Alloteau et al. 2019, 2021) in the context of the conservation of heritage objects in soda–lime glass.

13.4. Conclusion

Lead crystal glass has been manufactured at the industrial scale for several hundred years. Its implementation is a unique know-how materialized in exceptional pieces. Industrial glassmakers adapt the compositions and perform surface treatments adapted to the uses of their products to comply with the evolving regulations. At the same time, the progress of scientific knowledge fuels discussions on the evolution of standards, testing protocols and regulations related to food contact for lead crystal products, refining the understanding of glass leaching mechanisms and empirically developed treatments that protect against alteration.

Recent studies show that the structural role of lead varies as a function of its content in the glass, which significantly influences its properties. During leaching of industrial crystal glasses, lead is released by a diffusive mechanism at a lower rate than that of the alkali. Crystal leaching experiments allow the precise quantification of the amounts of lead released in solution and their modeling. The stakes are high and simultaneously involve science, toxicology, heritage, law and technology.

13.5. References

Ahmed, A.A. and Youssof, I.M. (1997a). Effect of repeated leaching on the release of lead and other cations from lead crystal glass (24% PbO) by acid solutions and water. *Glass Technology*, 38(5), 171–178.

Ahmed, A.A. and Youssof, I.M. (1997b). Interaction between lead crystal glass (24% PbO) and acetic acid. *Glastechnische Berichte-Glass Science and Technology*, 70(6), 173–185.

Ahmed, A.A. and Youssof, I.M. (1997c). Reactions between water and lead crystal glass (24% PbO). *Glass Technology*, 38(1), 30–38.

Akai, T., Fukumi, K., Yamashita, M. (2020). Acid and heat treatment to reduce lead leaching from cathode ray tube funnel glass. *J. Ceram. Soc. Japan*, 128(3), 149–152.

Alderman, O.L.G., Hannon, A.C., Holland, D., Feller, S., Lehr, G., Vitale, A.J., Hoppe, U., Zimmerman, M.V., Waterphul, A. (2013). Lone-pair distribution and plumbite network formation in high lead silicate glass, $80PbO \cdot 20SiO_2$. *Phys.Chem. Chem. Phys.*, 15(22), 8506.

Alloteau, F. (2017). Contribution à la compréhension des mécanismes de l'altération atmosphérique des verres et étude d'un traitement de protection à base de sels de zinc. PhD Thesis, PSL Université, Paris.

Alloteau, F., Valbi, V., Majérus, O., Biron, I., Lehuédé, P., Caurant, D., Seyeux, A. (2019). Study of a surface treatment based on zinc salts to protect glasses from atmospheric alteration: Mechanisms and application to ancient glass objects in museum. In *Glass Atmospheric Alteration: Cultural Heritage, Industrial and Nuclear Glasses*, Biron, I., Alloteau, F., Lehuédé, P., Majérus, O., Caurant, D. (eds). Hermann, Paris.

Alloteau, F., Majérus, O., Valbi, V., Biron, I., Lehuédé, P., Caurant, D., Lefèvre, G., Seyeux, A. (2021). Efficacy of zinc salts to protect glass against atmospheric alteration. Part II: Possible passivation mechanisms. *J. Am. Ceram. Soc.*, 104(5), 2052–2065.

Angeli, F., Villain, O., Schuller, S., Ispas, S., Charpentier, T. (2011). Insight into sodium silicate glass structural organization by multinuclear NMR combined with first-principles calculations. *Geochim. Cosmochim. Acta*, 75(9), 2453–2469.

Angeli, F., Charpentier, T., Molières, E., Soleilhavoup, A., Jollivet, P., Gin, S. (2013). Influence of lanthanum on borosilicate glass structure: A multinuclear MAS and MQMAS NMR investigation. *J. Non Cryst. Solids*, 376, 189–198.

Angeli, F., Jollivet, P., Charpentier, T., Fournier, M., Gin, S. (2016). Structure and chemical durability of lead crystal glass. *Environ. Sci. Technol.*, 50(21), 11549–11558.

Apelblat, A. and Manzurola, E. (2001). Solubilities of manganese, cadmium, mercury and lead acetates in water from T=278.15 K to T=340.15 K. *J. Chem. Thermodyn.*, 33(2), 147–153.

Ashbrook, S.E. (2009). Recent advances in solid-state NMR spectroscopy of quadrupolar nuclei. *Phys. Chem. Chem. Phys.*, 11(32), 6892.

Ashbrook, S.E., Griffin, J.M., Johnston, K.E. (2018). Recent advances in solid-state nuclear magnetic resonance spectroscopy. *Annu. Rev. Anal.l Chem.*, 11(1), 485–508.

Baeyens, J., Brems, A., Dewil, R. (2010). Recovery and recycling of post-consumer waste materials. Part 2. Target wastes (glass beverage bottles, plastics, scrap metal and steel cans, end-of-life tyres, batteries and household hazardous waste). *Int. J. Sustain. Eng.*, 3(4), 232–245.

Bertoncello, R., Milanese, L., Negro, R., Saragoni, L., Barison, S. (2003). Sol–gel preparation of non-hygroscopic siliceous thin films enriched with alkaline-earth ions. *J. Non-Cryst. Solids*, 324(1–2), 73–78.

Bertoncello, R., Milanese, L., Bouquillon, A., Dran, J.-C., Mille, B., Salomon, J. (2004). Leaching of lead silicate glasses in acid environment: Compositional and structural changes. *Applied Physics A*, 79(2), 193–198.

Bertoncello, R., Milanese, L., Dran, J.-C., Bouquillon, A., Sada, C. (2006). Sol–gel deposition of silica films on silicate glasses: Influence of the presence of lead in the glass or in precursor solutions. *J. Non Cryst. Solids*, 352(4), 315–321.

Bonnet, C., Bouquillon, A., Turrell, S., Deram, V., Mille, B., Salomon, J., Thomassin, J., Fiaud, C. (2003). Alteration of lead silicate glasses due to leaching in heated acid solutions. *J. Non Cryst. Solids*, 323(1–3), 214–220.

Charpentier, T., Kroll, P., Mauri, F. (2009). First-principles nuclear magnetic resonance structural analysis of vitreous silica. *J. Phys. Chem. C*, 113(18), 7917–7929.

Charpentier, T., Menziani, M.C., Pedone, A. (2013). Computational simulations of solid-state NMR spectra: A new era in structure determination of oxide glasses. *RSC Advances*, 3(27), 10550.

Cordie, P. (1994). Procédé pour le traitement de surface d'articles en cristal et articles ainsi obtenus. Patent, EP 0 609 116 B1.

Dal Bianco, B., Bertoncello, R., Bouquillon, A., Dran, J.-C., Milanese, L., Roehrs, S., Sada, C., Salomon, J., Voltolina, S. (2008). Investigation on sol–gel silica coatings for the protection of ancient glass: Interaction with glass surface and protection efficiency. *J. Non Cryst. Solids*, 354(26), 2983–2992.

Damodaran, K., Rao, B., Rao, K. (1990). A molecular dynamics study of a $PbO.SiO_2$ glass and melt. *Phys. Chem. Glasses*, 31(6), 212–216.

Dumas, J. (1989). Récipient en cristal traité pour empêcher la migration du plomb vers son contenu. Patent, EP0330792A1.

Eckert, H. (1992). Structural characterization of noncrystalline solids and glasses using solid-state NMR. *Prog. Nucl. Magn. Reson. Spectrosc.*, 24(3), 159–293.

Eckert, H. (2018). Spying with spins on messy materials: 60 years of glass structure elucidation by NMR spectroscopy. *Int. J. Appl. Glass Sci.*, 9(2), 167–187.

Edén, M. (2012). NMR studies of oxide-based glasses. *Annu. Rep. Prog. Chem., Sect. C: Phys. Chem.*, 108, 177–221.

Edén, M. (2015). Al-27 NMR studies of aluminosilicate glasses. In *Annual Reports on NMR Spectroscopy, volume 86*, Webb, G.A. (ed.). Elsevier Academic Press Inc., San Diego.

Fajans, K. and Kreidl, N.J. (1948). Stability of lead glasses and polarization of ions. *J. Am. Ceram. Soc.*, 31(4), 105–114.

Fayon, F., Bessada, C., Massiot, D., Farman, I., Coutures, J.-P. (1998). ^{29}Si and ^{207}Pb NMR study of local order in lead silicate glasses. *J. Non Cryst. Solids*, 232–234, 403–408.

Feller, S., Lodden, G., Riley, A., Edwards, T., Croskrey, J., Schue, A., Liss, D., Stentz, D., Blair, S., Kelley, M. et al. (2010). A multispectroscopic structural study of lead silicate glasses over an extended range of compositions. *J. Non Cryst. Solids*, 356(6–8), 304–313.

Geueke, B., Groh, K., Muncke, J. (2018). Food packaging in the circular economy: Overview of chemical safety aspects for commonly used materials. *J. Clean. Prod.*, 193, 491–505.

Gin, S., Jollivet, P., Fournier, M., Angeli, F., Frugier, P., Charpentier, T. (2015). Origin and consequences of silicate glass passivation by surface layers. *Nat. Commun.*, 6, 6360.

Graziano, J.H. and Blum, C. (1991). Lead exposure from lead crystal. *Lancet*, 337(8734), 141–142.

Graziano, J.H., Blum, C.B., Lolacono, N.J., Slavkovich, V., Manton, W.I., Pond, S., Moore, M.R. (1996). Human in vivo model for the determination of lead bioavailability using stable isotope dilution. *Environ. Health Perspect.*, 104(2), 176–179.

Guadagnino, E., Gambaro, M., Gramiccioni, L., Denaro, M., Feliciani, R., Baldini, M., Stacchini, P., Giovannangeli, S., Carelli, G., Castellino, N., Vinci, F. (2000). Estimation of lead intake from crystalware under conditions of consumer use. *Food Addit. Contam.*, 17(3), 205–218.

Guadagnino, E., Verità, M., Geotti-Bianchini, F., Shallenberger, J., Pantano, C.G. (2002). Surface analysis of 24% lead crystal glass articles: Correlation with lead release. *Glass Technol.*, 43(2), 63–69.

Helebrant, A. and Pekarkova, I. (1996). Kinetics of glass corrosion in acid solutions. *Berichte Der Bunsen-Gesellschaft-Physical Chemistry Chemical Physics*, 100(9), 1519–1522.

Hemesath, E. and René Corrales, L. (2005). Sensitivity of structural results to initial configurations and quench algorithms of lead silicate glass. *J. Non Cryst. Solids*, 351(18), 1522–1531.

Hight, S.C. (1996). Lead migration from lead crystal wine glasses. *Food Additives and Contaminants*, 13(7), 747–765.

Hong, N.V., Lan, M.T., Hung, H.V., San, L.T. (2019). Microstructure of lead silicate melt under compression: Insight from computer simulation. *Eur. Phys. J.B*, 92(12), 268.

Lee, S.K. and Kim, E.J. (2015). Probing metal-bridging oxygen and configurational disorder in amorphous lead silicates: Insights from ^{17}O solid-state nuclear magnetic resonance. *J. Phys. Chem. C*, 119(1), 748–756.

Lee, S.K. and Stebbins, J.F. (2003). Nature of cation mixing and ordering in Na-Ca silicate glasses and melts. *J. Phys. Chem. B*, 107(14), 3141–3148.

Mennig, M., Schmidt, G., Maetze, G., Fink, C., Betzholz, S. (1991). Decreasing lead exposure from lead crystal by sol-gel coating. In *Fundamentals of the Glass Manufacturing Process*, European Society of Glass Science and Technology (ed.). Society of Glass Technology, Sheffield.

Mizuno, M., Takahashi, M., Takaishi, T., Yoko, T. (2005). Leaching of lead and connectivity of plumbate networks in lead silicate glasses. *J. Am. Ceram. Soc.*, 88(10), 2908–2912.

Patrick, L. (2006). Lead toxicity, a review of the literature. Part 1: Exposure, evaluation, and treatment. *Altern. Med. Rev.*, 11(1), 2–22.

Rahimi, R.A. and Sadrnezhaad, S.K. (2012). Effects of ion-exchange and hydrolysis mechanisms on lead silicate glass corrosion. *Corrosion*, 68(9), 793–800.

Rahimi, R.A., Sadrnezhaad, S.K., Raisali, G., Hamidi, A. (2009). Hydrolysis kinetics of lead silicate glass in acid solution. *J. Nucl. Mater.*, 389(3), 427–431.

Rifqi, F. (1997). Procédé de traitement de substrats en verre. Patent, WO/1997/029058.

Rigato, V., Mea, G.D., Pantano, C.G. (1994). Hydrogen profiles in the surface of reduced lead-silicate glasses. *Surf. Interface Anal.*, 21(2), 144–149.

Rullier, R. (1975). Procédé de traitement des surfaces de verre. Patent, FR 2269500A1.

Rybarikova, L., Dvorska, L., Hradecka, H., Jiricek, P. (2001). Surface treatment of lead glasses for reducing the leaching of lead. *Ceramics-Silikaty*, 45(1), 31–34.

Sadrnezhaad, S.K., Rahimi, R.A., Raisali, G., Foruzanfar, F. (2009). Mechanism of deleading of silicate glass by 0.5N HNO_3. *J. Non Cryst. Solids*, 355(48–49), 2400–2404.

Schultz-Münzenberg, C., Meisel, W., Gütlich, P. (1998). Changes of lead silicate glasses induced by leaching. *J. Non Cryst. Solids*, 238(1–2), 83–90.

Shrikhande, V.K., Sudarsan, V., Kothiyal, G.P., Kulshreshtha, S.K. (2001). ^{29}Si MAS NMR and microhardness studies of some lead silicate glasses with and without modifiers. *J. Non Cryst. Solids*, 283(1–3), 18–26.

Smith, M.E. and van Eck, E.R.H. (1999). Recent advances in experimental solid-state NMR methodology for half-integer spin quadrupolar nuclei. *Prog. Nuc. Magn. Reson. Spectrosc.*, 34(2), 159–201.

Spierings, G.A.C.M. (1993). Wet chemical etching of silicate glasses in hydrofluoric acid based solutions. *J. Mater. Sci.*, 28(23), 6261–6273.

Yamashita, M., Matsumoto, S., Akai, T., Sugita, H., Imoto, Y., Komai, T., Sakanakura, H. (2014). Leaching behavior of cathode ray tube (CRT) glasses. *J. Ceram. Soc. Japan*, 122(1432), 1020–1027.

14

Lead in Glass: Standards and Regulations

Denis LALART[1] and Xavier CAPILLA[2]
[1] *Legal Expert, Aire-sur-la-Lys, France*
[2] *Institut du verre, Paris, France*

14.1. Lead uses in glassworks

14.1.1. *Crystal glass*

Lead-containing glass has interesting properties, in terms of elaboration processes and for the properties of the manufactured objects (see Chapter 1). Lead lowers the melting point and its working temperature, while increasing its workability range period of malleability (working time). Moreover, lead brings chemical stabilization to the glassy network. It renders glass brighter ("rainbow" effect). Furthermore, due to its higher density, lead glass has a specific sonority. Finally, lead glass is softer than the majority of other glasses and therefore easier to cut or polish.

These properties have been known and used since the beginnings of glassworks, in the Egyptian or Roman Antiquity, or at Venice in the Middle Ages, as explained in the previous chapters. But Lord Ravenscroft's filing of a patent in 1674 in the United Kingdom marked the official birth of lead glass, which is nowadays referred to as "crystal glass" (see Chapter 7). Lead is introduced in the composition in the form of litharge or lead minium.

Lead in Glassy Materials in Cultural Heritage,
coordinated by Anne BOUQUILLON and Patrice LEHUÉDÉ.

Since 1969, the term "lead crystal glass" has been defined and protected in Europe by Directive 69/493/CEE ("Council Directive of 15 December 1969 on the approximation of the laws of the Member States relating to crystal glass"). This Directive was transposed in the national regulations of all the European Union member states, and it was resumed by several standards. In France, the decree of 11 January 1972, mandating NF B 30-004 standard, transposes this European text. Compliance with Directive 69/493/CEE therefore guarantees the authenticity and the quality of the articles launched on the market under the "lead crystal glass" designation. To be referred to as "lead crystal glass", a glass item must meet various criteria related to the concentration of metallic oxides (particularly lead oxide PbO), density and refractive index. The items falling short of these measures are referred to as "crystal glass" or "crystallin".

Designation	Metallic oxides	Density	Refractive index
Full-lead crystal glass 30%	PbO ⩾ 30 %	⩾3.00	⩾1.545
Lead crystal glass 24%	PbO⩾ 24 %	⩾2.90	⩾1.545
Crystal glass, crystallin	ZnO, PbO, BaO, K_2O (together or separately) ⩾ 10 %	⩾2.45	⩾1.520
Crystal glass, crystallin	PbO, BaO, K_2O (together or separately) ⩾ 10 %	⩾2.40	–

Table 14.1. *Designation according to the Directive of 1969*

In the United States, there is no specific regulatory definition for the internal market, but lead crystal glass is defined as containing 24% lead oxide in customs regulations. The situation is the same worldwide.

Crystal glass is essentially used for decoration, lighting and nowadays, to a lesser extent, for tableware and jewelry.

14.1.2. *Other domains*

Besides crystal glass, lead is used in a variety of glass applications, for its specific properties, such as optical glass, X-ray shielding in radiology, solder glass with contents that may reach very high values – up to 80% in lead monoxide mass in the composition (INERIS 2015). It was also used in the cathode tubes of old TV sets with contents of about 22/23%. This was the most significant use in terms of volume among glassworks uses (INERIS 2015).

Besides glassworks, strictly speaking, lead is also used to formulate frit used to prepare ceramic enamels or glazes of enameled metallic articles.

Altogether, these glassworks uses represented about 12% of the amounts of lead used in 1990 (INERIS 2015), but a smaller proportion in 2012, since at this date 90% of the tonnages were used for batteries (ILA 2012).

14.2. Regulations related to lead

14.2.1. *Lead toxicity according to WHO*

Until 2000, lead toxicity was essentially associated with saturnism, which is chronic intoxication due to severe and prolonged exposure. World Health Organization (WHO 1999) had determined an exposure limit, expressed as a provisional tolerable weekly intake of 3 mg/person or 430 μg/day per ingestion. A data meta-analysis conducted in 2005 revealed that lead had also a toxic effect even at a low exposure level (WHO 2011; ANSES 2013). Lead accumulates in the body and impacts several organic systems. For example, an exposure of 72 μg/day results in a 1 mmHg increase in blood pressure. Lead is particularly harmful to small children. Indeed, exposure of 36 μg/day reduces the child's IQ by one point.

Lead is stored in the bones and can be released in the body under certain physiological conditions. It also impacts (male and female) fertility, kidneys, etc.

Considering these new observations, WHO (2019) now considers that there is no limit below which exposure to lead is not toxic.

14.2.2. *United States*

The American federal regulation, in the Consumer Product Safety Improvement Act (CPSIA), promulgated on 14 August 2008, requires the quasi-systematic elimination of lead from all the consumer items susceptible to come into contact with children under 12.

Companies commercializing consumer goods in California (United States) must know and make sure that their products comply with the requirements of Proposition 65, officially known as *Safe Drinking Water and Toxic Enforcement Act of 1986*. Californian consumers should be prevented from getting exposed to a chemical product listed by the law without prior information on the potential risk. Lead is one of the chemical substances particularly considered. This regulation gave rise to a certain number of legal cases involving lead crystal glass manufacturers. Crystal glass articles are now quasi-systematically sold with a "clear and reasonable warning" (warning label Proposition 65), as follows:

> Warning. Use of leaded crystal is known to the State of California to cause cancer, birth defects and other reproductive harms.

Considering online selling, the Californian law is also frequently referred to in other American states.

14.2.3. *Europe*

In 2002, the European Directive RoHS (2002/95/CE) was adopted. RoHS stands for "Restriction of the use of certain Hazardous Substances in electrical and electronic equipment". Its objective is to limit the use of several substances known as being dangerous. Lead is top of the list, with a maximal concentration of 0.1%. This Directive was revised in 2011, under the name Directive 2011/65/UE, more commonly referred to as RoHS-2. There is an exemption for components containing lead in glass or ceramic materials. Renewed until July 2021, this exemption is now under study. It corresponds to one of the known cases for which lead replacement or substitution is currently unfeasible in scientific and technical terms. This exemption is particularly important for the lighting market, especially for crystal chandeliers.

The Regulation CE no. 1907/2006, "Registration, Evaluation and Authorization of Chemicals" (REACH), became effective on June 1, 2007. All glasses, including crystal glass, result from high temperature melting of various inorganic raw materials. Atoms are organized in a glassy network via oxygen bridges forming a new substance (see Chapter 3). Like most types of glass, crystal glass benefits from a registration exemption under REACH regulation, provided it meets certain conditions.

Appendix XIV of the regulation provides a list of substances (54 substances by mid-2020) whose usage is subjected to authorization. Lead monoxide is at this stage identified as a substance of very high concern (SVHC) since 2012 without being listed in appendix XIV.

The usage of these substances to produce crystal glass is a matter of intermediate status and does not require applying for authorization, like boron components for manufacturing borosilicate glass, according to the response 1 218 of ECHA in 2017 (ECHA 2017).

Appendix XVII of the regulation lists the restrictions applicable to manufacturing, launch on the market and use of certain dangerous substances and of certain dangerous mixtures and items, among which are lead and its compounds (entry 63). The usage of lead and its compounds is thus limited for jewelry. Moreover, as new scientific data became available, the Commission decided in 2015 to extend this restriction to all the items that may be put in the mouth by children, modifying appendix XVII. Crystal glass is exempted from the restrictions stipulated in appendix XVII, as the "expected migration level is low (crystal glass, enamels and gemstones and semi-precious stones) or acceptable" (reference: Regulation (EU) 2015/628 of the Commission from 22 April 2015, considering 8). This exemption was partly based on the conclusions of a voluntary assessment of the risks posed by lead, known as VRAL, Voluntary Risk Assessment Report on Lead and Some Inorganic Lead Compounds, conducted in 2008.

Moreover, a regulation aimed at protecting employees exposed to lead was implemented. It stipulates an exposure limit value (VLEP 8h of 0.1 mg/m^3, article R.4412-149 of the Labor Code) as well as a biological limit value of lead in the blood (in France: 400 μg/L of blood for men and 300 μg/L for women – article R.4412-152 of the Labor Code; in Europe: 700 μg/L of blood – directive 98-24-CE). A reinforced individual follow-up must be implemented (article R. 4412-160 of the Labor Code) if the exposure in air is above 0.05 mg Pb/m^3 or if blood lead level is above 200 μg/L of blood for men (100 μg/L for women).

Following the development of these regulations, many actions have been implemented in order to protect employees, particularly in collaboration with INRS (Coates et al. 2001; Pale et al. 2014) to identify the sources of exposure and limit it.

However, studies over several years having shown effects at lower exposure values, these values are currently being revised with lower propositions: 180 μg/L of blood, (ANSES 2019), or 150 μg/L of blood (ECHA 2019).

Concerning the general population, the High Council for public health recommended in 2014 a reduction at the lowest level of the exposure to lead in order to take into account the effects without limits and lower the limit for defining saturnism in children from 100 to 50 μg/L of blood (HCSP 2014), which was validated by the decree from June 8, 2015 modifying the notification sheet model presented in appendix 27 of the decree from 22 August 2011 related to the compulsory notification of infectious diseases and other diseases mentioned in article D. 3113-7 of the public health code.

The measures adopted in order to limit the prevalence of lead in the general population bear fruit, as the prevalence of children saturnism in France decreased between the mid-1990s and mid-2000s (INVS 2008; INSERM 2019). The effects without limits of lead compounds, however, led the authorities to adopt still tougher regulations in order to limit the sources of exposure, as proven by the numerous actions dedicated to this subject in the National Plan for environmental health (2015–2019). This compels manufacturers to adapt their manufacturing process to the new requirements (modification of composition, surface treatment, etc.).

14.3. Food contact

14.3.1. *United States*

There is no specific food regulation related to glass-food contact safety, but the Food and Drug Administration (FDA) regulation for ceramics is applied in practice. The test is based on the ASTM C-738-94 standard: 24 h contact with 4% acetic acid.

Limits (ppm)	Flat articles	Small hollow articles	Large hollow articles	Cups and goblets	Decanters
Pb	3.0	2.0	1.0	0.5	0.5

Table 14.2. *Lead release limits according to FDA*

There was a debate in the United States, following an article (Graziano 1991) published in the renowned journal *The Lancet*. In a study on the elution of lead (Pb) from crystal decanters and glasses, port containing 89 μg Pb/L was placed in decanters and the Pb content of the wine rose steadily to 3,518 μg/L after 4 months. Wines and spirits stored in crystal decanters for a long (indeterminate) time contained Pb at concentrations up to 21,530 μg/L. In a short-term experiment, white wine eluted small amounts of Pb from crystal glasses within minutes. This article was widely disseminated in the mainstream media.

Then various studies were conducted in the United States (Appel et al. 1992) or in Europe (Guadagnino et al. 2000). None of them yielded results similar to that of Graziano. After a proprietary study (Hight 1996), and the implementation of a follow-up of crystal glass products launched on the American market, the FDA did not change its regulation.

14.3.2. *Europe*

The regulatory situation in Europe resembles that in the United States. There is no specific regulation related to the food-glass contact safety, but in practice a regulation aimed at ceramics is applied: Directive 84/500. The latter stipulates the maximal amounts of cadmium and lead released by ceramic objects into foodstuff with which they are in contact.

The limit values are lower than the ones of the FDA regulation (Table 14.3), and the test is similar. The same limits and the same test are stipulated by the international standards ISO 6486 and ISO 7086.

Following the American debate, the case of crystal glass has also been subjected to specific examination by the Council of Europe in 2004 (CoE 2004). Relying on exposure values that were at that time still considered permissible, the Committee of experts has made this conclusion:

> The studies on the migration of lead from the table glassware show that lead absorption due to migration is not dangerous for the consumers.

The Committee also recommended the application of the limit values and acceptance criteria stipulated by ISO 7086-2 for hollow glass articles and by ISO 6486-2 for flat articles.

Limits	Flat articles	Small hollow articles	Large hollow articles	Very large hollow articles
Pb	0.8 mg/dm^2	1.5 mg/L	0.75 mg/L	0.5 mg/L

Table 14.3. *Limits according to ISO 6486-2 and 7086-2 standards*

Considering the new toxicological data on lead, in 2010 the European Food Safety Authority (EFSA) called into question the reference data, and in particular:

– the disappearance of the previously established PTWI (Provisional Tolerable Weekly Intake);

– the impossibility to establish a new limit value to protect consumers;

– the definition of a guiding value corresponding to the level below which 1% of the population would be affected by exposure to lead;

– the previous estimation of the level of exposure through food, updated with a maximum of 146 μg/day.

EFSA insisted on the necessity to reduce the European population's exposure to lead.

In response, the European Commission initiated in 2012 the revision of Directive 84-500. It was announced that the new regulation would cover a larger field of application, including glass, as well as lead crystal.

The project is being elaborated both for the establishment of new limits and for the definition of appropriate tests.

Concerning the limits, the value known as *discussion starting value* (DSV) is 10 μg/L for hollow articles. This would correspond to a reduction by at least 2 orders of magnitude compared to the current limits.

Concerning the test, the European Commission initiated a research program involving a dozen European laboratories, coordinated by JRC, Joint Research Center. According to the studies conducted, there is a need to define testing conditions adapted to cases such as short and repeated usage (simulation of meals) or unique but long-term usage (Simoneau 2017). These results were the basis for the revision in 2019 of ISO 6486-1 and ISO 7086-1 standards.

14.4. Conclusion

Since the 2000s, a better understanding of lead toxicology raised awareness on the risks posed even by a low level of exposure of the population. This led to the promulgation of an increasing number of regulatory texts to reduce the set of lead usages in the world, and particularly in Europe.

Lead has been used for centuries in glassworks, and no specific public health problems have been observed. This can be explained by the chemical nature of glass, which retains lead in a macromolecular network. The current problem is to have the specificities of glass, as a particular substance, taken into consideration in the regulations and standards being developed.

14.5. References

ANSES (2013). Expositions au plomb : effets sur la santé associés à des plombémies inférieures à 100 µg/L. Saisine no. 2011-SA-0219 [Online]. Available at: https://www.anses.fr/fr/content/les-effets-du-plomb-sur-la-sant%C3%A9-associ%C3%A9s-%C3%A0-des-plomb%C3%A9mies-inf%C3%A9rieures-%C3%A0-100-%C2%B5gl [Accessed 4 February 2021].

ANSES (2019). Valeurs biologiques d'exposition en milieu professionnel. Le plomb et ses composés inorganiques. Agence nationale de sécurité sanitaire de l'alimentation, de l'environnement et du travail, Maison-Alfort [Online]. Available at: https://www.anses.fr/fr/system/files/VLEP2013SA0042.pdf [Accessed 4 February 2021].

Appel, B.R., Kahlon, J.K., Ferguson, J., Quattrone, A.J., Book, S.A. (1992). Potential lead exposures from lead crystal decanters. *Am. J. Public Health*, 82(12), 1671–1673.

Coates, L., Hasni-Pichard, H., Laborde-Castérot, H. (2014). Exposition au plomb chez les vitraillistes : évaluation et prévention. Références en santé au travail no. 139, 41-56 [Online]. Available at: https://www.inrs.fr/media.html?refINRS=TF%20220 [Accessed 4 February 2021].

CoE (2004). Conseil de l'Europe, Comité de santé publique, Comité d'experts sur les matières destinées à entrer en contact avec les denrées alimentaires. Déclaration de politique générale concernant la migration du Plomb de la vaisselle en verre dans les denrées alimentaires du 22 septembre 2004 [Online]. Available at: https://www.edqm.eu/sites/default/files/matieres_et_articles_en_papier_et_carton_en_contact_avec_des_denrees_alimentaires_122002.pdf [Accessed 4 February 2021].

ECHA (2017). Guidance on requirements for substances in articles. European Chemical Agency, Helsinki [Online]. Available at: https://echa.europa.eu/guidance-documents/guidance-on-reach [Accessed 4 February 2021].

ECHA (2019). ECHA scientific report for evaluation of limit values for lead and its compounds at the workplace. European Chemical Agency, Helsinki [Online]. Available at: https://echa.europa.eu/documents/10162/68cf7011-9c04-2634-efa6-b712f1b34a85 [Accessed 4 February 2021].

EFSA (2010). Scientific opinion on lead in food. *EFSA Journal*, 8(4), 1570 [Online]. Available at: https://efsa.onlinelibrary.wiley.com/doi/10.2903/j.efsa.2010.1570 [Accessed 4 February 2021].

Graziano, J.H. and Blum, C. (1991). Lead exposure from lead crystal. *The Lancet*, 337(8734), 141–142.

Guadagnino, E., Gambaro, M., Gramiccioni, L., Denaro, M., Feliciani, R., Baldini, M., Stacchini, P., Giovannangeli, S., Carelli, G., Castellino, N. et al. (2000). Estimation of lead intake from crystalware under conditions of consumer use. *Food Additives & Contaminants*, 17(3), 205–218.

HCSP (2014). Détermination de nouveaux objectifs de gestion des expositions au plomb. Synthèse et Recommandations, Collection Avis et Rapports [Online]. Available at: https://www.hcsp.fr/explore.cgi/avisrapportsdomaine?clefr=444 [Accessed 4 February 2021].

Hight, S.C. (1996). Lead migration from lead crystalwine glasses. *Food Additives & Contaminants*, 13(7), 747–76.

ILA (2012). International lead association. Uses statistics – Principal uses worldwide [Online]. Available at: https://www.ila-lead.org/lead-facts/lead-usesstatistics [Accessed 23 December 2020].

INERIS (2015). Données technico-économiques sur les substances chimiques en France : Plomb, INERIS –DRC-15-136881-10941A, 92 [Online]. Available at: https://substances.ineris.fr/fr/substance/getDocument/2611 [Accessed 23 December 2020].

INSERM (2019). Comprendre le saturnisme [Online]. Available at: https:/ww.inserm.fr/information-en-sante/dossiers-information/saturnisme [Accessed 23 December 2020].

INVS (2008). Intérêt d'une limitation des usages du plomb dans certains produits de consommation. Note technique, 26 [Online]. Available at: https://www.santepubliquefrance.fr/maladies-et-traumatismes/maladies-de-la-mere-et-de-l-enfant/saturnisme-de-l-enfant/documents/rapport-synthese/interet-d-une-limitation-des-usages-du-plomb-dans-certains-produits-de-consommation.-note-technique [Accessed 4 February 2021].

Pale, S., Aubrège, B., Lecossois, C., Pierre, F., Rouyer, J., Vallayer C. (2001). Exposition au plomb de salariés travaillant le cristal à froid. Documents pour le médecin du travail, 86, 201–205.

Simoneau, C., Beldi, G., Jakubowska, N., Peltzer, M. (2017). Towards suitable tests for the migration of metals from ceramic and crystal tableware: Work in support of the revision of the Ceramic Directive 84/500/EEC, EUR 28872 EN, Publications Office of the European Union, Luxembourg [Online]. Available at: https://publications.jrc.ec.europa.eu/repository/bitstream/JRC108092/final_report_ceramics_final.pdf [Accessed 4 February 2021].

WHO (1999). Joint FAO/WHO Expert Committee on Food Additives – Evaluation of certain food additives and contaminants: Fifty-third report of the Joint FAO/WHO Expert Committee on Food Additives [Online]. Available at: https://apps.who.int/iris/bitstream/handle/10665/42378/WHO_TRS_896.pdf?sequence=1&isAllowed=y [Accessed 4 February 2021].

WHO (2011). Safety evaluation of certain food additives and contaminants. Prepared by the Seventy-third meeting of the Joint FAO/WHO Expert Committee on Food Additives (JECFA). World Health Organization Food Additive Series, 64, Geneva [Online]. Available at: https://apps.who.int/iris/bitstream/handle/10665/44521/9789241660648_eng.pdf?sequence=1&isAllowed=y [Accessed 4 February 2021].

WHO (2019). Preventing disease through healthy environments – Exposure to lead: A major public health concern [Online]. Available at: https://www.who.int/publications/i/item/WHO-CED-PHE-EPE-19.4.7-eng [Accessed 4 February 2021].

List of Authors

Frédéric ANGELI
CEA
DES
ISEC
DE2D
University of Montpellier
Marcoule
France

Isabelle BIRON
C2RMF
IRCP
Paris
France

Anne BOUQUILLON
C2RMF
IRCP
Paris
France

Léa BRUNSWIC
CEA
DES
ISEC
DE2D
University of Montpellier
Marcoule
France

Xavier CAPILLA
Institut du verre
Paris
France

Daniel CAURANT
IRCP, CNRS
Chimie ParisTech
France

Thibault CHARPENTIER
NIMBE, CNRS
CEA
Paris-Saclay University
Gif-sur-Yvette
France

Laurent CORMIER
IMPMC, CNRS
Sorbonne University
MNHN
IRD
Paris
France

Patrick DEGRYSE
KU Leuven
Belgium

Stéphane GIN
CEA
DES
ISEC
DE2D
University of Montpellier
Marcoule
France

Marie GODET
IMPMC, CNRS
Sorbonne University
MNHN
IRD
Paris
France

Bernard GRATUZE
IRAMAT-CEB, CNRS
University of Orléans
France

Denis LALART
Legal Expert
Aire-sur-la-Lys
France

Patrice LEHUÉDÉ
C2RMF
IRCP
Paris
France

Odile MAJÉRUS
IRCP, CNRS
Chimie ParisTech
France

Adèle MUNOZ-BONGRAND
IRCP, CNRS
Chimie ParisTech
France

Daniel R. NEUVILLE
IPGP, CNRS
University of Paris
and
USTV
Paris
France

Isabelle PALLOT-FROSSARD
Foundation for Cultural Heritage Sciences
Cergy-Pontoise
France

Gauthier ROISINE
BUCT-Paris Curie Engineering School
Beijing
China

Nadine SCHIBILLE
IRAMAT-CEB, CNRS
University of Orléans
France

Alicia VAN HAM-MEERT
KU Leuven
Belgium

Marco VERITÀ
LAMA-Iuav
University of Venice
Italy

Gilles WALLEZ
IRCP
Sorbonne University
Paris
France

Index

E, F

G, H, I

L, M

N, O

P, R

S, T

U, V, W, X

Printed and bound by CPI Group (UK) Ltd, Croydon, CR0 4YY
13/09/2026
14960996-0007